W9-BTL-355

HANDBOOK OF NEURAL COMPUTING APPLICATIONS

HANDBOOK OF NEURAL COMPUTING APPLICATIONS

Alianna J. Maren, Ph.D.
The University of Tennessee Space Institute, Tullahoma, TN
Craig T. Harston, Ph.D.
Computer Applications Service, Chattanooga, TN
Robert M. Pap
Accurate Automation Corporation Chattanooga, TN

With Chapter Contributions From:

Paul J. Werbos, Ph.D.
  National Science Foundation, Washington, D.C.
Dan Jones, M.D.
  Memphis, TN
Stan Franklin, Ph.D.
  Memphis, TN
Pat Simpson
  General Dynamics, Electronics Div., San Deigo, CA
Steven G. Morton
  Oxford Computing, Inc., Oxford, CT
Robert B. Gezelter
  Columbia University, New York, NY
Clifford R. Parten, Ph.D.
  University of Tennessee at Chattanooga, Chattanooga, TN

ACADEMIC PRESS, INC.
Harcourt Brace Jovanovich, Publishers
San Diego   New York   Boston
London   Sydney   Tokyo   Toronto

This book is printed on acid-free paper. ∞

Copyright © 1990 by Academic Press, Inc.
All Rights Reserved.
No part of this publication may be reproduced or transmitted in any form or
by any means, electronic or mechanical, including photocopy, recording, or
any information storage and retrieval system, without permission in writing
from the publisher.

Academic Press, Inc.
San Diego, California 92101

United Kingdom Edition published by
Academic Press Limited
24–28 Oval Road, London NW1 7DX

Library of Congress Cataloging-in-Publication Data

Handbook of neural computing applications / Clifford R. Parten ... [et
al.].
        p.   cm.
    Includes bibliographical references.
    ISBN   0-12-546090-2
    1. Neural computers.   I. Parten, Clifford R.
QA76.5.H35443   1990
006.3--dc20   89-48104
                                                              CIP

Printed in the United States of America
90 91 92 93   9 8 7 6 5 4 3 2 1

ACKNOWLEDGMENTS

An endeavor such as this is possible only with the support, encouragement, and active involvement of many people. I want to thank, in order, my co-authors, contributing authors and editors, the marvelous support team at the University of Tennesee Space Institute, the reviewers of this book, and my family and friends. In addition, this work was undertaken while both I and my co- authors were supported by many research contracts which allowed us to investigate neural network capabilities for many applications areas. As with any undertaking, the time was too short for everything which we wanted to include. We both hope and anticipate that the next edition will be longer and more detailed, and will cover some of the applications which we did not have the opportunity to address in this first edition.

Thanks also go first to Robert M. Pap, who came up with the idea for this book and who organized the contractual arrangements. He provided numerous insights and ideas for the structure and content of the book. Craig Harston cheerfully contributed his chapters on time and with enthusiasm. Both Bob and Craig were instrumental in involving some of our contributing authors.

Each of our contributing authors—Stanley P. Franklin, Robert L. Gezelter, Dan Jones, Steven G. Morton, Patrick K. Simpson, Harold Szu, and Paul J. Werbos—were outstanding and a joy to work with. Each provided high-quality initial chapters, and cheerfully accomodated requests for revisions. More important, they contributed their own suggestions for revisions and passed on important papers which contributed to the overall quality of this document.

Without several key people, this book simply could not have been given the shape, form, and quality which it now has. I'd like to thank Alan Rose, President of Intertext Publications, who facilitated the production of this book, and Jeanne Glasser, our copy editor, who did a heroic job of wrestling the chapters from multiple authors into a coherent whole. In addition, Ms. Glasser coordinated artwork and proofs, helped us obtain permission for reproduction of previously published material, and facilitated a smooth transition to page layout and design. Her unfailing cheerfulness and support, and amazing amount of hard work within a short time, have made this book what it is.

Dr. Robert E. Uhrig, Director of the Tennessee Center for Neural Engineering and Applications, was tremendously helpful in his support and encouragement throughout the long process. I also appreciate his willingness to use this book as the text for his course in neural networks at the University of Tennnessee/Knoxville, in order to obtain student feedback and review.

Dr. Wesley L. Harris, Vice President of UTSI, and Mr. Ray Harter, Director of Finance and Adminstration at UTSI, helped create the environment which made this book possible.

Lisa Blanks, Sherry Veazey, and Larry Reynolds of UTSI's graphics department did an outstanding job in producing the beautiful graphics which are used throughout the book. The library staff of UTSI; Mary Lo, Marge Joseph, Carol Dace, and Brenda Brooks, helped immeasurably in obtaining needed books and resources and in checking on some of the references. I owe each of these people a great debt.

Many people have reviewed this book and offered helpful comments. I'm deeply appreciative of each of them for helping me see the forest as well as the trees. These individuals include Dr. Robert Uhrig, Dr. Harold Szu, Awatef Gacem, Bryan and Gail Thompson, Guy Lukes, Gary Kelly, and Michael Loathers. Dr. Cliff Parten and Mary Rich also provided valuable assistance.

While working on this book, I and my co-authors Robert Pap and Craig Harston were supported by several contracts for research and development in neural networks. We would like to thank each of our sponsoring organizations and contract officer technical representatives for their support. These include Richard Akita at the Naval Ocean Systems Command (N66001-89-C-7063), Dr. Joel Davis at the Office of Naval Research (N00014-90-C-0155), Ralph Kissel and Richard Dabney at the National Aeronautics and Space Agency—Marshall Space Flight Center (NAS8-38443), George Booth at the Department of Transportation/Transportation Systems Center (DTRS-57-90-C00057), Dr. Kervyn Mach at the Air Force National Aerospace Plane Joint Project Office (F33657-90-C-2206), Dr. Frank Huban at the National Science Foundation (ECS-8912896), and John Llewellen at the Department of Energy (DoE-DEFG-88-ER-12824). I and the co-authors have also been supported at various times by contracts with General Dynamics, UNISYS, and Sverdrup Technology, Inc.

We wish to thank Ms. Veronica Minsky, Dr. Stephen Lemar, Dr. Francois Beausays, and Dr. Bernard Widrow for allowing us advance publication of their research within this book. We further acknowledge and appreciate permission for reproduction of material published by Ablex Corporation, Donald I Fine, Inc., HarperCollins Publishers, the Institute of Electrecial and Electronics Engineers, Inc. (IEEE), the Optical Society of America, Pergamon Press, Tor Books, Springer-Verlag, and William Morrow & Co.

We also wish to express appreciation for permission to reproduce previously printed and/or copyrighted material by Dr. Robert Hecht-Nielson Dr. Kunihiko Fukushima, Dr. Teuvo Kohonen, Dr. Richard Lippmann, Ms. Anne McCaffrey, Dr. M.M. Menon and Dr. G.K. Heineman.

I also wish to thank Ms. Lucy Hayden, editor of the Journal of Neural Network Computing (Auerbach) for sponsoring my columns which have been published in that journal on a regular basis. These gave me my first opportunity to extend an in-depth exploration into the field of neural networks.

Finally, I'd like to thank my father, Dr. Edward J. O'Reilly, my sister, Ann Marie McNeil, and my friends Jean Miller, Cornelia Hendershot, Joni Johnson, and Marilyn Everett for their support and encouragement through the many hours of work on this book. Thank you, one and all.

Alianna J. Maren

PREFACE

In 1986, when neural networks made their first strong reemergence since the days of Rosenblatt's Perceptron, there was only one comprehensive book available on neural networks—*Parallel Distributed Processing,* edited by David Rumelhart, James McClelland, and the PDP Research Group. Now, as I write this preface, I have eight additional broad-sprectrum neural networks books in front of me, including this *Handbook.* These books, published within the last two years, reflect the extraordinary growth of interest in neural networks which has developed in recent years. Each book has something special to offer. What is unique, special, and useful about this one?

First, this book allows the reader to more clearly and cohesively organize information, currently available, on neural networks. With the number of neural networks growing at nearly an exponential rate, this is absolutely necessary. As an example of this growth, we can chart the descriptions of different types of networks gathered in a single publication over the past five years. In *Parallel Distributed Processing,* Rumelhart, McClelland, et al. describe four major types of networks. In a 1987 review paper, Richard Lippmann described seven. Later that year, Robert Hecht-Nielson wrote a review which summarized the characteristics of thirteen different networks. In 1988, Patrick Simpson wrote a pair of review papers (published in 1990 as *Artificial Neural Systems*) which described about two dozen networks. This book likewise describes two dozen specific network types and they are different from the ones described by Pat Simpson.

This almost exponential growth in the number of known networks was due at first to a coalescence of knowledge which had been painstakingly developed over a more than twenty-year period. More recently, it is due to the wide variety of networks which have been developed within just the past two years. As the number of diverse types of neural networks grow, we need a cohesive framework, a context within which to view and work with these different network types.

In addition to the number of networks available, there are new developments in both implementations and applications alternatives. Although such luminaries as Carver Mead have a long-standing commitment to neural network hardware realizations, most neural network implementations (analog electronic, digital electronic, optical, and hybrid) have been developed within the last five years.

Neural networks have been applied to a wide variety of applications. A glance at the proceedings of any one of the major neural networks conferences will reveal an astonishing variety of applications areas: from speech and hand-written test recognition, sensor data fusion, robotic control, signal processing, to a multitude of classification, mapping, and optimization tasks. While many of these applications use well-known networks such as the back-propagating Perceptron network or the Hopfield/Tank network, there are many other, less well-known networks which are growing steadily in significance. These include the Adaptive Resonance Theory network, the Brain-

State-in-a-Box network, the Learning Vector Quantization network, the Neocognitron, the self-organizing Topology-Preserving Map, and many others. All of these networks are finding practical use in a wide variety of applications areas. Many are either implemented or scheduled for implementation in real systems.

Not only has the number of applications grown, but also our knowledge of the ways in which various networks can be used for applications has increased. For example, the back-propagating Perceptron network has long been used as a classifier. However, this network can also perform signal segmentations, noise filtering, and complex mappings (in some cases, measurably better than algorithmic and/or rule-based systems). As another example, the Topology-Preserving Map has been used for sensory mapping. More recently, it has also been used for optimization tasks and for sensor data fusion. This extension of our awareness of the potential uses for different neural networks is one of the motivating factors for writing this book.

The *Handbook of Neural Computing Applications* may be conceptually divided into five parts. The first part (Chapters 1-6), forms a background and context for the exploration and study of different neural networks. In addition to chapters on the history and biology of neural networks, we cover the three key aspects which define each neural network type and function; structure, dynamics, and learning. We make the distinction between micro-, meso-, and macro-levels of structural description. These different levels of descriptive detail permit us to focus our attention on the neuronal, network, and system levels of structural organization respectively. By identifying various structural classes of networks (e.g. multilayer feedforward networks), we create an overall structure, or topology for organizing our knowledge of networks; for comparing, say, a back-propagating Perceptron with a Boltzmann machine, or a Hopfield network with a Brain-State-in-a-Box network. This allows people to organize their knowledge about the many neural networks more clearly and cohesively than ever before.

In Chapters 7-13, which form the second conceptual part of this book, we delve into specific neural network types. We describe each network in terms of its underlying concept, structure, dynamics, learning, and performance. We point out some of the most significant applications of each major network type. For certain key networks (e.g. the Hopfield, and later, the back-propagating Perceptron), we identify ways in which researchers have suggested improvements to the basic network concept. This allows network developers to build networks which have better storage, or which learn faster, or which yield more accurate results than was provided by the original network models.

Neural network implementations are a special issue; distinct from discussion of network types. In the third part of this book, we offer four chapters on selecting, configuring, and implementing neural networks. Chapter 14, contributed by Drs. Dan Jones and Stanley P. Franklin, discuss how to determine whether a neural network might be the right tool for a given application, and which networks to consider for various applications tasks. Chapter 15 deal with how to configure and optimize the back-propagation network, which is still the network of choice for 70% (or more) of current neural networks applications. Steven G. Morton, president of Oxford Computer, Inc. and developer of the Intelligent Memory Chip, provides a comparative discussion of analog and digital electronic implementation alternatives in Chapter 16. Dr. Harold L. Szu, inventor of the Cauchy machine and a long-time researcher in optical neural networks, discusses optical implementation possibilities in Chapter 17.

In the remaining chapters of this book, we address specific applications issues. While some of the applications chapters have been written by myself or my co-authors, we are especially privileged to have several contributed chapters written by outstanding experts in their applications domain. Dr. Paul J. Werbos, inventor of the back-propagation method for network learning (as well as several more advanced methods, such as the back-propagation of utility through time), has contributed a chapter on neurocontrol which is accessible to both control engineers who want an introduction to neural networks, and to neural networks researchers who want to learn about the potential of using a neurally-based approach for control. To my knowledge, this is the best introduction to neurocontrol available.

Patrick K. Simpson is probably the world's leading expert in using neural networks for sonar signal processing, and has contributed a chapter on that topic. Many of the processes which he describes could be applied to other types of spatio-temporal pattern recognition as well. Dr. Dan Jones, a physician with a background in electrical engineering and signal processing, has explored how neural networks can be used to aid medical diagnoses. Robert L. Gezelter, who specializes in system design with an emphasis on test and verification issues, has contributed (with Robert Pap) a unique and timely chapter on neural networks for fault diagnosis, which is prefaced by an introduction by Dr. Werbos. In addition to these guest chapters, Craig Harston, Robert Pap, and I have written chapters dealing with neural networks for spatio-temporal pattern recognition, robotics, business, data communications, data compression, and adaptive man-machine systems.

The last chapter in this book is a look to the future— "Neurocomputing in the Year 2000." As we approach the end of the second millennium, we are witnessing an accelerated degree of technical and societal change. This chapter explores the possibilities—from the mundane to the outrageous—of how our future will be influenced by neural computing technology.

Alianna J. Maren, Ph.D.
Visiting Associate Professor of Neural Network Engineering
The University of Tennessee Space Institute
September, 1990

CONTENTS

1

INTRODUCTION TO NEURAL NETWORKS

Alianna J. Maren

1.0 OVERVIEW

Neural networks are an emerging computational technology which can significantly enhance a number of applications. Neural networks are being adopted for use in a variety of commercial and military applications which range from pattern recognition to optimization and scheduling.

Neural networks can be developed within a reasonable timeframe and can often perform tasks better than other, more conventional technologies (including expert systems). When embedded in a hardware implementation, neural networks exhibit high fault tolerance to system damage and also offer high overall data throughput rates due to parallel data processing. Many different hardware options are being developed, including a variety of neural network VLSI chips. These will make it possible to insert low-cost neural networks into existing and recently developed systems, and facilitate improved performance in pattern recognition, noise filtering and clutter detection, process control and adaptive control (e.g. of robotic motion).

There are many different types of neural networks, each of which has different strengths particular to their application. The abilities of different networks can be related to their structure, dynamics, and learning methods.

1.1 PRACTICAL APPLICATIONS

Neural networks offer improved performance over conventional technologies in areas which include:

- Robust pattern detection (spatial, temporal, and spatio-temporal)
- Signal filtering
- Data segmentation
- Data compression and sensor data fusion
- Database mining and associative search
- Adaptive control

1

- Optimization, scheduling, and routing
- Complex mapping and modelling complex phenomena
- Adaptive interfaces for man/machine systems

The following subsections briefly discuss these neural networks' applications areas. In Part II, we explain in detail the different neural networks mentioned for these applications, and Part IV of this book is devoted to applications. (See Figure 1.1.)

1.1.1   Robust Pattern Detection

Many neural networks excel at pattern discrimination. This is important, because other available pattern recognition techniques (statistical, statistic, and artificial intelligence) are often not able to deal with the complexities inherent in many important applications. These applications include identifying patterns in speech, sonar, radar, seismic, and other time-varying signals, as well as interpreting visual images and other two-dimensional types of patterns. Some patterns stand out readily, and others require complex and extensive preprocessing before initiating pattern recognition.

Feedforward neural networks excel at pattern classification. Several variants of the basic back-propagation network have been developed to allow it to deal with temporal and spatio-temporal pattern recognition. These variants have been applied to interpreting patterns in speech, sonar, and radar returns. The *Adaptive Resonance Theory* (ART) network and the *Neocognitron* are also capable of classifying patterns. The ART network creates new pattern categories when it encounters a novel pattern, and the Neocognitron recognizes complex spatial patterns (such as handwritten characters) even when distorted or varied in location, rotation, and scale.

1.1.2   Signal Filtering

Neural networks are useful filters for removing noise and clutter from signals, or for constructing patterns from partial data. The back-propagation network is often used to create a noise-reduced version of an input signal. The MADALINE network has been in commercial use for about two decades in enhancing signals for telephone transmission.

Three autoassociative networks, the Hopfield network, the Brain-State-in-a-Box network, and the *Learning Vector Quantization* (LVQ) network, have each been used to create clean versions of a pattern when presented with noisy or degraded patterns. Demonstrations of the LVQ network as early as the 1970s showed that it could regenerate entire stored pictures when given only a portion of it as a cue.

1.1.3   Data Segmentation

Data segmentation, whether in temporally varying signals (e.g., seismic data) or in images, is a demanding task. Most current segmentation algorithms do not yield completely desirable results. For example, in images, the major boundary lines of regions are often missed due to noise or other small-scale interferences. Similar difficulties are found in segmenting meaningful data in speech, seismic returns, and other complex sensor data.

As an example, the back-propagation network has been successfully used for signal onset detection in seismic data, outperforming an existing automated picking program. There are many neural network approaches (including the Boundary Contour System) now available for image segmentation, most of which exhibit superior performance even when compared with some of the more complex image segmentation algorithms.

1.1.4   Data Compression and Sensor Data Fusion

In many fields, such as medicine, aerospace component testing, and telemetry, large amounts of data are produced, transmitted, and stored. Even with high-volume storage devices, the cost of data storage has become prohibitive. Nevertheless, the data must be kept. This puts a high priority on developing foolproof data compression methodologies.

Several different types of neural networks have been applied to this problem with promising results. The LVQ network has been used as a codebook for data compression. A variant use of the back-propagation network identifies compressed data representations stored in the hidden layers of this network. The self-organizing *Topology-Preserving Map* (TPM) is unique among networks because it is able to create a reduced-dimensionality data representation. For instance, it has been used to represent four-dimensionaal information (two x-y stereo *views* of a simulated robot end effecter) in a three-dimensional space.

The back-propagation network has also been used for creating fused representations of multisource data. More complex systems of networks can correlate information with different spatial or temporal response scales.

1.1.5   Database Mining and Associative Searching

A major problem which is beginning to surface in information retrieval is that explicit information can be easily accessed, whereas implicit information cannot. Explicit information can be found via keyword searches and direct queries. Implicit information is distributed across the patterns of data stored, either for a single event entered into a database or a group of related events. Neural networks are the most promising technology available for database mining. They also have the potential for finding data which is associated (via some metric) with a key datum.

1.1.6   Adaptive Control

Real-time adaptive control is crucial in many current and projected applications, ranging from process control to guiding robotic motion. Traditional control paradigms (e.g., Kalman filtering) work well within linear regions, or regions which can be given linear approximations. Neural networks, because of their inherent nonlinearity, can be used to control processes which are nonlinear. Further, neural networks can be used to build up a state model (by learning from example) which can be used for control. By using a neural network approach, it is possible to overcome the difficulties inherent in forming a process model. The adaptive quality of neural networks and their ability to generalize allows them to respond effectively in control situations which vary somewhat from those in which they were trained.

Some Applications of Well-Known Networks: I

Network	Year Introduced	Inventors/ Developers	Primary Applications	Advantages	Disadvantages	Most Relevant Chapter
ADALINE/ MADALINE	1960	B. Widrow	Adaptive signal filtering, adaptive equalization	Fast, easy to implement, can be done using analog or VLSI circuitry.	Linear relationship between input & output assumed. Only linear separable classification spaces possible.	7
Adaptive Resonance Theory	1983	G. Carpenter & S. Grossberg	Pattern recognition	Able to learn new patterns, form new pattern categories, and retain learned categories.	Nature of categorical exemplars may change with learning.	11
Back-Propagating Perceptrons: Basic	1974-1986	P. J. Werbos, D. Parker, D. Rumelhart	Pattern recognition, signal filtering, noise removal, signal/image segmentation, classification, mapping, adaptive robotic control, data compression	Fast operation. Good at forming internal represent- ations of features in input data or classification and other tasks. Well studied. Many successful applications.	Long learning time.	7
Recurrent	1987	Almeida, Pineda	Robotic control, speech recognition, sequence element prediction	Best network so far for classifying, mapping time-varying information.	Complex network, may be difficult to train and optimize.	17
Time-Delay	1987	D. W. Tank & J. J. Hopfield	Speech recognition	Performance equivalent to best conventional methods, faster operation.	Fixed window of temporal activity represented, responds awkwardly to differences in scale of input.	17
Functional-Link Network	1988	Y. H. Pao	Classification, mapping	Only two layers (input & output) needed; faster to train.	No clear way to identify functions for functional links.	15
Radial Basis Function Network	1987-1988	Multiple Researchers	Classification, mapping	Network with single hidden layer of RBF neurons performs equivalent to basic BP network with two hidden layers.	Not yet known.	15
Back-Propagation of Utility Function Through Time	1974	P. J. Werbos	Maximize performance index or utility function over time, neurocontrol (e.g. robotics)	Most comprehensive neural approach for model-based prediction and/or control.	Can use only after differentiable model identified, must adapt off-line if model is dynamic, and assumes model is exact.	22

Figure 1.1a Table of neural network applications.

Some Applications of Well-Known Networks: II

Network	Year Introduced	Inventors/ Developers	Primary Applications	Advantages	Disadvantages	Most Relevant Chapter
Bidirectional Associative Memory	1987	B. Kosko	Heteroassociative (content-addressable) memory	Simple, clear learning rule, architecture,&dynamics. Clear proof of dynamic stability.	Poor storage capacity, poor retrieval accuracy.	11
Boltzmann Machine, Cauchy Machine	1984, 1986	G. Hinton, T. Sejnowski, D. Ackley; H. Szu	Pattern recognition (images, sonar, radar), optimization	Able to form optimal representation of pattern features. Follows energy surface to obtain optimization minima.	Boltzmann machine-very long learning time. Cauchy machine offers faster learning	8
Boundary Contour System	1985	S. Grossberg, E. Mingolla	Low-level image processing	Biologically-based approach to excellent segmentation.	Complex, multilayered architecture.	12
Brain-State-in-a-Box	1977	J. Anderson	Autoassociative recall	Possibly better performance than Hopfield network.	Incompletely explored in terms of performance and applications potential.	9
Hopfield	1982	J. Hopfield	Autoassociate recall, optimization	Simple concept, proven dynamic stability, easy to implement in VLSI.	Unable to learn new states (fixed weights for discrete Hopfield), poor memory storage, many spurious states returned.	9
Learning Vector Quantization	1981	T. Kohonen	Autoassociative recall (pattern completion given partial pattern), data compression	Able to self-organize vector representations of probability distributions in data. Rapid execution after training is completed.	Unresolved issues in selecting numbers of vectors to use and length of time for appropriate training. Slow training.	10
Neocognitron	1975-1982	K. Fukushima	Recognition of hand-drawn characters and other linear-outline figures	Able to perform scale, translation and rotation invariant pattern recognition.	Requires many processing elements and layers, complex structures, scaling issues for real-world use still need to be resolved.	12
Self-Organizing Topology-Preserving Map	1981	T. Kohonen	Complex mapping (involving neighborhood relationships), data compression, optimization	Able to self-organize vector representations of data with a meaningful ordering among the representations.	Unresolved issues in selecting numbers of vectors to use and length of time for training. Slow training.	10

Figure 1.1b Table of neural network applications.

1.1.7   Optimization, Scheduling, and Routing

Large-scale optimization tasks often require expensive computational time. Some neural networks (including the Hopfield network, the Boltzmann machine, and the TPM network) have demonstrated the ability to provide near-optimal solutions to large-scale optimization problems in very short time periods. A neural network's capabilities are readily extended to scheduling and routing tasks. One practical application is in optimizing resources for major airlines.

1.1.8   Complex Mapping and Modelling Complex Phenomena

Neural networks can create mappings from one domain to another. Sometimes this takes the form of pattern classification, signal enhancement (noise reduction), or data compression — topics which we've just discussed. In some cases, neural networks are used as direct pattern heteroassociators. An example is using neural networks to guide parameter selection for automatic welding, which is being explored at Marshall Space Flight Center in Huntsville, AL. The inputs to the network system are eight parameters describing the desired weld (e.g., bead height, bead width, etc.). The outputs are eight parameters for setting the welding apparatus.

Neural networks can also be used to model continuous complex processes or behavior. An example of this exists at the Aerospace Medical Research Laboratory at Wright-Patterson AFB where it has been shown that neural networks can model human behavior in a simplified task involving collision avoidance and target positioning. The networks learned to duplicate the performance of the humans used to train the networks — even duplicating their mistakes! This type of study is a precursor to developing networks which can simulate human behavior (or other complex behavior) in demanding scenarios, such as modelling other fighter pilots in jet aircraft simulators.

1.1.9   Adaptive Interfaces for Man/Machine Systems

Man/machine interfaces are gaining recognition as one of the most significant aspects of any computer system. This is especially true if the system is designed to facilitate some human process, such as learning, data manipulation, or system control. Recent studies focus on how the workload should be divided between the person and the computer, and on the most effective ways in which the computer can aid the human user. Results indicate that the degree of assistance should vary depending on the task complexity. Other studies on human self-representation of information show that different people learn and think using different representational formats (e.g., visual, iconic, and linguistic). The man/machine interface can be significantly improved by adapting the interface to the user. This adaptation can be both user-specific and situation-specific. This type of task is a complex one, involving both multiscale pattern recognition and control. Neural networks are an ideal technology to facilitate such enhanced user interfaces to complex systems.

1.2   THE ADVANTAGES OF NEURAL NETWORKS

Neural networks offer specific processing advantages, which make it the technology of choice in multiple applications areas. These advantages include:

- *Adaptive learning.* An ability to learn how to do tasks based on the data given for training or initial experience.
- *Self-organization.* A neural network can create its own organization or representation of the information it receives during learning and operation.
- *Fault tolerance via redundant information coding.* Partial destruction of a network leads to the corresponding degradation of performance; however, some network capabilities may be retained even with major network damage.
- *Real-time operation.* Neural net computations may be carried out in parallel, and special hardware devices are being designed and manufactured which take advantage of this capability.
- *Ease of insertion into existing technology.* Near-term availability of neural networks on specialized chips, combined with their potential to offer improved performance on discrete tasks, will facilitate modular upgrades to existing systems and offer incremental design improvement options for systems under development.

1.2.1   Adaptive Learning

Adaptive learning is one of the most attractive features of neural networks; that is, they learn how to perform certain tasks by undergoing training with illustrative examples. Because neural networks can learn to discriminate patterns based on examples and training, we do not have to have elaborate a priori models, nor do we need to specify probability distribution functions.

For example, back-propagation networks can learn to distinguish straight lines from convex curved lines, and either of those from concave curved lines. Once trained, the network can make this discrimination when the lines are shifted up or down, if they are very noisy, and even if only a portion of the line is presented.

The designer's sole concern is the design of appropriate architecture. The designer does not have to be concerned with how the network will learn to discriminate among the possible choices; detailed network interactions need not be specified. It is necessary that the designer develop a good *learning algorithm* which will let the network learn to discriminate, and a set of *training patterns*.

Some networks continue to learn long after their initial training period is completed. These networks can modify their categories if the types of examples presented change, and they can create new classification categories if given an unfamiliar pattern. These capabilities offer long-term potential for using neural networks in environments where data or events are in flux.

1.2.2   Self-Organization

Neural networks use their adaptive learning capabilities to self-organize the information they receive during learning and/or operation. When the network self-organizes, it can (depending on the network type) create representations of distinct features in the presented data. Even when networks are taught to recognize certain classes of patterns, they self-organize the information used for pattern recognition. For example, the back-propagation network will create its own *feature representation* by which it can recognize certain patterns. This self-organization leads to generalization, which allows the neural network to respond appropriately even when presented with novel data or situations.

1.2.3   Fault Tolerance via Redundant Information Encoding

Neural networks are the first computational method available which are inherently fault tolerant. There are two distinct aspects of fault tolerance. First, networks can learn to recognize patterns which are noisy, distorted, or even incomplete. This is fault tolerance with regard to the data. Second, they can continue to perform (with graceful degradation) even when part of the network itself is destroyed. This is fault tolerance to damage within themselves. This was illustrated in the first test of 32 in × 32 out MOSES neural network chip. The test system required ten chips, and of the original wafer, eight chips were known to be defective. Ten chips (including eight defective ones) were used, the network was trained using a MADELINE 3 learning rule, and the network system was still able to perform pattern recognition. Software simulations of network performance with degraded networks show similar graceful degradation of network response.

The reason that neural networks are fault tolerant in this manner is that they have distributed (or redundant) information encoding. Most computer algorithms and data retrieval systems store each piece of information in a single, localized, addressable space. When neural networks store information, it is often not localized. Instead, the many interconnections between the nodes of the network will have taken on values so that when the network is given the right stimulus, it will generate an output pattern that represents the stored information. When given a different stimulus, the same network — with the same weights — will produce a different output.

1.2.4   Real-Time Operation

One of the top priorities in many application areas is the need to process large amounts of data very fast. Neural networks are well-suited for parallel implementation. Their structure is such that only a few steps need to be performed per neuron. For most networks operating in a real-time environment, the need for training or changing connection weights is minimal. This is the only time-consuming aspect of neural network operations. Thus, of all methodologies available, *neural networks are the best alternative for low-level, real-time pattern recognition and classification.*

As the result of a 1988 study, the Defense Advanced Research Projects Agency (DARPA) produced a report on neural networks which concluded that the computational units of neural networks can best be understood and characterized in terms of interconnects between units (storage) and interconnects-per-second (speed). The report estimates that today's computer tools, with simulation capabilities of roughly 10 million interconnects per second, fall far short of the computational capabilities of even modest biological networks (e.g., a fly). However, they see these limitations as hardware-imposed. Major research efforts are underway to improve simulation and hardware resources, which should result in major improvements in capabilities over the next decade.

1.2.5 Ease of Insertion into Existing Technology

An individual network can be trained to perform a single, well-defined task. (Complex tasks, such as making multiple pattern discriminations, will require systems of interacting networks.) Because a network can be rapidly prototyped, trained, tested, verified, and translated into a low-cost hardware implementation, it is easy to insert neural networks for specific purposes into existing systems. In this manner, neural networks can be used for incremental system improvement and upgrades, and each step can be evaluated before committing to further development.

1.3 A DEFINITION OF NEURAL NETWORKS

Neural networks are computational systems, either hardware or software, which mimic the computational abilities of biological systems by using large numbers of simple, interconnected artificial neurons. Artificial neurons are simple emulations of biological neurons; they take in information from sensor(s) or other artificial neurons, perform very simple operations on this data, and pass the results on to other artificial neurons. Neural networks operate by having their many artificial neurons process data in this manner. They use both logical parallelism (for all neurons in the same layer), combined with serial operations (as information in one layer is transferred to neurons in another layer). The three main characteristics which describe a neural network, and which contribute to its functional abilities are: structure, dynamics, and learning.

1.4 SUMMARY

Neural networks are useful for pattern recognition, signal filtering, data segmentation, compression and fusion, database mining, and adaptive control, to name a few. They offer the advantages of learning from example, self-organization, fault tolerance (graceful degradation), fast data processing, and ease of insertion into existing and newly developed systems. Neural networks achieve these abilities by using large numbers of simple, interconnected processing units which operate in logical parallelism.

REFERENCES

Anza Research Inc. (1986). *Neuralbase* (a neural network bibliographic database). Cupertino, CA.

Hammerstrom, D. (1986). "A Connectionist/Neural Network Bibliography." *Oregon Graduate Center Tech. Rpt.* CS/E-86-010.

Hammerstrom, D. (1987). "A Connectionist/Neural Network Bibliography, Volume II." *Oregon Graduate Center Tech. Rpt.* CS/E-87-008.

Klimasauskas, C.C. (1989). *The 1989 Neuro-Computing Bibliography.*, MIT Press, Cambridge, MA.

Suggested Reading

Articles and Books

Anderson, J.A. and Rosenfeld E. (1989). *Neurocomputing: Foundations of Research.*, MIT Press, Cambridge, MA.

Anderson, J.A., Pellionisz, A., and Rosenfeld, E. (1990). *Neurocomputing 2: Directions for Research* , MIT Press, Cambridge, MA.

Churchland, P. Smith. (1986). *Neurophilosophy.*, MIT Press, Cambridge, MA.

Churchland, P.S., and Sejnowski, T.J. (1988). "Perspectives on Cognitive Neuroscience," *Science, 242* , 741-745.

DARPA Neural Network Study. (October, 1987-February, 1988). Final Report. (Originally published in limited editions from MIT Lincoln Laboratory. Currently available from AFCEA Internat'l. Press (AIP), Dept. IE, 4400 Fair Lakes Court, Fairfax, VA, 22033-3899. Contact: (703) 631-6190 or (800) 336-4583, ext. 6190.

Grossberg, S. (1988). "Nonlinear neural networks: Principles, mechanisms, and architectures," *Neural Networks, 1* , 17-62.

Hinton, G.E., and Anderson, J.A. (1981, revised edition 1989). *Parallel Models of Associative Memory.*, Lawrence Erlbaum, Hillsdale, N.J.

Kohonen, T. (1988). "An introduction to neural computing," *Neural Networks, 1* , 3-16.

Lippman, R.P. (April, 1987). "An introduction to computing with neural nets," *IEEE ASSP Magazine,* 4-22.

McClelland, J. L., Rumelhart, D. E. and the PDP Research Group (Eds.) (1986). *Parallel Distributed Processing: Explorations in the Microstructure of Cognition: Volume 2: Psychological and Biological Models.* MIT Press, Cambridge, MA.

McClelland, J. L., Rumelhart, D. E. (1988). *Explorations in Parallel Distributed Processing: A Handbook of Models, Programs, and Exercises,*MIT Press, Cambridge, MA.

Rumelhart, D.E., McClelland, J.L., and the PDP Research Group (Eds.) (1986). *Parallel Distributed Processing: Explorations in the Microstructure of Cognition. Volume 1: Foundations* ,MIT Press, Cambridge, MA.

Schwartz, E.L. (1990). *Computational Neuroscience* , MIT Press, Cambridge, MA.

Sejnowski, T.J., Koch, C., and Churchland, P.S. (1988). "Computational Neuroscience," *Science, 241* , 1299-1306.

Simpson, P.K. (1990). *Artificial Neural Systems: Foundations, Paradigms, Applications, and Implementations,* Pergamon, New York.

Vemuri, V. (Ed.) (1988). *Artificial Neural Networks: Theoretical Concepts,* IEEE Computer Society Press, Washington, D.C.

Major Neural Networks Journals

IEEE Trans. on Neural Networks (research journal). Monthly since spring, 1990.

Journal of Neural Network Computing: Technology, Design, and Applications (an applications-oriented journal). Quarterly since Summer 1989 (Auerbach).

Neural Computation (research journal). Quarterly since Spring 1989 (MIT).

Neural Networks (the official journal of the International Neural Networks Society). Quarterly in 1988, bimonthly in 1989 and following years (Pergamon).

Neural Network Review (a critical review journal). Quarterly since 1989 (Lawrence Erlbaum).

Biological Cybernetics also publishes many articles on neural networks.

Conference Proceedings and Special Journal Editions

Advances in Neural Information Processing Systems I and II (1989 and 1990, respectively). Ed. by D.S. Touretzky (San Mateo, CA: Morgan Kaufmann).

Neural Networks for Computing (1986 and every following year, held in Snowbird, UT. 1986 proceedings published as AIP Conference Proceedings 151, ed. by J. S. Denker).

Proceedings of the IEEE First International Conference on Neural Networks, Volumes 1-4, (M. Caudhill and C. Butler, Eds.). (Obtain from IEEE Service Center, Piscataway, N.J.)

Proceedings of the Second International Conference on Neural Networks, Volumes 1-2. (Obtain copies of the Proceedings through the IEEE Service Center in Piscataway, N.J.).

Abstracts of Proceedings of the First Annual Conference of the International Neural Network Society, (M. Cauhill and C. Butler, Eds.) (Boston, MA; September 6-10, 1988; issued as the first supplement to the Neural Networks journal).

Proceedings of the (First) International Joint Conference on Neural Networks (June 18-22, 1989, Washington, D.C.; co-sponsored by IEEE and INNS).

Proceedings of the (Second) International Joint Conference on Neural Networks (January 15-19, 1990, Washington, D.C.; co-sponsored by IEEE and INNS).

Proceedings of the (Third) International Joint Conference on Neural Networks (June 17-21, 1990, San Diego, CA; co-sponsored by IEEE and INNS).

Proceedings of the International Neural Network Conference (July 9-13, Paris, France, co-sponsored by INNS and IEEE Neural Network Council).

Proceedings of Neuro-Nimes '90 — The Third International Workshop on Neural Networks and Their Applications (November 12-16, 1990, Nimes, France; sponsored by ERIEE and Multipole Technologique Regional de la Region Languidoc-Roussillon).

Proceedings of the Workshop on Neural Networks for Automatic Target Recognition (May 11-13, 1990; Boston, MA; sponsored by Boston University's Wang Institute, Center for Adaptive Systems, and Graduate Program in Cognitive and Neural Systems, along with AFOSR).

Proceedings of the Society of Photo-Optical Instrumentation Engineers (SPIE), *Vol. 634, Optical and Hybrid Computing.* Ed. by Harold Szu (SPIE, Bellingham, Washington, 1987; multiple articles on neural networks).

Computer, Special Issue on Neural Networks (March, 1988).

AI Expert, Special Issue on Neural Networks (August, 1988).

AI Expert, Special Issue on Neural Networks (June, 1990).

2

HISTORY AND DEVELOPMENT OF NEURAL NETWORKS

Craig Harston and A. J. Maren

2.0 OVERVIEW

Neural network simulations appear to be a recent development. However, this field was established before the advent of computers, and has survived at least one major setback and several eras. The first period established some basic foundations and the field emerged as a new technology. During the following period, there was a lack of interest by many until some fundamental limitations were overcome. The most recent period has seen considerable interest, funding, and technical progress. The most important aspect is that successful applications have emerged, and these applications have proven the value of the neural network computing paradigm.

This chapter provides a brief overview of the developmental history of the neural network field. This historical account is limited; however, there are additional publications available — many written by the participants in this evergrowing field. An important chronicle of the history of neural networks was edited by Anderson and Rosenfeld [1988]. Not only did they republish many of the early (1943-1987) neural network papers, but they introduced each article with an insightful and concise analysis of its significance.

We divide the history into several periods, including early enthusiasm, lack of support, and the re-emergence of the field. The early progress, which was accompanied by some excitement, was followed by a period of disinterest. This lack of enthusiasm was strengthened by the exposure of important limitations of the neuronal models of the time. Although there was little public support for the field, a few researchers persevered and made important headway. These include: Steve Grossberg, Teuvo Kohonen, Paul Werbos, Kunihiko Fukushima, Shun-Ichi Amari, Bernard Widrow, A. Harry Klopf, and James Anderson. Their work overcame the early objections of the field and opened the way for renewed interest and growth of neural network simulations.

2.1   EARLY FOUNDATIONS

In the Nineteenth Century, William James wrote: "When two brain processes are active together or in immediate succession, one of them, on re-occurring tends to propagate its excitement into the other" [James, 1890]. James defined a neuronal process of learning as early as 1890 and a diagram in his work indicates that multiple connections were considered important. He defined re-integration as a process which reconstructed the missing information and provided total recall. This work did not inspire any simulations that we know of, but it does establish the opinion of the time concerning the brain.

Inspired by neurophysiologists such as Donald Hebb and Karl Lashley, work in the neural network field began in the 1940s [Hebb, 1949; Lashley, 1950]. Early simulations were done with paper and pencil. The first computer simulations were not done until the mid-Fifties.

2.1.1   Initial Simulations

Using formal logic, McCulloch and Pitts [1943] developed models of neural networks based on their understanding of neurology. These models made several assumptions about how neurons worked. Their networks were based on simple neurons which were considered to be binary devices with fixed thresholds. Connecting synapses had identical weights and the inhibitory synapses blocked a transmission completely. Their models also included the effects of synaptic delay.

The results of their model were simple logic functions such as "a or b" and "a and b". The limitations of these models were recognized by Rosenblatt [1958] when he referred to the McCulloch-Pitts' models as "logical contrivances." Nevertheless, McCulloch and Pitts' models drew their importance from that of multiple connections and the potential power of simple interacting processors. The theoretical importance of their work remained with computer scientists and had little impact on neuroscience.

McCulloch and Pitts' work was not continued to a large extent. Their model allowed for complex looping. This meant that the behavior of the model was difficult to predict analytically. Also, the available computational resources of the day were not adequate to simulate the proposed systems. The McCulloch and Pitts model remains an important milestone for the neural network field, but has found little application today.

2.1.2   First Computer Simulations

Two groups published early computer simulations of neuronal models. Inspired by the neuroscience of the day, these groups used computers to determine the value of their ideas [Farley and Clark, 1954; Rochester, Holland, Haibt and Duda, 1956]. The group from IBM research laboratories [Farley and Clark, 1954] maintained close contact with Donald Hebb and Peter Milner, neuroscientists at McGill University. When their first simulations did not work, they consulted the neuroscientists. The IBM researchers then revised their models to include the latest information from the neuronal laboratories.

Such interaction with neuroscientists established a multidisciplinary trend which continues to the present day. Several concepts from Hebb's laboratory proved to be valuable. Initial simulations allowed connection weights to grow without bounds. This is not a physiological phenomenon and the simulated synaptic weights became unusably large. When the system was normalized to restrict the magnitude of the connection weights, the simulations worked better. Additionally, inhibition was introduced into the system. Inhibition modified the weights by decreasing the connection weights under appropriate circumstances. Early concepts of artificial neuronal action were binary, or on/off in nature. Hebb and Milner pointed out to the IBM research team that the frequency of firing was a more realistic picture of neuronal output than digital on/off patterns.

2.2   PROMISING AND EMERGING TECHNOLOGY

Not only was neuroscience influential in the development of neural networks, but psychologists and engineers also contributed to the progress of neural network simulations. Bernard Widrow of Stanford University and Frank Rosenblatt, a psychologist, initiated major techniques and trends in the field.

Rosenblatt [1958] stirred considerable interest and activity in the field when he designed and developed the *Perceptron*. The Perceptron had three layers with the middle layer known as the association layer. There was competitive process between units in the output layer, resulting from inhibitory connections between the output units that helped to exclusively classify patterns. This three-layer system could learn to connect or associate a given input to a random output unit. Rosenblatt's contribution was considered a milestone at the time, establishing the nature of relationships between input and output with the system thereby making separate and distinct classifications. The Perceptron was computationally precise and was a true learning machine. Although the system exhibited complex adaptive behavior, Rosenblatt recognized that it did have limitations.

The ADALINE (*ADA*ptive *LIN*ear *E*lement) network was developed shortly after the Perceptron. Development of the ADALINE, by Widrow and Hoff (1960) of Stanford University, was one of the most impressive happenings of this period. The ADALINE was an analog electronic device made from simple components. Both the ADALINE and the MADALINE (for *Many ADALINE*s) employed a more sophisticated learning procedure than the Perceptron technique. The Widrow-Hoff technique is called the *Least-Mean-Squared* (LMS) learning rule, also known as the Delta Rule, because it works with minimizing a *delta* or difference between the observed and desired output. Unfortunately, continued work with neural network simulations was not rewarded by universities, so Widrow distinguished himself with an alternate line of research early during his career.

2.3   DISENCHANTMENT

Researchers became active in the field and often focused on the Perceptron system developed by Rosenblatt. His system functioned well in a limited domain, and learning was evidenced (Block, 1962). The proof of the *Perceptron Convergence Theorem* illustrated that if a solution existed, a Perceptron with an error-correcting rule could be successful. Perceptrons, with a single layer of association units, could learn to discriminate between linearly separable patterns. When these systems were tested with complex patterns that could not be discriminated with a hyperplane, little progress was seen. Thus, researchers became frustrated and increasing skeptical about the potential and value of neural network systems.

In 1969, Minsky and Papert published a book which has been described as "... brilliant...simply and clearly written in an elegant and informal style" [Anderson and Rosenfeld, 1988]. The Minsky and Papert analysis summed up a general feeling of frustration among researchers, and was thus accepted by most without further analysis. The fundamental error made by Minsky and Papert was to generalize the limitations of single layer Perceptrons to multilayered systems. Their statement "... our intuitive judgment that the extension (to multilayer systems) is sterile" (p. 232) was later demonstrated to be inadequate. This concept was refuted when multilayered neural networks proved to have powerful nonlinear discriminative abilities. Multilayered systems, using a more powerful learning rule, have solved parity and many other nonlinear problems [Rumelhart and McClelland, 1986].

The significant result of Minsky and Papert's book was to eliminate funding for research with neural network simulations. The conclusions supported the disenchantment of researchers in the field. As a result, considerable prejudice against this field was activated and unfortunately continues to some degree yet today.

2.4 INNOVATION

Although public interest and available funding were minimal, several researchers continued working to develop neuromorphically based computational methods for problems such as pattern recognition. This work provided the basis for later popularization and expansion. We believe that several paradigms were generated which modern work continues to enhance. Grossberg's influence founded a school of thought which explores resonating algorithms. Anderson and Kohonen developed associative techniques independent of each other. Klopf investigated models for learning. Werbos developed and used the back-propagation learning method, however several years passed before this approach was popularized. Amari was involved with theoretical developments, while Fukushima developed a step-wise trained multilayered neural approach for complex character recognition.

2.4.1 Self-Organizing Neural Networks

In 1967, Shun-Ichi Amari published a paper which established a mathematical basis for a learning theory (error-correction method) dealing with adaptive pattern classification. In 1972, he expanded on his work to show how a self-organizing network could form a representative pattern from a set of stimulus patterns, and fix this new pattern as a stable state of the network. He considered this self-organizing network to be a model for associative memory [Amari and Masina, 1988].

2.4.2 Associative Memory Neural Networks

Some of the first heteroassociative networks were devised by Shun-Ichi Amari, John Anderson, Teuvo Kohonen, and Steve Grossberg. Amari took a very mathematical approach; Anderson was interested in replicating certain forms of biological/psychological associative behaviors; Kohonen was investigating mathematically described associative networks; and Grossberg was trying to make neural networks that would model the classic conditioned-response association.

Initial work in associative memory was published in 1972 [Anderson, 1972; Kohonen, 1972a] by independent workers unaware of each other's contribution. Anderson was a neurophysiologist and Kohonen had an electrical engineering background. This approach is often described as being Hebbian in nature although other neural network techniques also have Hebb-like concepts and features. James Anderson began to develop his neural network, called the *Brain-State-in-a-Box*, in the 1970s. One of his first comprehensive articles was written in 1977 with Jack Silverstein, Stephen Ritz, and Randall Jones. Anderson's articles are particularly interesting because of his neurophysiologically oriented stance.

Teuvo Kohonen's self-organizing *Adaptive Vector Quantization* (AVQ) and *Learning Vector Quantization* (LVQ) networks organize their own representation of categories among input data. The Kohonen net has a layer of highly interconnected neurons which have a cooperative relationship. Teuvo Kohonen has enhanced his basic concepts to create the self-organizing *Topology-Preserving Map* (TPM) [Kohonen, 1972b]. In this network, neurons have positive (excitatory) connections to their nearest neighbors, and negative (inhibitory) connections to neurons which are further away.

Kohonen has written several classic books, including *Self-Organization and Associative Memory* which includes the technical basis for his LVQ network, as well as addressing some substantially larger issues. For anyone who has a solid grounding in linear algebra, the book is self-contained.

2.4.3 Multilayered Feedforward Neural Networks

Back-propagation nets are probably the most well-known and widely applied of the neural net options today. They are an outgrowth of earlier work on Perceptrons, with the addition of a hidden layer and use of the Generalized Delta Rule for learning. In essence, the back-propagation network is a Perceptron with multiple layers, a different threshold function in the artificial neuron, and a more robust and capable learning rule.

Several people published the back-propagation technique independently, apparently without knowledge of the other's work. The earliest was published by Paul Werbos who began with a strong interest in how the human mind processed information. The back-propagation method was a small, albeit essential element in a larger model which he proposed in 1974. As Werbos described it, "the only thing new about back-propagation is that it is an efficient way of calculating derivatives." In order to convince the skeptical members of his Ph.D. committee that the derivatives calculated by this method were exact and correct, he had to prove a generalized theorem which he called the *chain rule for ordered derivatives*. As a practical application, he applied his model to socio-economic forecasting. The proof and the application are found in his Ph.D. thesis, recently rewritten and published [Werbos, 1989]. This article explains how the back-propagation method can be applied to dynamic processes, which is the application for which this method was originally invented. Most current applications of the back-propagation method (even those that pertain to dynamic systems) are static in time. Therefore, this article provides the reader with a basis for using the back-propagation in a more general context. Werbos also developed a neurally-based forecasting approach called Heuristic Dynamic Programming [Werbos, 1977].

2.4.4   Resonating Neural Networks

For some time now, Gail Carpenter and Steve Grossberg [1988] have worked together developing *Adaptive Resonance Theory* (ART) networks based on biologically plausible models. ART networks are radically different from anything we have encountered up until now. During the twenty-year development period, Grossberg and his colleagues were actually investigating very different issues from those researchers who developed the feedforward networks.

The feedforward network developers wanted to make classifiers. They used only a few inspirations from neurophysiological systems. Specifically, they drew on the neurally-based architectural concept of using layers of neurons, which were known to exist in the visual cortex. They based the learning strategies of some of their networks on very simple applications of the Hebb rule. Others, taking more of an engineering approach, designed learning algorithms that minimized the Least Mean Squared (LMS) error of an expected output pattern with the actual one.

In contrast, Grossberg and his colleagues set out to model a much more comprehensive set of neurophysiological phenomena. By this time, Gail Carpenter and Steve Grossberg were collaborating regularly on work addressing this problem. Other researchers at Boston University, such as Michael Cohen, were joining forces with Grossberg and Carpenter in this area. Their work led to specification of the requirements of a neural network heteroassociator that would have some of the performance characteristics that neurophysiological systems displayed.

2.4.5   Cooperative/Competitive Networks

Kunihiko Fukushima of the NHK Science and Technical Research Laboratories of Japan developed a multilayered neural network system capable of interpreting handwritten characters. The original network, patterned after the visual system, was published in 1975 and called the *Cognitron*. His group also published extensively during the Eighties about their enhanced system called the *Neocognitron*.

2.4.6   Reinforcement Learning, Temporal Difference Learning, and Adaptive Critics

A. Henry Klopf has developed a basis for learning in artificial neurons based on a biological principle for neuronal learning called "heterostasis" [Klopf, 1972]. This is related to his hypothesis that each neuron is "hedonistic" in that it adapts to maximize efficiency of its excitatory synapses when firing (pleasure) and minimize the activity of its inhibitory synapses (pain). Klopf's work formed the basis on which he has developed an enhanced principle of learning in artificial neurons called "drive-reinforcement learning," which accounts for a wide range of classical conditioning phenomena [Klopf, 1986, 1987].

Richard Sutton and Andrew Barto, studying with Klopf, developed an understanding of artificial neural processes which led them to develop (with Charles Anderson) a theory of reinforcement learning and a control model called the Adaptive Critic [Sutton, Barto, and Anderson, 1983]. Sutton has extended the early work to develop the temporal difference model [Sutton, 1984, 1987], and Barto has developed with others an associative reward-penalty model for reinforcement learning [Barto & Anandan, 1985; Barto & Jordan, 1987].

2.5   RE-EMERGENCE

Progress during the late 1970s and early 1980s was important to the re-emergence of interest in the neural network field. Several factors influenced this movement. For example, comprehensive books and conferences provided a forum for people in diverse fields with specialized technical languages, and the response to conferences and publications was quite positive. The news media picked up on the increased activity and tutorials helped disseminate the technology. Academic programs appeared and courses were introduced at most major universities throughout the United States. Attention is now focused on funding levels throughout Europe, Japan, and the United States Significant funding is becoming available and several new commercial ventures are active with applications in industry and financial institutions.

As an example of the pace at which this field has emerged, Figure 2.1 shows the number of neural networks identified in the comprehensive review articles which appeared between 1987 and 1988 [Lippman, 1987; Hecht-Nielson, 1988; Simpson, 1988]. They illustrate an exponential growth, not so much in the number of networks which were developed, but in the public's awareness of the existence of neural networks developed over the previous decade. Figures 2.2a and 2.2b illustrates the actual historical development of the major neural network systems discussed in the previous section. It is clear that significant work was done in the 1960s and 1970s, and that a mushrooming has taken place in neural network concepts and major system designs since about 1984.

2.5.1   Conferences and Seminars

In 1979, Geoffrey Hinton, James Anderson, and Don Norman organized a small conference in La Jolla. The people whom they brought together came from diverse fields: neurophysiology, cognitive psychology, artificial intelligence, mathematics, and electrical engineering. As the conference progressed, the area in which their research interests overlapped became focused, and synergy began to emerge. The vision which formed was the prospect of machine-based intelligence and perception. Although topics such as neural plasticity, distributed representation, and parallel processing had been discussed as far back as the 1940s (e.g., with the work of McCulloch and Pitts on neural networks and Lashley on engrams), this was perhaps the first coherent resurgence of interest in neural networks since the days of Rosenblatt's work on the Perceptron.

How many neural networks are there?

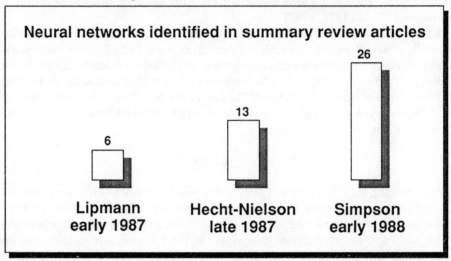

Figure 2.1   The number of neural networks identified in comprehensive review articles which appeared between 1987 and 1988 [Lippman, 1987; Hecht-Nielson, 1988; Simpson, 1988].

As a result of this conference, Hinton and Anderson [1989] orchestrated a set of edited papers, *Parallel Models of Associative Memory*. Even though this book was first published in 1981, it is still of substantial interest and has recently been revised. Because the field of neural networks was embryonic, the researchers were involved in questions that helped identify the main attributes characterizing neuromorphic computing. The publication of the *Parallel Distributed Processing* book by Rumelhart, McClelland, and the PDP research group clearly defined and defended the potential of an interdisciplinary approach (Rumelhart et al., 1986; McClelland et al., 1986). DARPA sponsored a review of the neural network field which was published in 1988. This report was generally optimistic and provided an objective evaluation of the potential of the neural network field. The DARPA report legitimized the field to Department of Defense and other agencies so funding could be improved.

A workshop of core neural network researchers at Snowbird, Utah in 1986 was a watershed event, and the impetus for planning a public conference in San Diego. The IEEE group in San Diego was surprised at the turnout to their meeting which was the called *IEEE First International Conference on Neural Networks*. It was at this June, 1987 meeting that the International Neural Network Society was organized with Stephen Grossberg as the first President. *Neural Networks*, the official journal of this international society, soon followed. Annual meetings of the IEEE conference and the International Society merged in 1989 as the International Joint Conference on Neural Networks. Several journals and newsletters dedicated to the neural network field soon emerged. Available journals include: *Neural Networks, Journal of Neural Network Computing, Neural Network Review, Neural Computation, The International Journal of Neural Networks Research and Applications,* and *the IEEE Trans. on Neural Networks.* (Publishers of these journals are listed at the end of Chapter 1.)

2.5.3 Technical Advancements

A selection of recent advancements reviewed below include: the back-propagation network, the Hopfield network, and the *Bidirectional Associative Memory* (BAM) network.

- Popularization of Back-propagation
  Although Werbos initially published the technique of propagating errors, later and apparently independently Parker (1985) and Le Cun (1986) published the back-propagation technique. In 1986, as an important part of the re-emergence of the neural field, David Rumelhart and James McClelland of the University of Southern California, popularized this method in their famous book, *Parallel Distributed Processing,* discussed in detail later in this book.
- Hopfield Network
  Of all the auto-associative networks, the Hopfield network [Hopfield, 1982, 1984] [Hopfield and Tank, 1986; Tank and Hopfield, 1987] network is the most widely known today. The major concept underlying both the Hopfield network and Anderson's Brain-State-in-a-Box network (1972) is the same. This concept is that a single network of interconnected, binary-value neurons can store multiple stable states. The primary reason that the Hopfield network has attracted so much attention is twofold. First, he illustrated how the stability of network dynamics could be mathematically described by a Lyapunov function. He was able to demonstrate that the network described was globally stable. Second, his network was relatively easy to build, using current VLSI technology.
- BAM Neural Network
  Bart Kosko has developed several lines of research with his hetero-associative networks, which include the BAM, the Fuzzy Cognitive Maps, and the Fuzzy Associative Memory. His research has led to multiple versions and evolutions of these networks [Kosko, 1986, 1987a, 1987b, 1987c].

2.6 CURRENT STATUS

Significant progress has been made in the field of neural networks — enough to attract a great deal of attention and fund further research. Advancement beyond current commercial applications appears to be possible, and research is advancing the field on many fronts. Neurally based chips are emerging and applications to complex problems developing [Mead, 1989]. Clearly, today is a period of transition for neural network technology.

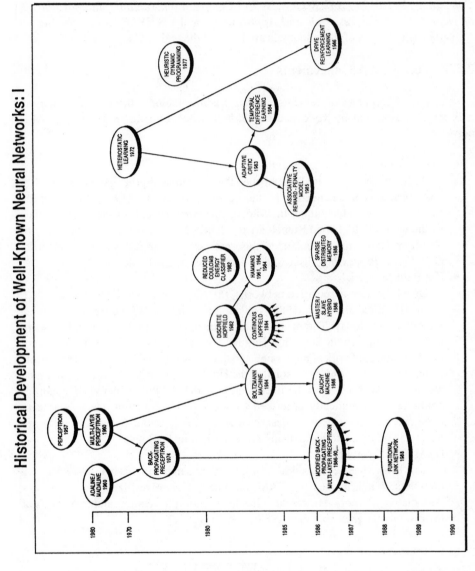

Figure 2.2a. Illustrates the actual historical development of the major neural network systems.

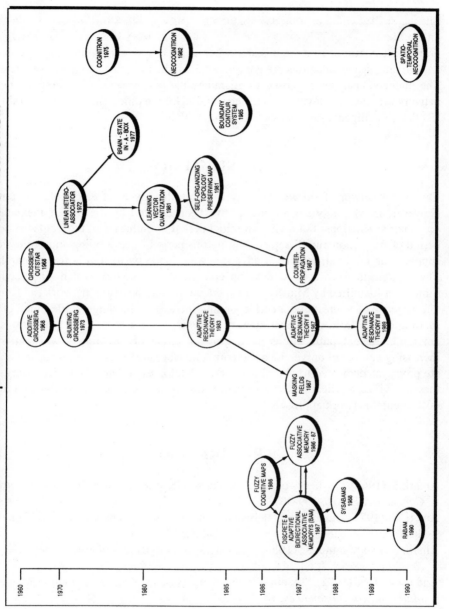

Figure 2.2b Illustrates the actual historical development of the major neural network systems.

2.6.1 Major Issues and Concerns

The major issues of concern today are the scalability problem, testing, verification, and integration of neural network systems into the modern environment. Neural network programs sometimes become unstable when applied to larger problems. The defense, nuclear and, space industries are concerned about the issue of testing and verification. The mathematical theories used to guarantee the performance of an applied neural network is still under development. The solution for the time being may be to train and test these intelligent systems much as we do with humans.

2.7 SUMMARY

The neural network field was conceived before the advent of electronic computers. However, many important advances have been boosted by the use of inexpensive computer simulations. Following an initial period of enthusiasm, the field survived a period of frustration and disrepute. During this period when funding and professional support was minimal, important advances were made by relatively few researchers. These pioneers were able to develop convincing technology which surpassed the limitations identified by Minsky and Papert. Currently, the neural network field enjoys a resurgence of interest and a corresponding increase in funding. While much of this funding comes from the Department of Defense, some commercial applications are emerging. Recent publications by a growing circle of researchers working in an increasing number of universities and corporate laboratories have surfaced, and graduate programs dedicated to neural network technologies are emerging. The success of these activities, applications, and current research will determine the value and future of the neural network approach.

REFERENCES

Amari, S. (1967). "A theory of adaptive pattern classifiers," *IEEE Trans. on Electronic Computers, EC-16,* 299-307.

Amari, S. (1972). "Learning patterns and pattern sequences by self organizing nets of threshold elements," *IEEE Trans. Computers, C-21,* 1197-1206.

Amari, S., and Maginu, K. (1988). "Statistical neurodynamics of associative memory," *Neural Networks, 1,* 63-74.

Anderson, J.A. (1972). "A simple neural network generating interactive memory," *Mathematical Biosciences, 14,* 197-220.

Anderson, J.A. and Rosenfeld, E. (1988). *Neurocomputing Foundations of Research,* MIT Press, Cambridge, MA.

Anderson, J.A., Silverstein, J.W., Ritz, S.A., and Jones, R.S. (1977). "Distinctive features, categorical perception, and probability learning: some applications of a neural model," *Psych. Review, 84,* 413-451, and reprinted in J.A. Anderson and E. Rosenfeld (Eds.) (1988). *Neurocomputing* , MIT Press, Cambridge, MA.

Barto, A., and Anandan, P. (1985). "Pattern recognizing stochastic learning automata," *IEEE Trans. on Systems, Man, and Cybernetics,* SMC-15, 360-375.

Barto, A., & Jordan, M. (1987). "Gradient following without back-propagation in layered networks," *Proc. IEEE First Int'l. Conf. Neural Networks* (San Diego, CA; June 21-24, 1987), II-629-II-636.

Bartok A., Sutton, R., & Anderson, C. (1983). "Neuron-like adaptive elements that can solve difficult learning control problems," *IEEE Trans. on Systems, Man, and Cybernetics, SMC-13,* 834-846.

Block, H.D. (1962), "The Perceptron: a model of brain functioning. I," *Reviews of Modern Physics, 34,* 123-135.

Carpenter, G.A., and Grossberg, S. (March, 1988). "The ART of adaptive pattern recognition," *Computer,* 77-88.

DARPA Neural Network Study, (October, 1987-February, 1988). *Final Report.* MIT Lincoln Laboratory, AFCEA International Press Fairfax VA.

Farley, B. and Clark, W.A. (1954). "Simulation of self-organizing systems by digital computer," *IRE Transaction of Information Theory, 4,* 76-84.

Fukushima, K., Miyake, S. and Ito, T. (1975). "Neocognitron: a neural network model for a mechanism of visual pattern recognition," *IEEE Transactions on Systems, Man, and Cybernetics,* SMC-13, 826-834.

Grossberg, S. (1976). "Adaptive pattern classification and universal recoding: II. feedback, expectation, olfaction, illusions," *Bio. Cybernetics, 23,* 187-202.

Hecht-Nielsen, R. (1988). "Neurocomputing: Picking the human brain," *IEEE Spectrum,* March, 36-41.

Hinton, B. and Anderson, T. (Eds.) (1989, 1981). *Parallel Models of Associative Memory,* Lawrence Earlbaum, Hillsdale, NJ.

Hebb, D.O. (1949). *The organization of behavior,* Wiley, New York.

Hopfield, J.J. (1982). "Neural networks and physical systems with emergent collective computational abilities," *Proc. of the National Academy of Sciences, USA, 79,* 2554-2558.

Hopfield, J.J., and Tank, D. (1986). "Computing with neural circuits: A model," *Science, 233,* 625-633.

Hopfield, J. J. (May, 1984), "Neurons with graded response have collective computational properties like those of two-state neurons," *Proc. Nat'l. Acad. Sci. USA, 81,* 3088-3092.

James, W. (1890). *Psychology,* Holt, New York, Chapter XVI, 253-279.

Klopf, A. (1972). "Brain function and adaptive systems: A hterostatic theory," *Air Force Research Laboratories Technical Report,* AFCRL-72-0164.

Klopf, A. H. (1986). "A drive-reinforcement model of single neuron function: An alternative to the Hebbian neuronal model," in J. S. Denker (Ed.), "Neural Networks for Computing," *AIP Conference Proceedings, 51,* Snowbird, UT; Am. Inst. Physics, New York, NY, 265-270.

Klopf, A. (1987). "A neuronal model of classical conditioning," *AFWAL Technical Report,* AFWAL-TR-97-1139.

Kohonen, T. (1987). *Content-Addressable Memories, 2nd. Ed.,* Springer-Verlag, New York.

Kohonen, T. (1989). *Self-Organization and Associative Memory,* 3rd Edition, Springer-Verlag, Berlin, NY.

Kohonen, T. (1972). "Correlation matrix memories," *IEEE Trans. Computer* C-21, 353-359.

Kohonen, T. (1972). "Self-organized formation of topologically correct feature maps," *Biological Cybernetics, 43,* 59-60.

Kosko, B. (1987). "Constructing an Associative Memory," *BYTE,* Sept., 137-144.

Kosko, B. (1987). "Adaptive bidirectional associative memories," *Applied Optics, 26,* 4947-4960.

Kosko, B. (1987). "Fuzzy associative memories," *Fuzzy Expert Systems,* Addison-Wesley, Reading, MA.

Kosko, B. (1986). "Fuzzy cognitive maps," *International Journal of Man-Machine Studies, 24,* 65-75

Lashley, K.S. (1950). "In search of the engram," *Society of Experimental Biology Symposium, No. 4: Psychological Mechanisms in Animal Behavior,* Cambridge University Press, Cambridge, MA., 454-480.

LeCun, Y. (1986) "Learning processes in an asymmetric threshold network," *Disordered Systems and Biological Organization,* E. Beinenstock, Fogelman Souli, and G. Weisbuch (Eds.), Springer, Berlin.

Lippmann, R.P. (April, 1987). "An introduction to computing with neural nets," *IEEE ASSP Magazine,* 4-22.

Mead, C. (1989). *Analog VLSI and Neural Systems,* Addison-Wesley Publishing Co.

Minsky, M.L., and Papert, S. (1969, 1987). *Perceptrons ,* MIT Press,Cambridge, MA. and reprinted in *Neurocomputing,* ed. by J. Anderson and E. Rosenfeld (1987). MIT Press, Cambridge, MA., 161-170.

McCulloch, W.S. and Pitts, W. (1943). "A logical calculus of the ideas immanent in nervous activity," *Bulletin of Mathematical Biophysics, 5,* 115-133.

McClelland, J.L, and Rumelhart, O.E., and the PDP Research Group. (1986). "Parallel distributed processing: explorations on the microstructures of cognitron, Vol. 11," *Psychological and Biological Models,* MIT Press, Cambridge, MA.

Parker, D. (1985). "Learning logic," Technical report TR-87, *Center for Computational Research in Economics and Management Science,* MIT Press, Cambridge, MA.

Rochester, N., Holland, J.H., Haibt, L.H. and Duda, W.L. (1956). "Tests on a cell assembly theory of the action of the brain, using a large digital computer," *IRE Transactions of Information Theory,* IT-2, 80-93.

Rosenblatt, F. (1958). "The Perceptron: A probabilistic model for information storage and organization in the brain," *Psych. Review, 65,* 386-408, and reprinted in *Neurocomputing,* J. Anderson and E. Rosenfeld (Eds.) (1988).MIT Press, Cambridge, MA., 92-114.

Rumelhart, D.E., Hinton, G.E. and Williams, R.J. (1986). "Learning internal representations by error propagation," *Parallel Distributed Processing,* Editors Rumelhart, D.E., McClelland, J.L. and the PDP Research Group (Eds.), Chapter 8, 318-364.

Rumelhart, D.E. and McClelland, J.L. and the PDP Research Group (1987). *Parallel Distributed Processing: Exploration in the Microstructure of Cognition, Vol.1: Foundations,* MIT Press, Cambridge, MA.

Simpson, P.K. (1988). "A review of artificial neural systems. I: Foundations and II: Paradigms, Applications and Implementations," Preprint privately circulated, and later published as *Artificial Neural Systems: Foundations, Paradigms, Applications, and Implementations,* (1990). Pergamon, New York, NY.

Sutton, R. (1984). "Temporal credit assignment in reinforcement learning," *University of Massachusetts, Computer and Information Systems Technical Report,* COINS-TR-84-02.

Sutton, R. (1987). "Learning to predict by the methods of temporal differences," *GTE Laboratories TEchnical Report,* TR87-509.1.

Tank, D.W., and Hopfield, J.J. (1987). "Collective computation in neuronlike circuits," *Scientific American, 257,* 104-114, 158.

Werbos, P. (1974). *Beyond Regression: New Tools for Prediction and Analysis in the Behavioral Sciences.* Ph.D. dissertation, Harvard University, Cambridge, MA.

Werbos, P. (1977). "Advanced forecasting methods for global crisis warning and models of intelligence," General Systems Yearbook.

Werbos, P. (March-April, 1989). "Maximizing long-term gas industry profits in two minutes in Lotus using neural network methods," *IEEE Trans. on Systems, Man, and Cybernetics.*

Widrow, B., and Hoff, M.E. (1960). "Adaptive Switching Circuits," *Institute of Radio Engineers, Western Electronic Show and Convention, Convention Record, Part 4),* 96-104, and reprinted in *Neurocomputing,* J. Anderson and E. Rosenfeld (Eds.), MIT Press, Cambridge, MA 126-134.

3

THE NEUROLOGICAL BASIS FOR NEURAL COMPUTATIONS

Craig T. Harston

3.0   NEUROSCIENCE AS A MODEL

There are several reasons to advocate using the nervous system as a model for computer simulations. Of course, the first is because the simulations should work well if a successful example like the brain is used. Much of the progress in neural network modeling is derived from improved neurological knowledge. When early neurological models were updated with concepts fresh from the laboratories of Hebb and Milner at McGill University, the simulations demonstrated improvement [Rochester, Holland, Haibt and Duda, 1956]. Not only did the neurology improve the simulations but, these results helped establish neurological thinking as a source of workable and tenable ideas.

A second reason is that the brain can be better understood as we develop computer simulations. New concepts are integrated into the simulations and, in a sense, productive ideas are proven out [Churchland and Sejnowski, 1988]. The fact that adaptive, parallel, distributed computer networks can learn, recognize complex patterns, generalize, and discriminate similar to humans and animals in behavioral experiments is a powerful rationale for believing that the brain also accomplishes these behaviors by the same mechanisms. Research suggests that the brain incorporates parallelism and distributed information with adaptive connections in order to learn, recognize, generalize, and discriminate. These successful systems verify the importance and plausibility of parallel distributed knowledge in the brain.

A third reason to use the brain as a model is the potential of applying workable ideas from the nervous system to solve difficult problems. Industry and government agencies have problems which have not been automated by computer processes. Specifically, identification of aircrafts could be used by the military and civilian aircraft control. Intelligent monitoring of chemical processes could be integrated with expert process control. Effective identification of security threats would discriminate between false alarms and true breaches. Voice recognition and handwritten character recognition could be used widely to improve data entry efficiency.

While the neural network field incorporates computer simulation of aspects of the central nervous system, the field is not restricted to modelling the biological system. Many neural network research papers deal with mathematical models of optimization borrowed from physics or chemistry.

Biological approaches to neural network simulation include the CMAC, patterned on a model of the cerebellum, which is responsible for coordinated motor movement [Albus, 1979]. Other researchers offer theoretical developments based on biology for the computerized neural network field [Amari, 1977; Churchland and Sejnowski, 1989]. Baron [1987] has written a text which integrates computational and neurological concepts, and Part I of Churchland's *Neurophilosophy* [1988] contains a useful overview of neurophysiology and related material [Shepherd, 1988].

Some researchers use neural structures and function as the basis for neural network systems development. Parallel networks, one for recognizing shape and the other for identifying location, can be integrated by another network (Anzai and Shimada, 1988). The system is patterned after the visual system in the brain and similar to the proposed associative cortical model. Spinocerebellar, magnocellular (rubro-) and transcortical concepts were used to develop robotic neural networks (Miyamoto et al., 1988). The neurophysiology of sensorimotor components of gaze was used for simulated neural networks (Pellionisz, 1986). These reports are evidence that neurobiology can be used to develop successful computer systems.

3.1   THE SINGLE NEURON

The neuron is the fundamental building block of the nervous system. Neurons exist in many shapes, sizes and lengths. These attributes are important to the function and utility of the neuron. Classification of these cells into standard types has been done by many neuroanatomists.

3.1.1   Dendrites, Cell Bodies, Hillocks, and Axons

The neuron is a living cell and as such it comes complete with many common features of a biological cell. There is a nucleus with an internal nucleolus. The cell body contains the usual mitochondria, Golgi apparatus, rough endoplasmic reticulum and free ribosomes. The shape of the cell body can be round, triangular, drop-like, pointed on two ends or stellate with many points emanating to dendrites or an axon. Usually there is a mound on the surface of the cell which terminates into the axon. This is called the axon hillock and has special electrophysiological functions summarized below. The axon leaves the axon hillock and carries electrophysiologic pulses from the cell to other areas. There is extensive branching on the far end of the axon. The axon terminals end on the dendrites of the following neurons. Leading to the cell body, there are a multitude of tiny dendrites which come together forming larger branches and trunks which attach to the cell body.

3.1.2 Synapses

One of the unique things about neurons is the ability to communicate with other neurons. The axon terminals form a connection which almost touches the dendrites of other target neurons. There is a space between the two cells called the synaptic gap of about 50 to 200 Angstroms. One neuron communicates to another by transferring electrical energy along the axon to the axon terminals and from here to the next cell by releasing specific chemicals called neurotransmitters into the synaptic gap. This process takes about 0.1 to 0.2 milliseconds of time, which is a long time by electronic standards. These neurotransmitters depolarize the dendritic membrane and increase the chance that the target neuron will continue the transmission.

3.1.3 Bioelectric Nature of the Neuron

The nerve cell membrane maintains an electrical difference or voltage potential across the cell's wall. It does this by controlling the distribution of charged ions inside and outside the neuron. In the resting state, some ions are kept inside while other charged ions are pumped out of the cell. This takes energy to maintain these ionic concentrations and their associated voltage potential. Sodium, potassium, chlorine and calcium are some of the major ions involved. This information is easily available in the neurophysiological literature (Llinas, 1988).

Some neurotransmitters enhance the membrane potential difference by polarizing the membrane or increasing the ionic difference. The chlorine ion is principally involved with hyperpolarization. Neurotransmitters can also depolarize the nerve by opening channels in the membrane so the ions can pass through. The depolarization tends to spread from the synaptic area to surrounding membrane. As this activity travels away from the synapse, the effect decreases with time and distance. If there are enough synaptic transmissions occurring nearly simultaneously, the effects will summate and the depolarization will travel over a larger distance. This effect degrades or decreases over time and distance. Computer simulations may need to incorporate time based signal degradation to successfully process information which varies over time.

If enough synapses release transmitters simultaneously, the depolarization will spread over large areas of the nerve cell membrane. When the depolarization is widespread, it may encompass the axon hillock. When the axon hillock is depolarized, the axon depolarizes. Unlike the cell membrane, the axon depolarization is transmitted down the axon without degradation. This effect is called the axon potential and can occur 0 to 1500 times a second.

Early thinking suggested that single axon potentials correspond to information processed by the brain. However, the magnitude of the axon potential is constant and not related to the amount of stimulus. Many neurophysiologists realized that the most distinctive feature of axonal activity was their frequency. The neurons would respond to a stimulus with a machine gun-like barrage of activity when monitored and reported with a speaker. Given a stimulus, the axon's response was related to frequency. A resting neuron would pop off randomly or fire with a steady base line frequency. The neuron's response to a stimulus would be a change from the resting axon potential rate to a rapid-fire pattern.

The frequency-based thinking has influenced computerized neuronal models. When early computer simulations failed, computer researchers consulted with neuroscientists. One group from IBM (Rochester, Holland, Haibt and Duda, 1956) visited with Milner and Hebb to learn of more realistic ideas about the brain. One idea of benefit was the relationship of frequency to the information processes of a neuron (Anderson and Rosenfeld, pp.66, 1988). The IBM group used frequency related output to improve their models.

3.1.4   Multiple Neurons

The nervous system is comprised of millions of nerves, about 10 to the 12th power. While single neurons are interesting, it is the interaction of many neurons which makes learning, recognition, decision making, discrimination and generalization possible. Some neuronal relationships are simple. For example, the pain reflex is mediated by a sensory and a few motor neurons. One neuron senses a painful stimulus and carries the message to the spinal cord. The information is passed to motor neurons via a single set of synapses. The motor neurons transmit signals to muscles which in turn move the body away from painful stimulus. There are other neurons which transmit information to the brain but the reflex is simple in nature.

Most operations of the nervous system are much more complex than simple reflexes. Neuroanatomists have documented the variety of nerve tracts, neurons, and networks in the brain and spinal cord. This research provides a basic framework for understanding the brain, which has a rich diversity of neuronal architecture. There are systems within systems, general as well as specific subsystems, that are never inactive but always emitting many frequencies of activity.

3.2   EARLY RESEARCH

Two questions are particularly troublesome albeit fascinating. One, how do we learn; the other, exactly where are memories stored. Two early neural researchers have contributed ideas which have proven to be important for simulated neural intelligence. These researchers were Donald Hebb (1949) and K.S. Lashley (1950).

3.2.1   Hebb's Law

Donald Hebb formulated the theoretical basis for learning. He proposed that a synaptic activity facilitated the ability of neurons to communicate. Also if two events were active at the same time, the involved synapses would be facilitated such that activity from one of the events could activate the same subsystem previously activated by the other stimulus. These ideas established the basis for learning or association by the brain.

Neurochemists, physiologists, and physiological psychologists have been guided by Hebb's statement for years. His theory has stimulated untold efforts to discover the nature of the learned facilitation. Many feel that these changes must involve changes of protein synthesis. Researchers have investigated RNA and DNA, which were thought to mediate protein synthesis. Others have looked at different types of protein and changes of protein levels. Recent work suggests that calcium mediated translocation of proteins could be the result of associative conditioning (Alkon, 1989). An experience dependent on biochemical processes for early neuronal growth has been reported (Aoki and Siekevitz, 1988). The presence of cell wall protein phosphorylation near the synapse may facilitate synaptic activity on a long term or relatively permanent basis. As such, it would be the biochemical basis of learning. Such has been suggested by [Brown et al., 1988].

Synaptic facilitation is simpler in computer simulations. The effectiveness of the synapse is represented by a number. This number is called a weight in the neural network field. The larger the weight, the more effective the synapse. If the synapse is active or needs to be strengthened, then the weight can be increased. Alternatively, the simulated synapse can be weakened by decreasing the weight. Modification of the weights is considered to be learning in the neural network field. Computer simulators have found that limits must be placed on the magnitude of the weight so changes are often limited. The weight becomes, in effect, the memory of the computerized network.

3.2.2 Lashley's Work

Lashley began his research effort to find the engram, the location in the brain where memory is stored. He began by training rats in mazes, then areas of the cortex in the brain were removed during surgery. Following recovery from the operation, the rats were tested in the maze again. Lashley, a Harvard neuropsychologist, expected that some rats would not be able to remember the maze if the right part of the brain was missing. In that case, we would know in what part of the brain the memory for maze learning would be stored. Many rats were trained and many different parts of the brain were removed.

The result was difficult to understand at first. The rats could remember the mazes quite well. Their memory was so good that Dr. Lashley began to think that maze learning was not stored in the cortex. After much thought, Lashley was able to evaluate his work and make two important conclusions. He felt that the different cortical areas had equal potential for maze learning. This was called the *law of equipotential.* The second idea was called the law of mass action. It was based on the fact that the rats would make maze errors in proportion to the amount of cortical tissue removed.

From these laws, several things can be surmised, all of which have major implications for computer simulation of neural processing. There appears to be massive parallelism and duplication of function in the brain. Lashley's work suggests that many areas of the cortex are involved with memory. This occurs to the degree that one area can be inoperable, yet the other involved areas continue so that the animal functions almost without decrement. Functionally this redundancy provides resistance to faults. The parallelism also provides the ability to process massive amounts of data.

Recent research has found considerable functional specificity in the cortex, but this does not negate Lashley's findings. The cortex in the back of the head is most related to visual processes. There are sensory and motor areas across the top of the cortex. Right-left specificity of verbal and conceptual functioning has been reported. Even when functionality of a system is limited to an area, that area includes many neurons. In the cortex these areas are called columns as discussed below. It is the parallel processing by many functioning units which gives the nervous system its recognition speed, flexibility, and analytic abilities. While localized specificities do exist in the cortex, distributed learning and functionality over many neurons is also well established.

3.3   STRUCTURAL ORGANIZATION OF BIOLOGICAL NEURAL SYSTEMS

There are several major design principles which can be found underlying the structural organization of different areas of the brain. These may offer long-term potential as models for design of artificial neural systems. These principles are:

- Layers of processing elements
- Columns of processing elements
- Specialization of neural tissue into both specific and non-specific systems

3.3.1   Layers of Processing Elements

Sensors interface the world to the brain and from this point the data is passed through multiple levels of the nervous system. Through this transfer from low to high levels of the brain different levels of cerebral functioning such as abstraction, conceptualization and feature detection occur. For example, the nervous system can detect specific features yet it also deals with ambiguous information. These sensations may not be understood until it is processed by many layers in several areas of the brain.

Information sensed by the eyes, ears or, from touch is passed through many layers of nerves. In the retina alone, there are three or four layers of neurons in addition to the photo receptors. From here visual information passes to the lateral geniculate body and then to the visual cortex at the back of the head. In the cortex there are six layers of cortex which are involved in processing visual images. Visual processing does not stop at the visual cortex but is passed along to nearby areas of the cortex where it is further defined. Associations with other events occur in more distant cortical areas. Clearly, visual information is processed by many layers of neurons before it is used or consciously understood.

3.3.2   Columns of Processing Elements

The brain, including the cortex, is organized into smaller working subgroups. Although the cortex appears to be one large convoluted multilayered neural network with horizontal stripes made of six different neuronal layers, it does not function as a single six-layered network. Researchers have shown evidence that the cortex is divided into columns acting as functional units (Hubel and Wiesel, 1974; Hubel, Wiesel and Stryker, 1978). Input arrives through the bottom of the cortex terminating with neurons in the second and fourth layers; there appear to be collateral synapses with the third, fifth and sixth layers as well. This information is processed by the neurons in the cortex and leaves via neurons in the third and fifth cortical layers to other parts of the brain, cortex, or spinal cord.

The functional significance of the six cortical layers is beginning to emerge. There appears to be some advantage to the layers which can be tested empirically by computer simulations. As with the cortex, large computer simulated neural networks often do not function well, while small interrelated networks with separate but relevant functions can operate very well (Kawato et al., 1987; Nguyen and Widrow, 1989; Barto, Sutton and Anderson, 1983; Harston, Maren and Pap, 1989). It appears that the design of interrelated computer networks may be critical for the successful application of computer simulated neural network technology.

3.3.3 Specialization of Neural Tissue

The nervous system has both specific and non-specific systems for sensory and motor processing. This varies to the point where some areas of the cortex, called association areas, have general learning ability. Even these areas may specialize in a general way, being involved with speech which is a general and high level of sound processing.

Specific systems mediate exact information through the sensory system to the cortex. The reverse pathway from the motor cortex to the motor neurons in the spinal cord activates specific actions. These actions are often fixed or reflex-like.

The non-specific sensory system acts as a warning system to the brain. Along with specific sensory information, sensory information is distributed to non-specific areas of the brain. These non-specific areas control alertness and attention. If the incoming signal is novel or important, then these non-specific areas alert or activate additional areas of the brain. This non-specific area of the brain is called the reticular activation area and is located at the core of the brain stem. This activation process causes us to respond appropriately to certain stimuli.

The indirect motor system is activated following a goal-oriented movement. Many subsystems are activated in the brain stem and cerebellum, which modulate and control movements started by the motor cortex or reflexes. It is these indirect subsystems which provide for smooth coordinated movements. Superior performance, indeed learned movements, are absorbed and improved by these indirect subsystems.

It appears that features are often detected by a nervous system that is tuned specifically for the event. It appears that the nervous system is prewired to identify and respond to specific items in the environment before much learning can occur.

3.4 STRUCTURALLY LINKED DYNAMICS OF BIOLOGICAL NEURAL SYSTEMS

Several of the dynamic processes which occur in biological neural systems are integrally linked to the structures of these systems. These dynamics form the basis from which the higher properties of the system emerge. These structurally-linked dynamic processes include:

- Distributed representation of information
- Temporal encoding of information

- The role of inhibition
- Feedforward and feedback processing loops

3.4.1   Distributed Representation of Information

The information which is accessed by the various biological sensors is represented in a distributed manner across multiple neurons. Further, neural processes seem to involve activations of multiple neurons. This is true not only in a local sense, but also in terms of distributing processing across complementary subsystems.

It may be desirable to mimic these featurization and distribution processes for use in automated control systems. There are two sensory distribution systems in the brain. One is the specific sensor-thalamic-cortical system and the other is a nonspecific system used for attention and motivation. The specific system transfers information from the sensors through the thalamus and onto a sensory modality specific area of the cortex.

During the sensor input process, features are extracted from the input matrix, there may be additional refinement of the data during the input process. For example, there are cortico-thalamic connections as well as thalamo-cortical neural pathways. These may set up a reverberatory loop between the subcortical thalamus and the neuronal layers in the cortex. It can be postulated that these reverberatory circuits may refine the definition extracted from the sensory data. There may be some degree of cooperative/competitive pattern matching by these loops. An alternative explanation suggests that this reverberation serves as short term memory (STM) in the brain. Another reverberatory loop within the brain is the subcortical limbic system. The limbic system could also be involved with STM and object identification, except that it appears to be involved mostly with emotion and as such may be principally involved with motivation.

3.4.2   Temporal Encoding of information

Neural information is encoded as spike trains from axons. A given axon will typically have a constant spiking amplitude or potential, but its frequency of response will carry information content. Collections of neurons may take on a temporal pattern of spiking activity which is information-rich.

In most artificial neural systems, the input is from a single clock cycle, although some architectures do allow for some form of temporal feedback of information (e.g., recurrent and time-delay networks). It is understood that the "connection weight strength" encoded for many neural networks corresponds not to spiking amplitude, but to frequency.

3.4.3   The Role of Inhibition

Inhibition is the effect of one neuron blocking the action of another neuron. This inhibitive mechanism is common if not prevalent throughout the nervous system. Inhibition is so pervasive that we could say that as a baby grows, much of its learning is the result of learning what not to do. Most of the stimuli around us can safely be ignored. However what we ignore may be learned. Inhibitory learning by the nervous system is the result of neurons learning the inhibition of inappropriate activity.

Neural network simulations have also used the inhibition concept. One of the first computer simulations did not work initially (Rochester et al., 1956). When the developers added inhibitory decreases of synaptic influence (i.e., -1 as well as +1) to their model, the computer simulation of neurons functioned much better than in earlier trials. The idea of inhibition as used by downward adjustment of weights is found in the back propagation of error technique where over-sized results are used to decrease relevant weights in the network. Lateral inhibition also plays a role in the self-organizing Topology-Preserving Map and the multilayer cooperative / competitive networks which we discuss in Part II.

While inhibition is used for neural network simulations, the inhibitory concept probably is not used as much here as it is in the brain. There may be considerable untapped potential if more extensive inhibitory concepts were used for computer simulations. It is possible that components and functions in the nervous system are organized around and based on inhibitory mechanisms. This suggests that network simulations could be improved with inhibition by designing systems which are active prior to learning and training. This system would learn by inhibiting or decreasing unnecessary or irrelevant behaviors. The result would preserve the desired behaviors which would not be suppressed by the inhibition.

3.4.4   Feedforward and Feedback Processing Loops

It appears that the brain uses circular or reverberatory loops to process information. This looping occurs when one part of the brain processes input and passes information on to another area. This new area processes and passes the information back to the originating location or through other intermediate locations. In the end, the information is returned through the original brain area to reverberate through the structures again. Examples of this circular or reverberant processing are the cortico-thalamic loop and the limbic system.

The cortical-thalamic system has interconnections between the basal nuclei of the thalamus and the cortex. The connections are not randomly distributed but are topographically organized and connected. For example, input to the brain goes through sensory specific areas of the thalamus; from here the information is sent to sensory specific areas of the cortex. Presumably the information is processed in the cortex and circulates back to the thalamus. Information may be circulated or processed many times by this thalamic-cortico loop. Some computer simulated neural networks also use circular processing between the neural layers to define and refine input patterns. The most notable neural network with these characteristics is the Adaptive Resonance Theory (ART I and ART II) networks, developed by Gail Carpenter and Stephen Grossberg [Grossberg, 1988]. The ART networks specialize in refining and focusing the definition or categorization of different sets of input.

Short term memory (STM) is an active circular brain pathway which keeps a memory active for a few seconds or possibly minutes. This temporary memory is kept active by reactivating the appropriate neural pattern repeatedly. This process is required for recollection, and it may also be used to keep a memory alive long enough for the neurons to learn and create structural modifications in the neurons. These modifications become the basis for long term memories of important events. Such a mechanism of STM may or may not be important for simulated neural networks. Electronic memory can serve very well as STM, however reverberatory circuits—such as STM—might integrate more compatibly into computer neural network system's designs.

3.5   EMERGENT PROPERTIES ARISE FROM THE DYNAMICS OF BIOLOGICAL NEURAL SYSTEMS

From the basic structurally-linked dynamics described in the previous section, several system-level emergent properties arise. These relate to the higher-level abilities of an organism. These more global processes include:

- Processing goals corrolate with processing levels
- System activity emerging as associations between interacting subsystems

3.5.1   Processing Goals Corrolate With Processing Levels

There are three aspects of the input process delineated below which can be used by simulated networks. First, is the normalization and featurization of input. Second, there is a refinement or focusing of the information. Third, duly processed information is broadcasted to associate with other areas.

Input from the sensory organs is processed through several neurons before it gets to the cortex. There are two sensory systems; one is general and consists of widespread activation of lower levels of the brain when a new stimulus is available. This serves to alert or motivate the animal. The other is specific in nature. Input from pressure receptors in the finger are relayed through a synapse in the spinal cord to the thalamus in the center of the brain. This information is passed to a specific area in the cortex. Sound is processed by a similar number of synaptic relays, but instead of a synapse in the spinal cord, it has a relay in the brain stem. At every level of synapses additional signals are generated to various areas.

Another important function performed by sensory systems concerns transforming the intensity of a signal to one that is compatible with the nervous system. For example, the ear adjusts itself to be sensitive to very low sounds while the nerves are still active. However, loud sounds cause the ear to become less sensitive and the amount of neural activity is controlled. The total effect of the auditory control is to provide logarithmically increasing responses to sound. Thus, low sounds are important but loud sounds do not overburden the system. This example suggests that input to computer systems should be normalized to provide reasonable dynamics within a narrow range for simulated neurons.

3.5.2   Associations Between Interacting Subsystems

The output of the nervous system is the result of associated activity from many areas of the brain. For example, when the location of an object in the visual field and the motivational components are associated in the posterior parietal cortex, an animal will focus on the intended item. The shift of gaze requires many complex movements of the eyes, head, and body, resulting from activity in the association areas of the cortex linking with the motor command areas in the motor cortex.

The initial motor command activates the pyramidal motor system which directly synapses with the motor neurons in the spinal cord. These cause a prompt initial movement by the affected muscles. The pyramidal motor system is a direct motor command mechanism which ensures that body activity is responsive to cortically based decisions.

A second system which acts parallel to the pyramidal system, and is responsible for cortically commanded movement is the extrapyramidal system. When the cortex initiates the pyramidal system activity, it also activates the extrapyramidal system. This parallel system activates many additional circuits which continue to influence the spinal motor neurons and in turn the muscles long after the direct cortical-pyramidal activity has ceased. The extrapyramidal system involves feedforward and feedback systems so the intended movement is smooth, coordinated, and efficient.

3.5.3   Feature Abstraction and Association

Sensory input in and of itself has no meaning. It is the task of sensory neural systems to extract information from the sensory input. The visual system is a good example of this process. Color is separated into different systems because different color-specific receptors are activated in the retina; movement is detected in the retina. The nerves appear to be interconnected with inhibitory synapses in the retina so when a pattern moves across the visual field some neurons are activated by the movement. In fact, movement is detected in one direction but not the other (Poggio and Koch, 1987). Other features are determined as the information is processed from the retina to the lateral geniculate body to the visual cortex in the occipital lobe. Features such as edges, angles, and binocular disparity are developed in the cortical areas. These features appear to be processed in parallel systems. There are interconnections between some of the systems but most of these features are fused in association areas of the cortex.

The visual system does not appear to identify a dog or other object at the occipital cortical level. This area of the cortex is the first to receive visual information through the most direct route to the cortex, however objects are not recognized here. The type of information presented to this area of the cortex are called features. Examples of visual features are edges, color, and movement. Tones may be auditory features. Non-specific features might be attentional, signifying new or prominent information.

3.6   LEARNING IN BIOLOGICAL NEURAL SYSTEMS

The ability of biological neural systems to adaptively learn in response to experience and environment is one of the most remarkable aspects of these systems. There are many aspects associated with such learning. Among them are:

- Learning overlays hardwired connections
- Synaptic plasticity versus stability: A crucial design dilemna
- Synaptic modification provides basis for observable organism behavior

3.6.1   Learning Overlays Hardwired Connections

There is a continuum from hardwired development of the nervous system to adaptive openness. This includes intermediate states where the brain is adaptive early in life, yet develops a fixation or inflexibility with maturation. Even with some things seemingly frozen, there is a continued ability to adapt and learn new things as an adult.

This range of adaptability in neural tissue seems to be reflected in the range of available behaviors in complex biological organisms. While some behaviors seem to be established at birth, others are developed and molded after birth. Even so, some development is limited to the animal's maturation period. A few songbirds will learn to sing while young, but cannot learn songs when they are adult. Some species can only learn variations of songs known to their species. They cannot learn songs from other species even if that is the only song they ever hear. On the other hand, mockingbirds learn songs from several different species. Clearly, the nervous system in different species has diverse developmental abilities (Marler, 1976).

The brain and nervous systems establish many neurons prior to birth. Few, if any, new neurons appear following birth in the human. Many important functional neuronal connections are programmed by genetics and develop without learning. These connections are considered to be a type of hardwiring.

It is appears that new learning is superimposed over the basac hardwired architecture networks. Electronic memory can serve very well as STM, however reverberatory circuits—such as STM—might integrate more compatibly into computer neural network system's designs.

3.6.4   Conceptualization

The purpose of the cortical associative learning system is to establish conceptual meaning, making decisions that will instigate action [Wise & Desimone, 1988]. Once the features have been extracted from the sensory suite and the appropriate features have been fused, associative neural networks can be used to link output to form meaningful concepts. Computerized associative algorithms are powerful and understood to a considerable degree (Kosko, 1990; Simpson, 1990). Other useful associative algorithms may be based on other techniques discussed later in this book. The use of associative networks is important; however, the organization of these networks also must be developed. For this purpose, the cortex of the brain could be modeled to provide the necessary developmental ideas. This associative cortical model will be patterned on the biological success of the brain. Once input has been featured and fused by preprocessing subsystems of the brain, the cortex transfers the information to several cortical areas where it becomes associated with other preprocessed information. There are several areas of the cortex involved both serially and in parallel. While the information goes to different areas, it also jumps to additional cortical areas for further cortical association. This process must be directed and controlled to provide meaning and specificity. These associations should be designed and managed to link fused information that will form ideas and concepts.

3.7   FUNCTIONAL RESULTS OF NEURAL ARCHITECTURE

Unlike modern data processing and communications systems which transfer and process information along a single line, the nervous system uses many lines of communication (i.e., axons and dendrites) and many information processors (i.e., neurons). There are advantages to this redundancy.

Fault tolerance exists and is important in biological nervous systems. Some neurons and their associated axons die each day and are not replaced. Fortunately, the redundancy of many neurons per function allows the system to continue functioning even with considerable neuronal loss. As a corollary, computational systems with redundant components can provide fault tolerance. Fault tolerant systems are important for normal working conditions; however, they are critical for high-risk situations, such as those in the nuclear power industry, space exploration and warfare. In these high-risk environments nonredundant failure can be catastrophic.

Simulated neural systems can be naturally fault tolerant because of the distributed nature of the processing and redundant use of information. Neural networks are insensitive to unreasonable data values. It may be a natural result of neural networks to fill in gaps where information is missing. For example, we do not notice the blind spot in our retina because the perceptual system based on biological neural networks fills in the missing spot with context-similar patterns. The result of simulated neural network systems may be context adjusted information which represents the most appropriate input.

The major advantage to parallel systems is speed. Certain supercomputers, noted for their fast computational abilities, accomplish this speed by using parallel processors. Problems, which can be broken into separate subproblems, are handled by separate parallel processors and can be solved more quickly than when handled by serial computers. Animals, including humans, can solve some complex problems much faster than our fastest electronic computers. While humans do not perform mathematical calculations as quickly as computers, we do recognize patterns and solve complex motor manipulative tasks almost instantly.

Input from the sensory organs is processed through several neurons before it gets to the cortex. There are two sensory systems; one is general and consists of widespread activation of lower levels of the brain when a new stimulus is available. This serves to alert or motivate the animal. The other is specific in nature. Input from pressure receptors in the finger are relayed through a synapse in the spinal cord to the thalamus in the center of the brain. This information is passed to a specific area in the cortex (see Figure 3.1). Sound is processed by a similar number of synaptic relays, but instead of a synapse in the spinal cord, it has a relay in the brain stem. At every level of synapses additional signals are generated to various areas.

3.8   COMPUTER SIMULATIONS BASED ON THE BRAIN

There is a correspondence between various neural architectures and the major approaches to neurocomputing . For example, the hierarchical information flow from one layer to the next, is copied by the feedforward neural networks. These architectures are

Primary Sensory & Associative Cortical Areas

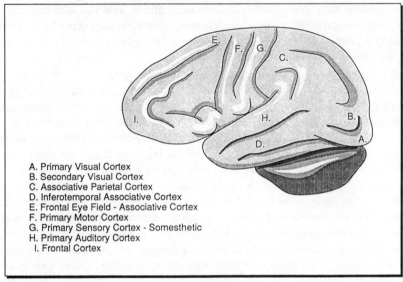

A. Primary Visual Cortex
B. Secondary Visual Cortex
C. Associative Parietal Cortex
D. Inferotemporal Associative Cortex
E. Frontal Eye Field - Associative Cortex
F. Primary Motor Cortex
G. Primary Sensory Cortex - Somesthetic
H. Primary Auditory Cortex
 I. Frontal Cortex

Figure 3.1 Illustrations of cortical areas of the brain.

capable of complex pattern identification. Often these feedforward networks use the error of performance of feedback to learn. The nervous system also uses feedback to learn. Looping between groups or levels of neurons has inspired the development of resonating neural network computational approaches. The use of inhibition, as well as excitation, has guided the development of cooperative/competitive actions found in autoassociative neural network simulations.

Promising areas of the brain to model and those that are currently areas of intense interest include the retina, audition, and olfaction. More research on orientation and habituation could focus on the reticula formation. Development of ideas and concepts should focus on associative relationships between cortical areas. Motor function for robotics and telerobotics could model the motor cortex, brain stem nuclei, and cerebellum. Short term memory might be simulated by looping circuits. The role for emotional motivation for electronic systems has not been examined.

REFERENCES

Albus, J.S. (1979). "Mechanisms of planning and problem solving in the brain," *Math Biosciences, 45,* 247-293.

Alkon, D.L. (1989). "Memory storage and neural systems," *Scientific American*, July, 42-50.

Amari, S. (1977). "Neural theory of association and concept-formation," *Biological Cybernetics, 26,* 175-185.

Anderson, J.A. and Rosenfeld, E. (1988). *Neurocomputing Foundations of Research,* MIT Press, Cambridge, MA, 65-67.

Anzai, Y. and Shimada, T. (1988). "Modular neural networks for shape and/or location recognition," *Abstracts of the Second International Neural Networks Conference* (Sept. 6-10, Boston, MA), 158.

Aoki, C. and Siekevitz, P. (1988). "Plasticity in brain development," *Scientific American,* December, 56-64.

Baron, R.J. (1987). *The Cerebral Computer An Introduction to the Computational Structure of the Human Brain,* Lawrence Erlbaum Publishers, Hillsdale, NJ.

Barto, A.G., Sutton, R.S. and Anderson, C.W. (1983). "Neuron like adaptive elements that can solve difficult learning control problems," *IEEE Transactions on Systems, Man, and Cybernetics, SMC-13*: 834-846

Brown, T.H., Chapman, P.F., Kairiss, E.W. and Keenan, C.L. (1988). "Long-Term Synaptic Potentiation," *Science, 242,* 724- 728.

Churchland, P.S. (1988). *Neurophilosophy,* MIT Press, Cambridge, MA

Churchland, P.S. and Sejnowski, T.J. (1988). "Perspectives on Cognitive Neuroscience," *Science, 242,* 741-745,

Harston, C.T., Maren, A.J. and Pap, R.M. (1989). "Neural Network Sensory Motor Robotics Application," *Proceedings of WESCON/89,* San Francisco, CA; Nov. 1989, 699-708.

Hebb, D.O. (1949). *The Organization of Behavior,* Wiley, New York, NY.

Hubel, D.H., and Wiesel, T.N. (1974). "Sequence regularity and geometry of orientation columns in the monkey striate cortex," *Journal of Computer Neurolology, 158,* 267-294.

Hubel, D.H., Wiesel, T.N. and Stryker, M.P. (1978). "Anatomical demonstration of orientation columns in macaque monkey," *Journal of Computer Neurology, 177,* 361-380.

Grossberg, S. *(Editor) (1988). Neural Networks and Natural Intelligence,* MIT Press, Cambridge, MA.

Kawato, M., Uno, Y., Isobe, M. and Suzuki, R. (1988). "Hierarchical neural network model for voluntary movement with application to robotics," *IEEE Control Systems Magazine,* April, 8-16.

Klopf, A.H. (1987). "A Neuronal Model of Classical Conditioning," *Interim Report, AFWAL-TR-87-1139,* Air Force Wright Aeronautical Laboratories.

Kosko, B. (1989). "Unsupervised learning in noise," *Proc. First Intl. Joint Conference on Neural Networks, I,* Washington, D.C.; June 18-22, 1989, 7-17.

Lashley, K.S. (1950). "In search of the engram, society of experimental biology symposium, No. 4," *Psychological Mechanisms in Animal Behavior,* Cambridge University Press, Cambridge, MA.

Llinas, R.R. (1988). "The Intrinsic Electrophysiological Properties of Mammalian Neurons: Insights into Central Nervous System Function," *Science, V. 242,* 1654.

Marler, P. (1976). "Sensory templates in species specific behavior," *Simpler Networks and Behavior,* J.E. Fentress, (Ed.), Sinauer, Sunderland, MA, 314-329.

Miyamoto, H., Kawato, M., Setoyama, T., & Suzuki, R. (1988). "Feeback-error-learning neural network for trajectory control of a robotic manipulator," *Neural Networks, 1,* 251-265.

Nguyen, D. and Widrow, B. (1989). "The truck backer-upper: An example of self-learning in neural networks," *PRoc. First Intl. Joint Conference on Neural Networks, II,* Washington, D.C.; June 18-22, 1989, 357.

Pellionisz, A.J. (1986). "Tensor network theory and its application in computer modeling of the metaorganization of sensorimotor hierarchies of gaze." *AIP Conference Proceedings 151, Neural Networks of Computing,* J.S. Denker (Ed.) (Proc. of Neural Information Processing Conference, Snowbird, Utah), 339-344.

Poggio, T. and Koch, C. (1987). "Synapses that compute motion," *Scientific American,* May, 46-52.

Rochester, N., Holland, J.H., Haibt, L.H. and Duda, W.L. (1956). "Tests on a cell assembly theory of the action of the brain, using a large digital computer," *IRE Transaction on Information Theory, IT-2,* 80-93.

Rossen, M.L., Niles, L.T., Tajchman, G.N., Bush, M.A. and Anderson, J.A. (1988). "Training methods for a connectionist model of consonant-vowel syllable recognition," *IEEE International Conference on Neural networks, I,* 239-246.

Shepherd, G.M. (1988). *Neurobiology,* Oxford University Press, New York, NY.

Simpson, P. (1990). "Higher-ordered and intraconnected bidirectional associative memories," preprint. *IEEE Trans. on Systems, Man, and Cybernetics,* 637-653.

Wise, S.P. and Desimone, R. (1988). "Behavioral neurophysiology: Insights into seeing and grasping," *Science, 242,* p. 736-741.

4

NEURAL NETWORK STRUCTURES: FORM FOLLOWS FUNCTION

Alianna J. Maren

4.0 OVERVIEW

The primary method used to accomplish different tasks using neural networks is to develop the appropriate structures. Throughout this chapter, network structure is related to network function. Also delineated are areas not yet taken advantage of by designers. *Form follows function* and *biology as inspiration* are the two themes which interweave throughout this chapter.

4.1 LEVELS OF STRUCTURAL DESCRIPTION

Even though neural networks are still a new technology, they are not simple. In order to understand their structures, we have to look at them at several different levels. (See Figure 4.1.)

- *Micro-Structure.* Tbe lowest level of network structure — the neural element itself.
- *Meso-Structure.* Form related to function.
- *Macro-Structure.* No single network can accommodate complex tasks — several networks are needed.

Let's briefly examine the organization of an artificial neural network, particularly at the micro- and meso-structural levels. A typical artificial neuron is usually just a record in an array. The record stores the activation of the neuron. This activation is bounded, usually between 0 and 1 or -1 and 1. Some neurons have only two possible states, which

Neural Network Micro-Structures

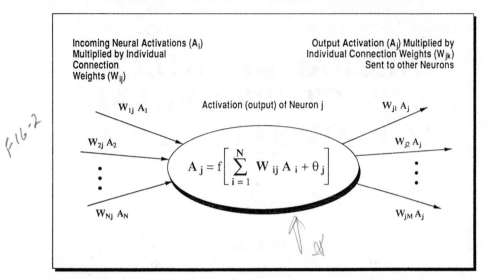

Figure 4.1   The different levels at which we can examine the structure of a neural network are its micro-, meso-, and macro-structures.

are the extreme or bounded values. Other neurons can have activations which vary between the bound limits.

The neurons are connected to each other, either positively or negatively and have a known connection strength. The values of these strengths are commonly called connection weights and are usually stored in an array. Typically, a neural network receives an input pattern, operates on it, and produces an output pattern. The input pattern is a way of selectively activating different neurons in the input layer of the network. Neurons which are activated above some defined threshold (which may vary) can output a signal to the connecting neurons. This process is usually accomplished in one cycle of network operation. The newly activated neurons can now send out their signals to the neurons to which they are connected. This process continues until a pattern of activations appears in the output layer of the network.

Although a single neuron may send out only one signal value from its end, the values which are received by the connected neurons may differ. This is because the strength of the signal which is sent out by a neuron is modified by the values of the connection weights. For example, consider the neuron that has two connections; the weights of these connections are -0.5 and 1.5. Suppose this neuron is activated, and sends out a signal of strength 1. Because the signal will be weighted (multiplied) by its connections, the signals that the two connecting neurons will receive will be -0.5 and 1.5, respectively. It is the purpose of these connection weights to send different signals to neurons; this allows neural networks to produce different and useful patterns of output in response to input stimuli.

Each layer of neurons in a network accesses and processes data simultaneously. Each operation by a layer of neurons is thought of as a cycle of operation by the network. A network with three layers would typically have three cycles of operation, accessing the input in the first layer, processing and passing the input to the second layer (resulting in a pattern distributed across the neurons in the second layer), and processing and passing this pattern to the third layer. The resulting pattern in the third layer would be thought of as the output of the neural network.

Conceptually, operations on a given layer happen in one step. However, keep in mind that all references to simultaneity need to be interpreted in terms of the implementation. A true parallel implementation can access and process data simultaneously. Most current neural networks, accessing data sequentially, are emulations running on personal computers, minicomputers, or mainframes. The concept of a single clock cycle, as though the network were implemented in parallel, is still applicable.

The input for a neural network may be analog or digital. Many interesting real-life network applications require that the network use analog input, or handle a digitized representation of multi-state or continuously variable input. If the selected network design is optimized for bi-state neurons, some preprocessing will be necessary to represent the input data in binary format.

The output of the network requires activation of one or more neurons, either for a single cycle of operation or (for more advanced networks) for several cycles. This output may be interpreted as pattern classification, as a pattern associated with the input (heteroassociation), as a completed or noise-cleaned version of the input pattern (autoassociation), or in a number of other ways.

The neurons in a neural network may be organized in a number of different ways. A typical neural network might have different layers of neurons, some which accept the input, others which process it, and yet another layer to store the output. Some neural networks combine all these functions into one or two layers. Some networks can pass information from one layer to another, but not back. Others allow backwards connections. Some networks allow information to move vertically between layers, but not between neurons in the same layer. Other networks will have one layer with lateral connections between the neurons in that single layer. Some neural networks allow a neuron to send information back to itself, in a recurrent or self-stimulating mode, others do not. The architectural structure of the network affects its performance and also the applications for which it is suited.

4.2   NEURAL MICRO-STRUCTURES

Let's summarize the operation of a typical artificial neuron, as illustrated in Figure 4.2. This neuron accepts an incoming signal or set of signals from other neurons or input devices, sums together the signals received over a single operating cycle, adds or subtracts a threshold value, and passes this summed value through an activation function or a transfer function. This yields a value which is functionally dependent on the earlier summed value. This value is called the *activation* of the neuron. If the transfer functions yielded a positive value for the neuron's activation (or a value above a certain predefined threshold), then the activated neuron outputs a signal which is gated (multiplied) by different values called weights, and sent on to other neurons.

The most significant ways in which artificial neurons differ from each other, and can be modified for new performance include: changing the transfer function, adding new

Neural Network Transfer Functions

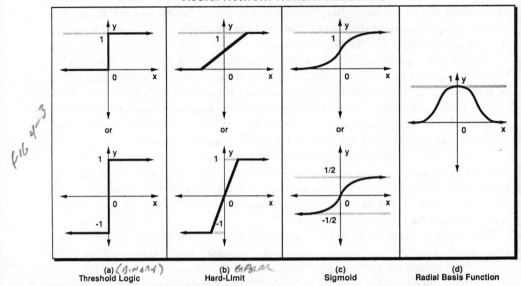

Neural Network Transfer Functions

(a) *(BINARY)*	(b) *BIPOLAR*	(c)	(d)
Threshold Logic	Hard-Limit	Sigmoid	Radial Basis Function

Figure 4.2   At the micro-structure level, neurons are typically characterized by their transfer functions, which operate on the incoming signals to produce a single output for the neuron. This output is weighted by the connection strengths to each of the individual connecting neurons, so that those neurons may each receive a different value, and may be either positive or negative, depending on the sign of the connection weight.

parameters or functions to the network (such as bias, gain, and adaptive thresholds), and tapping the potential for storing information for longer than a single cycle of operation. These choices are illustrated in Figure 4.2.

4.2.1   The Transfer Function

The most common distinguishing factor among most of the artificial neurons being used today is their transfer function. This function specifies how the neuron will scale its response to incoming signals, and produces the neuron's activation. If the activation is strong enough (i.e., passes some threshold criteria), then the artificial neuron will output a signal to the neurons to which it is connected.

Four typical transfer functions are:

- Threshold-logic nodes
- Hard-limit nodes
- Continuous-function (sigmoid) nodes
- Radial basis functions

4.2.1.1   Threshold-Logic Nodes   The easiest way to define the activation of a neuron is to consider it as being binary — either 0 or 1. Threshold-logic nodes create binary state neurons by applying a simple transfer function. If the summed input is greater than or equal to the neuron's threshold, the activation is 1; if it is less, the activation is 0. The neurons in the discrete Hopfield/Tank network are binary. Networks which have threshold-logic nodes are easy to implement in hardware (which accounts in part for the widespread early interest in the Hopfield/Tank network), but are often very limited in their capabilities.

4.2.1.2   Hard-Limit Nodes   In the hard-limit node, both an upper and a lower bound are set on the summed input from other neurons, plus the thresholds, called the total summed input. If this total summed input is less than the lower bound, the activation is defined to be 0 (or -1). If the summed input is greater than or equal to the upper bound, the activation is 1. If the summed input is somewhere between the upper and lower bounds, then the activation is defined as a linear function of the summed input. The equations below summarize the transfer function operation in a hard-limit neuron. These functions are not differentiable at transition points, which limit their use in applications which require sophisticated learning abilities.

Suppose that the upper bound on summed input is α, and the lower bound is $-\alpha$. Suppose that the upper and lower bounds on the activation are 1 and 0, respectively. Then, for a hard-limit node:

$$y = 0 \text{ for } x \ -\alpha,$$
$$y = x/(2*\alpha) +1/2 \text{ for } -\alpha \times \alpha, \text{ and}$$
$$y = 1 \text{ for } x = \alpha, \qquad\qquad\qquad \textbf{Eq. 4.1}$$

Where x is the sum of all the input activations affecting the neuron during the current cycle of operation

α is a constant

y is the output activation of the neuron

4.2.1.3   Continuous Function Nodes   One of the major reasons why work on early neural networks (e.g., the Perceptron) came to a halt was that the learning rule could not be substantially improved with the threshold-logic nodes that were used. In 1974, Paul Werbos presented a doctoral dissertation which proposed a new learning law for Perceptron-type neural networks. This was the *back-propagation* method for learning. In order to accomplish this, he had to change the transfer function of the neurons from a threshold logic function to a smoothly varying one. The reason was that his method required the differentiation of output activation as a function of its inputs. The derivative of a threshold-logic node is undefined at the transition point, and zero everywhere else. This doesn't help a learning method which uses derivatives.

Werbos suggested a transfer function which had a sigmoid shape, as shown in Figure 4.3. Although a sigmoid function has been widely used, it is not the only function which would work. Any function which is smoothly defined over the interval from minus to positive infinity or, (more practically) the interval of possible input values, and is monotonically increasing, and has both upper and lower bounds, will work just as well. For example, an arctangent function is also sometimes used. The equation for the sigmoid function is as follows:

FIG 1

**Neural networks can be described
in terms of their structures or architectures.**

• **Micro-Structure** -	What are the characteristics of each node in the network?
• **Meso-Structure** -	How is the network organized? • Number of Layers • Connection Patterns • Flow of Information
• **Macro-Structure** -	How can different networks be put together to accomplish different tasks or address large-scale problems?

Figure 4.3   The four classical types of neuron transfer function include threshold logic, hard limit, and sigmoid-type functions, and radial basis functions.

$$y = 1/(1 + \exp(-\alpha x)), \qquad y = \frac{1}{1 + e^{-\alpha x}} \qquad \text{Eq. 4.2}$$

Where x is the sum of all the input activations affecting the neuron during the current cycle of operation

α is a constant

y is the output activation of the neuron

The sigmoidal shape of the transfer function means that for most values of the total input stimulus (the independent variable), the value given by the transfer function is close to one of the asymptotic values. Usually the higher-valued asymptote is 1, and the lower one is 0 or -1. This allows the output value of the transfer function to usually be roughly grouped into one of two classes: high or low. (The interpretation of this is straightforward and similar to the previously described neural elements. For high values, the neuron fires, for low values, it doesn't fire.)

One important factor concerning the sigmoid function (or any other similarly shaped function), is that its derivative is always positive, and is close to zero for either large positive or large negative values of x. It is at its maximum value when x is 0. As we shall see later, this will prove to be important.

4.2.1.4   Radial Basis Functions   A final, and still fairly uncommon, transfer function is the radial basis function, which is typically a Gaussian function. This function is useful when creating a neural network for continuous function mappings. The centers and the widths of these functions may be adapted, which makes them a more adaptive function than sigmoid functions (which typically have an adaptive threshold or bias term added to them, resulting in a shift of the location of the "center" of the function). Mappings which may require two "hidden" layers of sigmoid function units can sometimes be accomplished by a single layer of neurons using radial basis functions. We discuss radial basis functions in more depth in Part III, in the chapter on optimizing the performance of feedforward networks.

4.2.2   Bias, Gain, and Threshold Functions

Some neural architectures specify a threshold function which is added to the summed input of signals coming into the neuron. This type of threshold is like a bias term (pre-transfer function threshold). This bias acts like another node in the layer below with a constant output, which is connected to the node of interest. It usually has an adjustable value. The imaginary node providing the bias term has no relationship to any of the nodes in the network. It provides a means of adding a constant value to the summed input, which can be used to scale the average input into a useful range. Use of a bias term is a way to take some of the burden off training the weights from the lower layer to the current node, so that they do not have to provide an appropriately scaled input value.

The usual input to a neuron is the sum of the inputs from other neurons times their connection weights. This entire value can be multiplied by a gain factor. A gain term sometimes appears in the transfer functions of certain networks (e.g., the back-propagating Perceptron). This gain is usually a fixed value. However, it can be adaptively modified in networks which use simulated condensation (a Monte Carlo-based training algorithm) or back-propagation (both described more fully in Part II). When this value is made into an adaptable parameter in back-propagation, it also affects learning rates.

An adaptable gain factor can also be applied to the output of the activated neuron, so that the neuron no longer produces a constant-valued output. This allows certain neurons to output signals that are stronger than others. This capability may be useful when a neuron learns to generalize and/or detect significant features.

When multiple adaptable parameters are used in a single neuron, the network designers need to decouple them as much as possible in order to minimize conflicts while training. For example, the input bias and input gain terms can be decoupled by having the gain applied to the summed input before the bias is added, rather than after.

4.2.3   Memory Kernels for Storing Activations

Most artificial neurons have extremely short-term memories, recalling only the input which they have received and summed during the current cycle of operation. This is clearly ineffective for certain applications, especially those in which some temporal sequences are found in the input patterns (e.g., speech recognition, processing sonar signals), or in which the network is controlling some time-varying process (e.g., robotic motion, process control). Because the temporal context of information is so important, it is useful to develop artificial neurons which have some form of memory. There are two easy ways to do this, and both are analogous to biological neuron behaviors. The first is to encode a memory function or memory kernel within the node, so that it sums information over a time interval which is greater than one cycle of operation. The second is to allow for exponential decay of previous activations. The second method can be implemented either internally, or as a recurrent connection. These methods differ in that the first sums in a temporal interval of input stimulus before the transfer function is applied, and the second sums in the post-transfer function activation from the previous cycle(s) into the current input stimulus.

4.2.4 Connection Weights with Temporally Dependent Behaviors

The previous discussion has focused on the neurons in the neural network, ignoring the potential for interesting behavior in the connection weights. Simple neural networks use simple values stored in matrix and updated via a learning algorithm for connection weights. Unless the network is actively learning, the signals passed by these weights do not change with time. When connection weights are allowed to have temporal dependencies, the entire network can exhibit interesting behaviors.

4.3 NEURAL MESO-STRUCTURES

The meso-structural level of a neural network is the level at which the physical organization and arrangement of the neurons in the network is considered. (See Figure 4.4). Meso-structural considerations are especially important in that they help us distinguish different classes or types of network architectures. These considerations help us make distinctions between the many different types and classes of networks

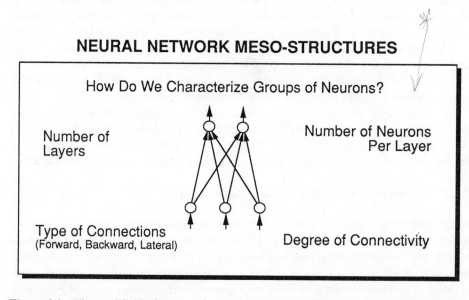

NEURAL NETWORK MESO-STRUCTURES

How Do We Characterize Groups of Neurons?

Number of Layers

Number of Neurons Per Layer

Type of Connections
(Forward, Backward, Lateral)

Degree of Connectivity

Figure 4.4 The modifiable features of a neural network at the meso-structure level include numbers and arrangements of layers, numbers of neurons per layer, and network connectivity.

which currently abound. To help bring some conceptual order into this field, we have made some basic distinctions which allow us to form categories of networks.

The first distinction is in the number of layers in a network. We classify networks as being single-layered, bilayered, or multilayered. Within this first distinction, we note that the type of connectivity allowed creates the possibility of different structures. We identify two major types of single-layered networks: those that are explicitly laterally connected, and those which have only implicit connectivity (as evidenced by a topological ordering of neural elements). Bilayered networks typically have both feedforward and feedback connections. They usually do not have lateral connections. There is a large class of multilayered networks which have strictly feedforward connections, and there are a number of multilayered networks with complex connectivities (feedforward, feedback, and lateral).

Based on these distinctions, we can identify five different structurally related classes of networks, as illustrated in Figure 4.5. These are:

- Multilayer, feedforward networks
- Single-layer, laterally-connected networks
- Single-layer, topologically ordered (vector matching) networks
- Bilayer, feedforward/feedback networks
- Multilayer, cooperative networks

Six Basic Typologies of Neural Network Meso-Structures

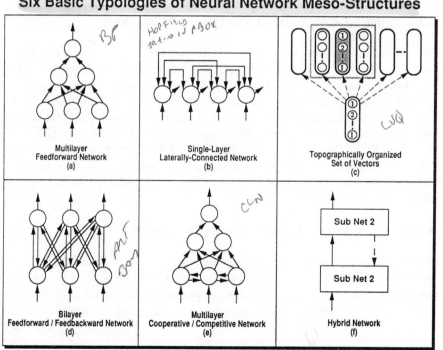

Figure 4.5   The six major types of neural network meso-structures.

The possibility of combining different architectures opens up a a sixth type of network structure, that of the hybrid network. In the following subsections, we briefly describe each of these major classes. In the next major subsection of this chapter, we consider the macro-structure of neural network systems, which involves combining these meso-structures into larger systems. Hybrid networks can be considered as borderline situation between meso- and macro-structures, and we postpone discussion to the section on macro-structures.

4.3.1 Multilayered Feedforward Neural Networks

In multilayered feedforward networks, all neural signals propagate in a "forward" direction through the network layers. There are no self-connections, lateral connections, or back-connections. Examples include the Perceptron and the back-propagating Perceptron network. (In back-propagation networks, even though the results of forward signal passing may be used to correct weights of previous neurons, the operations of the net are strictly feedforward.)

Feedforward networks are typically the network of choice for pattern classification applications. Some feedforward networks can learn to generalize about important distinguishing characteristics of their input patterns. Their ability to do this depends on the learning algorithm and specific architecture of the network. Feedforward networks, or networks with more than two layers, are the optimal choice when generalization as well as pattern recognition is desired.

4.3.2 Single-Layer, Laterally-Connected Networks

The second most widely-known type of network architecture uses a single layer of laterally-connected neurons. Because this network has only one layer, it can activate only one pattern at a time. The lateral (and sometimes recurrent) connections cause different patterns to appear in the single layer with each cycle of operation. Laterally-connected networks are typically used for pattern autoassociation. Autoassociative networks can store many patterns, but can only manifest one at a time. They are good for regenerating clean versions of patterns they have learned when they are given a noisy or incomplete pattern as a starting point. The Hopfield/Tank network and Anderson's Brain-State-in-a-Box are examples of single-layer, laterally-connected autoassociative networks.

4.3.3 Single-Layer, Topographically-Ordered Networks

Networks based on a single layer of topologically-ordered vectors form another class of networks, one which is growing in importance. These networks do not have explicit connections as are found in the other network classes. During learning, a measure of the vector distance between the different vector neurodes is used to adjust their relative positions in vector space. This class of network encompasses the Learning Vector Quantization and self-organizing Topology-Preserving Map networks, both developed by Teuvo Kohonen.

4.3.4   Bilayer Feedforward/Feedback Networks (Associative/Resonating Networks)

Some two-layer networks pass information backward as well as forward during processing. To do this, they have to have feedback connections as well as feedforward between the neural elements. This means that these networks will have two sets of connection weights, one going from the first layer to the second, and the other connecting the second layer back to the first. The values stored in these matrices are typically different, and not just the transposition of each other. This type of network structure is particularly good for associating a pattern in the first layer with another pattern in the second layer, which is called pattern heteroassociation. They can also be used for pattern classifiction.

The most recent networks of this type to emerge involve a dynamic called *resonance*, in which the patterns in the first and second layers repeatedly stimulate each other until the pattern in each layer settles to a stable state. This dynamic provides a way to get improved recall of patterns stored in each layer. The two most popular two-layer, feedforward/feedback networks are the Carpenter/Grossberg Adaptive Resonance Theory (ART) network and Kosko's Bidirectional Associative Memory (BAM) type of network.

4.3.5   Multilayer Cooperative/Competitive Networks

Some of the networks which we put into the *feedforward/feedback* category have lateral connections. The lateral connections are specifically designed so that the cooperative (excitatory, or positive) connections and the competitive (inhibitory, or negative connections) balance each other in a certain way. The special design considerations which go into making these networks causes us to label them *cooperative/competitive networks*. The cooperative processes used in these networks are often designed to explicitly mimic certain biological networks which have inspired their design. For example, Steve Grossberg and his colleages have developed a Boundary Contour System, which is a layered set of nets which finds the boundary line segmentations in an image. The architecture of this system is a deliberate emulation of biological vision processing, and has several performance characteristics which are very much like those of biological vision. Cooperative/competitive networks allow complex and subtle processes to be carried out.

4.3.6   Other Considerations

Other architectural considerations may affect the way a network operates. These include the design of input and output representations.

- Type of input. Some networks can work only with binary signals, others can work with real-valued or analog input. Some networks are inherently suited for input which can be represented as a one-dimensional list of input values, where others are well-suited for dealing with two (or higher) dimensional inputs. Some can handle only static inputs, others can work with time-varying sequences of information.

- Type of output. Some networks output only a classification of the input data, others can yield output patterns in response to input category to which the input belongs, or may create completions of the input. Some networks are associative, so that presenting such a network with a given input pattern yields a totally different output pattern. Some networks produce corrections or adjustments in an ongoing process. These distinctions can affect the designer's choice of a network.

4.4   THE MACRO-STRUCTURE

Sometimes, no known neural network architecture is adequate for a complex problem or task. In this case, we may wish to develop a system of interacting neural networks. This brings us to the consideration of neural network macro-structures — the design of systems of networks. There are two types of macro-structures: strongly coupled and loosely coupled networks. Strongly coupled networks can be treated as a single network, and are created by fusing together two or more networks into a single new structure. This process usually produces a new hybrid network. Loosely coupled structures connect networks which retain their structural distinctness. This approach allows us to create systems of interacting networks which can solve more challenging problems.

The goal in creating a strongly coupled hybrid network is to create a new network type which highlights the strengths and minimizes the weaknesses of the combined network systems. Examples of strongly coupled hybrid networks include the Hamming network and the counter-propagation network. Although these networks have some usefulness, the more exciting and potentially useful advances are being made in the area of creating interacting systems of networks.

True network macro-structures involve two or more interacting networks. The primary concern in designing a macro-system is to understand and decompose the applications problem in a modular way, so that the various tasks can be partitioned and their interactions isolated. Major concerns in designing such a system of networks include: specifying the number, type, and size of the component networks, and specifying their connection patterns. (See Figure 4.6)

4.5   SUMMARY

It is possible to conceptually organize the rapidly evolving field of neural networks by thinking of networks at three different levels of detail: micro-structure, meso-structure, and macro-structure. At the lowest level of detail (micro-structure), we consider the characteristics of individual neurons. The most important consideration is usually the transfer function for a neuron, which is typically threshold logic, hard-limit, or sigmoid.

At the meso-structural level, we consider the design and classification of individual neural networks. There are five major types of network classes: multilayer feedforward networks, single-layer autoassociation networks, single-layer vector matching networks, bilayer feedforward-feedback networks, and multilayer cooperative networks.

Neural Network Macro-Structures

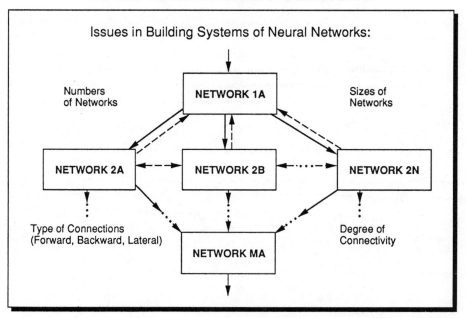

Figure 4.6   The modifiable features of a system of neural networks at the macro-structure level include the number and types of networks, the size of each network, and the connection configuration and dynamics between the different networks.

Hybrids are a possible sixth class. Each of these networks serves a different type of purpose. Feedforward networks are often used for pattern classification. Single-layer laterally connected networks are ideal for autoassociation. Feedforward-feedback networks are useful for heteroassociation, or associating one pattern with another. Multi-layer, cooperative/competitive networks are often useful for very sophisticated processes in which the interrelation of different types of information helps in making some sort of decision.

Sometimes, no single network can solve a tough problem. In that case, a system of interacting networks must be created. This is work at the macro-structural level. There are different types of macro-structures: those with tightly coupled known networks producing a new hybrid network, and systems created with two or more loosely interconnected networks.

5

DYNAMICS OF NEURAL NETWORK OPERATIONS

Alianna J. Maren

5.0 OVERVIEW

Network dynamics are important because they specify how a network will perform during its operation. Two examples, a back-propagating Perceptron and a Hopfield network, illustrate the differences in dynamics between two popular neural networks. For networks such as the Hopfield network (and many others), the stability of network dynamics is very important. Taken together, network structures and dynamics allow us to specify the functional abilities of a neural network, which relate to its applications potential.

5.1 TYPICAL NETWORK DYNAMICS

Every neural network will have a specific set of network dynamics, or way in which it processes data. These dynamics are related to the structure of a network. For example, in a multilayer feedforward network, all of the flow of information is forward, from one layer of neurons to another. There is no lateral communication, nor is there feedback information flow. In a laterally-connected network, there is only a single layer. Information flows from each neuron to the other in this layer. Unlike a feedforward network, in which information flows through a given neuron only once for a given task, the process in a laterally-connected (or a bilayer resonating) network will repeat several times.

This brings up a major issue in network dynamics — that of network stability. In a feedforward network, such as the back-propagating Perceptron, stability is not an issue. That is because information flows from the first layer through the middle ones to the last layer, and the process stops (for that particular calculation). In a laterally-connected network (Hopfield or Brain-State-in-a-Box) or a bilayer resonating network (Bidirectional Associative Memory or Adaptive Resonance Theory network), the process iterates several times. Theoretically, it would be possible for a process to be never-ending. Thus, when a network has the potential for recurrent processing, network stability (or convergence to a stable solution) is a very important concern.

When we consider network dynamics converging to a stable state, that is distinct from the learning process in which the connection weights converge to stable values. The first issue concerns stability of network dynamics; the second concerns stability in network learning. In this chapter, we focus on stability of network dynamics. For feedforward networks and some others, convergence of weight connection values to a stable state is an important issue. We address that issue elsewhere in this book.

Let's begin our consideration of network dynamics by examining two typical networks, the back-propagating Perceptron and the Hopfield network.

5.1.1   A Back-Propagation Network Example

The back-propagating Perceptron has the micro- and meso-structures shown in Figure 5.1. In operation, all information flows forward from the input neurons to the neurons in the hidden or middle layer(s), and from there to the output neurons. No information is passed backward (or back-propagated) during actual network operation; the back-propagation refers strictly to the learning stage [Werbos, 1974].

As an illustration, let's take a simple back-propagation network which has been configured to solve a specific problem; the X-OR (exclusive-OR) problem [Rumelhart et al., 1986]. This means that patterns of (0,0) or (1,1) should produce a value close to zero in the output node (the top node of the network), and input patterns of (1,0) or (0,1) should produce a value near one in the output node. Finding a set of connection weights for this task is not easy; it requires application of the back-propagation learning rule for (usually) several thousand iterations to achieve a good set of connection weights and neuron thresholds.

The Back-Propagation Network

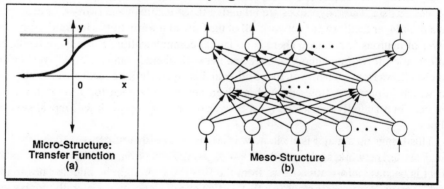

Figure 5.1   The architecture of the back-propagating Perceptron. (a) The "sigmoid" transfer function is used at the micro-structural (neuronal) level. (b) The meso-structure of the back-propagation network is the same as the meso-structure of the multilayered Perceptron.

A Back-Propagation Network to Solve the "X-Or" Problem

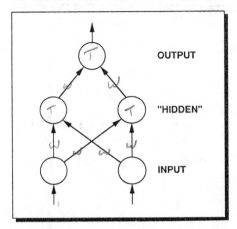

Figure 5.2   A simple back-propagation network architecture for solving the X-OR problem. The inputs will be 0's or 1's in either of the input nodes, and we want the output to be 1 if the input pattern is (0,1) or (1,0), and 0 if the input pattern is either (0,0) or (1,1). In addition to adapting the connection weights, the thresholds for activating both the hidden and output nodes can also undergo adaptation, again using the back-propagation learning rule to find optimal values.

The basic architecture for this problem has two input nodes, two hidden nodes, and a single output node, as shown in Figure 5.2. The structure shown here also has variable thresholds on the two hidden and one output node. This means that there are a total of nine variables in the system: four weights connecting the input to hidden nodes, two weights connecting the hidden to output nodes, and three thresholds. (The back-propagation network will also work when just the connection weights are varied.) The weight values could be very hard to come by unless some learning algorithm (e.g., the back-propagation method) is used to drive the collection of weights toward a useful set of values.

Just for fun, and as an example of how challenging this seemingly easy problem is, try to come up on your own with some possible combinations of weights that will make this network produce the desired classifications for the X-OR inputs.

If you took the time to try this, even for ten or fifteen minutes, you probably agreed that it would be a good idea to have a *learning law* to find these weights. We discuss the back-propagation learning law in Part II. Figure 5.3 shows a set of weight and threshold values for this network which was obtained using this law. Let's use them to illustrate how this network works.

Suppose we put in a pattern, say (0,1). That mean that there is 0 activation in the left-hand neuron on the first layer and an activation value of 1 in the neuron on the right.

Now, we move our attention to the next layer up. For each neuron in this layer, we calculate an input which is the weighted sum of all the activations from the first layer. The weighted sum is achieved by vector multiplying the activations in the first layer by

**A Back-Propagation Network with
Connection Weights and Thresholds
Adapted to Solve the "X-Or" Problem**

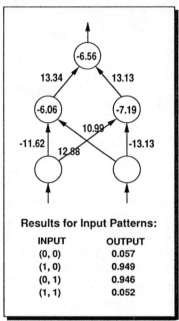

Results for Input Patterns:

INPUT	OUTPUT
(0, 0)	0.057
(1, 0)	0.949
(0, 1)	0.946
(1, 1)	0.052

Figure 5.3   An example set of weights and thresholds for this X-OR configuration. There are several "sets" of similar weights and thresholds that will work, and infinitely many values within each set.

a "connection weight matrix". For this specific example, we get a value of 0*(-11.62)+1*(10.99) = 10.99 for the neuron on the left in the second layer, and 0*(12.88)+1*(-13.13) = -13.13 for the neuron on the right.

These are not the activation of these neurons, though. To obtain the activations, we add a threshold value (which is found for each neuron using the back-propagation rule), and apply a transfer function. The transfer function is defined for each different network. In the case of the simple back-propagating Perceptron, it is the sigmoid function shown in Figure 5.1.

In the specific example we have shown, the activation of the neuron on the left side of the hidden (middle) layer is the transfer function applied to the difference (10.99-6.06), or 4.94. Applying the transfer function yields an activation value close to 1. The activation of the neuron on the right is the transfer function applied to (-13.13+7.19), or -5.14. Applying the transfer function yields a value close to 0.

Approximating the next step, we use a value of 1 for the activation of the neuron on the left, and 0 for the neuron on the right, multiply each activation by its appropriate connection weight, and sum the values as input to the topmost neuron. This is approximately 1*(13.34)+0*(13.13) = 13.34. We add the threshold of -6.56 to obtain a value of 6.78. Applying the transfer function to it will yield a value close to 1 (0.946), which

The Discrete and Continuous Hopfield Network

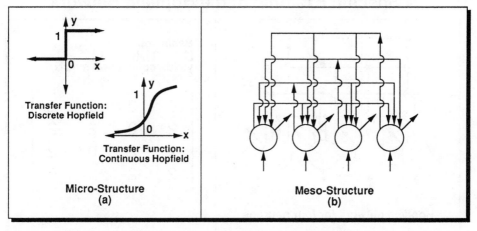

Micro-Structure
(a)

Meso-Structure
(b)

Figure 5.4   The architecture of the Hopfield network. (a) At the micro-structural level, the discrete transfer function is simple bi-state. (b) At the meso-structural level, the network is completely laterally connected, but has no recurrent (self) connections.

is the desired result. Using the other three binary input patterns, we can similarly show that this network yields the desired classification within an acceptable tolerance.

In general, our formula for the activation of the jth neuron in a layer is:

$$y\_activ_j = f(w_{ij} \ input\_activ_i + theta_j) \qquad \textbf{Eq. 5.1}$$

Where y_activ_j is the activation of the jth neuron in the receiving layer

    f is the transfer function
    w_{ij} is the connection weight between the ith neuron in the sending layer and the
        jth neuron in the receiving layer
    $input\_activ_i$ is the input to the ith neuron in the sending layer
    $theta_j$ is the threshold of the jth neuron in the receiving layer

5.1.2   A Hopfield Network Example

The Hopfield network is very different from the back-propagation network [Hopfield and Tank, 1986; Tank and Hopfield, 1974]. Its laterally-connected structure is shown in Figure 5.4. Unlike the back-propagation network, a single neuron can be activated many times. In fact, the rule for operating a Hopfield network is that successive neuronal activations don't stop until the network pattern stabilizes; that is, all of the values stored in each neuron remain constant, even after processing inputs from the other neurons.

Consider a specific Hopfield network configured to store two orthogonal patterns, (1 1 0 0) and (0 0 1 1). This network is illustrated in Figure 5.5, along with the symmetric connection weight matrix associated with this network. The rule for the dynamic processes of the Hopfield network is that we input a pattern to the network, select a

A Specific Example of the Hopfield Network

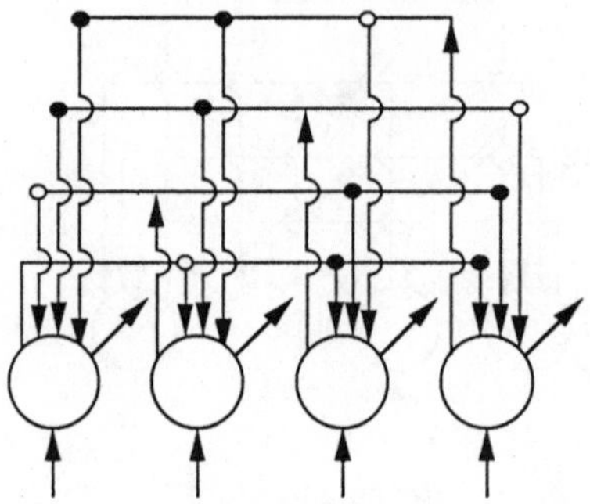

Basis set:
[1 1 0 0], [0 0 1 1]
yields connection matrix:

$$\begin{bmatrix} 0 & 2 & -2 & -2 \\ 2 & 0 & -2 & -2 \\ -2 & -2 & 0 & 2 \\ -2 & -2 & 2 & 0 \end{bmatrix}$$

Hopfield network for four neurons
(a)

(b)

Figure 5.5　An example of a specific Hopfield network for four neurons. The symmetric connection weight matrix, shown as a matrix on the right, is graphically illustrated on the left, where dark circles indicate negative values for connection weights, and open circles indicate positive values.

neuron at random, and operate on that neuron by replacing its value with that of the sum of all of its different neuron activations multiplied by their corresponding weight values. We then apply the binary-valued transfer function to this sum. We must keep note of whether or not the new value is different from the old. We continue this process, selecting neurons at random, until none of the neuronal values change.

To make this illustration easier, we will carry out these operations in batch mode, by multiplying the input vector by the connection weight matrix. For each neuron, or element in the input vector, the transfer function indicates that if a resulting value is negative, we reset it to 0 (since the inputs are bi-state 0 and 1). If the resulting value is 0, we leave it at 0. If the resulting value is positive, we reset it to 1.

Figure 5.6 (a and b) illustrates how either of the initial patterns, when given to the network, produce themselves. In these two cases, the autoassociative process correctly regenerates the two original patterns — when they are presented as input in clean, uncorrupted form. Figure 5.6 (c) illustrates how a corrupted version of the second original pattern (the second bit is changed from 0 to 1) yields a totally spurious result — (0 0 0 0). This happens often with the simple Hopfield network. In Part II, we discuss some ways to overcome this deficiency.

5.2　ENERGY SURFACES AND STABILITY CRITERION

In networks such as the Hopfield network, the Brain-State-in-a-Box network, and the Bidirectional Associative Memory (BAM) network, dynamic stability is a very important consideration. In order to have confidence in a network's performance, we need

Examples of Hopfield Network Autoassociative Recall

$$[1\ 1\ 0\ 0] \begin{bmatrix} 0 & 2 & -2 & -2 \\ 2 & 0 & -2 & -2 \\ -2 & -2 & 0 & 2 \\ -2 & -2 & 2 & 0 \end{bmatrix} = [2\ 2\ -4\ -4] \Rightarrow [1\ 1\ 0\ 0]$$

stable recall
of stored pattern

(a)

$$[0\ 0\ 1\ 1] \begin{bmatrix} 0 & 2 & -2 & -2 \\ 2 & 0 & -2 & -2 \\ -2 & -2 & 0 & 2 \\ -2 & -2 & 2 & 0 \end{bmatrix} = [-4\ -4\ 2\ 2] \Rightarrow [0\ 0\ 1\ 1]$$

stable recall
of stored pattern

(b)

$$[0\ 1\ 1\ 1] \begin{bmatrix} 0 & 2 & -2 & -2 \\ 2 & 0 & -2 & -2 \\ -2 & -2 & 0 & 2 \\ -2 & -2 & 2 & 0 \end{bmatrix} = [-2\ -4\ 0\ 0] \Rightarrow [0\ 0\ 0\ 0]$$

new pattern

$$[0\ 0\ 0\ 0] \begin{bmatrix} 0 & 2 & -2 & -2 \\ 2 & 0 & -2 & -2 \\ -2 & -2 & 0 & 2 \\ -2 & -2 & 2 & 0 \end{bmatrix} = [0\ 0\ 0\ 0] \Rightarrow [0\ 0\ 0\ 0]$$

stable recall
of spurious
pattern

(c)

Figure 5.6   (a) and (b). When the two original vectors are input into Hopfield matrix, the connection strengths lead to complete recovery of the initial vectors in an autoassociative process. However, the Hopfield network is particularly susceptible to noise, as is illustrated in (c), in which a degraded version of the second original pattern leads to a spurious result.

assurance that the network will converge to a solution when given an input pattern. (Whether or not the solution is correct is another issue.) This concern about the stability of network dynamics has prompted a number of theoretical considerations.

Most of the investigations into network stability are based on an energy function approach to modelling network dynamics. In this context, the Lyapunov Theorem has proved to be very useful. The essence of this theorem is that if we can define an energy function for a network, and if this function meets certain criteria, then the system will be globally stable. This means that whatever its starting point, it will converge during its processing to a stable state. These stable states (there may be several of them for a given network configuration) correspond to minima in the energy surface.

The Lyapunov Theorem states that if we can define a continuous vector function $f(\vec{u})$ and a scalar function $E(\vec{u})$, then $E(\vec{u})$ is a Lyapunov function if the following conditions are true:

- $E(\vec{u}) = 0$ for all values of $\vec{u}$
- If $\vec{u}$ (the vector norm of $\vec{u}$) approaches infinity, then $E(\vec{u})$ approaches infinity
- The derivative of $E(\vec{uL}) \neq 0$ *for all values of* $\vec{u}$
- The derivative of $E(\vec{u}) = 0$ for all singular points of $f(\vec{u})$

If these conditions are met, the energy equation can then be called a Lyapunov function. The system is then globally stable, which means that its dynamic processes will converge.

The challenge in using the Lyapunov function approach is to find an energy function which corresponds to the actual nature of a network. This is not an easy thing to do. John Hopfield made a significant contribution when he developed an energy state equation for a collection of neural-like elements, because his energy equation was a Lyapunov function, and because it was based not only on the properties of a collection of simplified neurons, but could also be realized in hardware [Hopfield, 1982, 1984]. Cohen and Grossberg [1983] have developed a Lyapunov function which establishes the basis for the global stability of the Adaptive Resonance Theory network. Kosko (1988) has published a Lyapunov function which establishes stability of his Bidirectional Associative Memory (BAM) network. Cohen and Grossberg, in particular, have shown how the dynamic equations of several other well-known neural networks (e.g., the Brain-State-in-a-Box) can be recast in the formalism of their equations, thus demonstrating that those networks are also globally stable. All of these equations are discussed in some detail in Appendix A.

When we use the energy surface approach to modeling, the network dynamics correspond to traversing the energy surface from a high point (an input configuration) to a low point (the final pattern configuration). This is illustrated in Figure 5.7, which shows the energy surface as a function of two variables. (Since the dimensionality of an energy surface corresponds to the dimensionality of the number of variables in the system, this illustrative surface has a lower dimensionality than the four-dimensional surface which would be needed to illustrate the simple Hopfield example given earlier, and has a lower dimensionality than any practical neural network system.) As can be seen in this example, there are two local minima. Suppose the starting point (initial configuration) is the one marked with an "x" in the figure. The system will follow a path to the nearest (most accessible) energy minimum. In the illustration shown, this will not be the global minimum for the surface. However, many network dynamics (such as the Hopfield network) can allow the network to become stabilized in a local

Energy Surface "Contour Map"

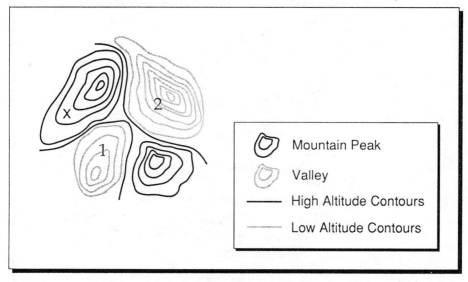

Figure 5.7   An illustration of an energy surface which contains two local minima, one of which is lower than the other. Most network dynamics would cause an energy configuration initialized at "x" to converge to point 1, and not reach the deeper minimum at point 2.

minimum. There is substantial research underway centering on finding ways for networks to avoid falling into local minima. The work on the Boltzmann machine is an example of such an approach, as are the modifications to the Hopfield network which we discuss in Part II.

5.3   NETWORK STRUCTURES AND DYNAMICS

Figure 5.8 illustrates how we can create a taxonomy of network types which are related by their structures and dynamics. This is especially useful because we can make a direct correlation between the structures and dynamics of networks to their functional abilities, and hence to their different applications possibilities.

Beginning at the left side of Figure 5.8, we note that the single-layer networks are good for autoassociation and the bilayer and multilayer networks are good for heteroassociation. This makes sense, since a single-layer network can store at any moment a single pattern. For a fully laterally-connected, single-layer network, the dynamics of operation change on the original input pattern until it reflects a pattern corresponding to one of the energy minima. If this resulting energy-minimum pattern matches one of the those stored in the network as a template, we have achieved autoassociation. In the case of the vector-matching networks (such as the Learning Vector Quantization network), an input vector is matched to the closest of the stored vectors, again resulting in a form of autoassociation. (This can also be used for data compression, as in a codebook.)

A Topological Organization of Major Networks

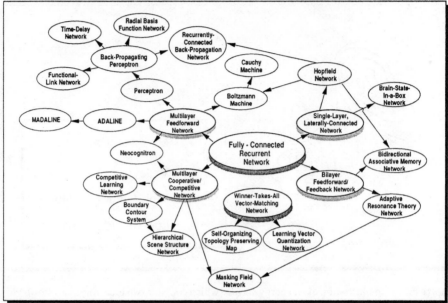

Figure 5.8   The different types of networks can be grouped by similarities in their structure and dynamics. The structures and dynamics of different classes of networks influence their functional abilities, which allows them to be used for similar applications. See text for further explanation.

The bilayer networks immediately suggest that two patterns can be correlated with each other, yielding pattern heteroassociation. Pattern association and pattern recognition are facilitated by multilevel networks, in which the middle layers act as feature extractors and generalizers. Networks such as the back-propagation network and the Boltzmann machine have at least one middle layer which is a feature generalization layer. This layer gives them powerful pattern recognition, signal filtering, and data compression abilities. The Boltzmann machine learns by a statistical method which traverses an energy surface, allowing it to function as an optimizer as well as a pattern associator.

Networks such as the Neocognitron have multiple layers with complex connection patterns. The combination of lateral connectivity with feedforward information flow (and sometimes feedback flow) allows them to take pattern recognition to a very high level. The Neocognitron, for example, can recognize hand-drawn characters. Its complex multilayer structure allows it to recognize patterns, even with shift, rotation, and size variance.

Figure 5.9 illustrates how the functional abilities of different networks corrollate to their structures and dynamics.

Network Structures and Dynamics Relate to Their Functional Abilities and Applications Uses

STRUCTURE:	Single Layer, Laterally-Connected	Topological Arrangement of Vectors	Bilayer, Feedforward/ Feedback	Multilayer, Feedforward	Multilayer, Feedforward &/or Feedback &/or Laterally-Connected
DYNAMICS:	Recurrent *FEEDBACK*	Vector Matching	Recurrent	Feedforward	Cooperative / Competitive
NETWORK TYPE:	Hopfield Brain-State-in-a-Box	Adaptive Vector Quantization Learning Vector Quantization Self-Organizing Topology - Preserving Map	Bidirectional Associative Memory Adaptive Resonance Theory	Basic Perceptron ADALINE / MADALINE Back-Propagating Perceptron Boltzmann Machine Functional-Link Net	Competitive Learning Net Masking Field Neocognitron Boundary Contour System Hierarchical Scene Structure
APPLICATIONS USES:	Autoassociation Optimization (Hopfield)	Autoassociation Data Compression Optimization (LVQ)	Heteroassociation Pattern Recognition	Heteroassociation Pattern Recognition (spatial, temporal, & spatio-temporal) Data Compression Signal Filtering Optimization (Boltzmann Machine) Image Processing	Pattern Recognition (spatial, temporal, & spatio-temporal) Image Processing

Figure 5.9 Corrollation of structures, dynamics, and applications for major classes of networks.

REFERENCES

Cohen, M., and Grossberg, S. (1983). "Absolute stability of global pattern formation and parallel memory storage by competitive neural networks," *IEEE Trans. Systems, Man, and Cybernetics, SMC-13*, 815-825.

Hopfield, J. J. (April, 1982). "Neural networks and physical systems with emergent collective computational abilities," *Proc. Nat'l. Acad. Sci. USA, 79*, 2554-2558.

Hopfield, J. J. (May 1984). "Neurons with graded response have collective computational properties like those of two-state neurons," *Proc. Nat'l. Acad. Sci. USA, 81*, 3088-3092.

Hopfield, J.J., and Tank, D.W. (1986). "Computing with neural circuits: A model," *Science, 233*, 625-633.

Kosko, B. (1988). "Feedback stability and unsupervised learning," *Proc. Second IEEE Int'l. Conf. on Neural Networks* (San Diego, CA; July 24-27, 1988), I-141–I-142.

Rumelhart, D.E., Hinton, G.E., and Williams, R.J. (1986). "Learning internal representations by error propagation." In D.E. Rumelhart, J.L. McClelland, and the PDP Research Group (Eds.), *Parallel Distributed Processing, Vol. I.*,, MIT Press, Cambridge, MA, 318-362.

Tank, D., and Hopfield, J. (1987). "Collective computation in neuronlike circuits," *Scientific American, 257, 6*, 104-114.

Werbos, P. (1974) *Beyond Regression: New Tools for Prediction and Analysis in the Behavioral Sciences*. Ph.D. dissertation, Harvard University, Cambridge, MA.

6

LEARNING BACKGROUND FOR NEURAL NETWORKS

Craig T. Harston

6.0 OVERVIEW

There are two topics relevant to learning and neural network technology. One is about learning techniques used in the neural network field, and the other is a review of psychological learning theory and associated behavioral technology. Both topics of network and behavioral learning are important. Interestingly, these two fields are only vaguely similar; however, the field of neural network learning could gain much from the behavioral/psychological learning field.

Psychology is a broad field with diverse theoretical points of view. Simply put, the field is dichotomized with cognitive approaches at one end and behavioristic technology at the other end. The field of neural networks encompasses contributions from both of these different views. This chapter reviews theory and techniques from the behavioristic point of view. Behavioral techniques can and have been applied to the neural network field [Wieland and Leighton, 1988; Harston and Martinez, 1988]. In the future, we expect behaviorism to contribute useful albeit basic concepts and techniques to computerized neural networks. The cognitive approach is also important and involves many researchers, however, it is beyond the scope of this chapter. Other resources may be helpful [Churchland, 1986; Simon, 1987; Edelman, 1989].

The relevance of behavioristic training techniques becomes evident during computer neural network training. When simulated neural networks are trained one trial at a time as directed by a person, the computer system responds much like an animal does when trained by behavioristic technology. Behaviorists use a technique called shaping to progressively teach an animal to do tricks. Our experience has shown that computerized neuronal systems can also be taught through similar progressive training. Not only does such a computer system learn, but it appears that more difficult problems and a larger quantity of problems can be taught to the network through shaping procedures. During such training experiences, we have noticed that neural network computer systems will forget in a manner similar to animals. That is, they seem to forget when they learn new material, but they quickly respond appropriately when reminded with just one training trial. It is significant that the computer system learns to generalize and even progress in a way similar to animals through behavioristic training. Such experiences have been published [Wieland and Leighton, 1988; Harston and Martinez, 1988]. It should not surprise us that computer simulations, because they are based on the operations of the brain, behave similarly.

Behavioristic learning techniques and theory are important for computer simulated nervous systems. Psychological learning technology can be used for the development and training of neural network simulations. With this important relationship in mind, we have included a primer of behavioristic technology. It is hoped that these concepts will be directly useful to your work with functioning neural networks.

6.1 INTELLIGENCE: AN OPERATIONAL DEFINITION

Intelligent systems are defined by what they do. Learning is an intelligent process. Trick horses appear to count by pawing the ground. In fact, they do not count. They have learned to watch the trainer and stop counting when the trainer shows subtle signs of relief. Some trainers may be unaware that the horse has not learned to count but has learned to watch the trainer. Regardless of what the horse learns, the learning indicates intelligence. While humans and animals show various degrees of intelligence by what they do and learn, machines can also be intelligent. Computer chess games must be intelligent because they play the game with considerable skill. Some computer programs play the role of the psychologist by asking everyday questions. They may ask, "How are you today?" or "How do you feel about that?". If the conversation involves certain key words such as "depressed," "mother" or "sex," the computer will use these words in a question. A degree of intelligence is demonstrated by these computer programs.

Computers have achieved success in the professional arena, being applied to such applications as: diagnosing medical problems, maintaining locomotive engines, running a canning machine, and advising oil drilling crews. These computer systems have been programmed to use rules and facts to make conclusions. They use the same rules used by the experts to do the job. These systems are called expert systems [Feigenbaum, McCordic and Nii, 1990]. Their performance indicates a useful degree of intelligence.

Though learning is an unique and important indication of intelligence, rarely has it been achieved through computer programming. Successful learning programs have been developed by simulating the biological model of the brain. This is the field called neural networks.

6.2 LEARNING AND CONDITIONING

Learning can be defined as a change in behavior due to training procedures. This is a before and after process, where performance is measured before and after training. The difference indicates how much is learned. Behaviorists often refer to learning as conditioning. Behavioral definitions of learning similar to this one can be studied in further depth in Kimble's revision of Hilgard and Marquis's (1961) book *Conditioning and Learning*.

Behaviorists describe two types of conditioning. The first is classical conditioning which was established by the Russian physiologist, Pavlov. The second type was developed as the law of effect by Thorndike. The law of effect is often called operant or instrumental conditioning. These two types of conditioning seem to be fundamentally different as discussed below.

6.2.1 Classical Conditioning

Classical conditioning was first described by Pavlov and later by Bekhterev. Pavlov presented dogs with pairs of stimuli which became associated. Following the association, the dogs responded to both stimuli in similar ways. This type of learning is illustrated in the well-known example of the dogs' salivation in response to a ringing bell. Pavlov measured the salivation of dogs when given meat powder. When the dogs were given the powder, they produced saliva. This response is a normal reflex which is called the *unconditioned response* (UCR). The meat powder is called the **unconditioned stimulus (UCS)**. During conditioning, Pavlov rang a bell just prior to giving the dog the meat powder. After this pair of stimuli were given together, the dog would salivate when the bell was rung as well as when the meat powder was given. The bell is called the *conditioned stimulus* (CS) and salivation to the bell is the *conditioned response* (CR). This is conditioning, and the classical example of learning (see Figure 6.1).

The learning is due to pairing the conditioned and unconditioned stimuli. The association of these two stimuli is so strong that they both cause the response of salivation. Learning can be defined as an association of two stimuli. This definition of learning suggests how to simulate learning with computers. Extensive computer simulation of the classical model was done by A. Harry Klopf [1987].

6.2.2 Operant Conditioning

Another type of learning was published by Thorndike called the Law of Effect. Others have called it instrumental or operant conditioning. B. F. Skinner was a strong proponent of this type of learning and tried to explain all of human behavior with this concept. He popularized this topic through widespread publications [Skinner, 1971]. Operant learning involves three components: feedback, training based on previously known responses, and shaping or progressive training.

Instrumental learning is dependent on feedback. When an animal does something correctly, it receives a reward. If it does something incorrectly, it is punished. Behaviorists refer to rewards as reinforcements. These are stimuli which increase the frequency of a given behavior. On the other hand, punishment is something which decreases the probability of the behavior. Reinforcement and punishment serve as feedback and are the basis for learning (see Figure 6.2).

Shaping, the behavioral training technique which progressively teaches new responses, is accomplished by reinforcing increasingly more approximate behaviors. Note that there is a backwards order of the training procedure. The last behavioral require-

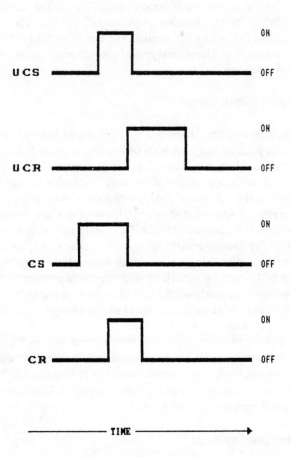

Figure 6.1 Classical conditioning paradigm. The UCS instigates a reflexive UCR. When the CS is presented just prior to the UCS, a conditioned response (CR) is learned.

ment is shaped or taught first. After the last behavior is taught, the next to the last behavior is taught. As each trick is learned, the previous behavioral requirement is taught until the entire sequence of movements is learned. Complex behaviors are taught with this technique. (Notice that operant learning is based on previously known responses.)

6.2.3 Implications for Computers

The three components of operant conditioning have important implications for computer simulated learning. Neural network technology also uses feedback as a learning tool. Often this takes the form of difference learning. As seen later in this book, some

	POSITIVE	NEGATIVE
REINFORCEMENT	* Present reward * Performance up * Obtain goal	* Withdraw pain or threat * Performance up * Escape/avoidance
PUNISHMENT	* Apply threat/pain * Performance down * Avoid response	* Remove reward * Performance down * Extinction

Figure 6.2   Operant conditioning paradigm. Actions, results and goals grouped by combinations of presentations and removals with the resulting operant behavior. Reinforcement results with increased behavior, and punishment with decreased frequency of behavior.

neural networks learn by using the desired response as feedback. Comparisons between the computed response and the correct response becomes the basis for learning. The difference between the two is fed back into the system and the computer system is adjusted toward making the correct response.

The concept of reinforcement is used in some neural networks to indicate when the computer neural network connections should be updated. This kind of reinforcement occurs when two events occur together and must be associated with one another. The association is formed because network connections which are active are updated or strengthened at the same time. This sets up the network to respond when either of the two original events are active [Kosko, 1989; Klopf, 1987].

The second component is that learning must start with known behaviors. This suggests that an animal cannot learn to respond with behavior which is not in its behavioral repertoire. This is obvious and does not present difficult problems for computer learning. It is true that the computer system will not be able to do anything which is not part of its mechanisms. More importantly, it does suggest that learning can be associated with previously acquired response mechanisms. Possibly newly learned responses should be built upon systems which are well established. For animals, learning is not an arbitrary process. Associations are not random, but new material is attached or associated with mechanisms which are important and meaningful to the animal. This is germane to computerized systems. Possibly neural network systems should build one function based on another which is more fundamental. This would be similar to biological systems, so highly intelligent systems would be nothing more than a coordinated conglomerate of limited functional subsystems.

Progressive training, the third component, appears to make learning more efficient. This process, called shaping, allows animals to learn more quickly. Also animals learn more complex behaviors which could not be learned otherwise. There are important implications for computer learning. Shaping, as used with animals, cannot be used in computer learning because, unlike animals, computer systems do not emit spontaneous responses to be reinforced. If shaping cannot be used, then something similar will be helpful. Progressive training embodies the essential elements of shaping and has proven to be useful. Wieland and Leighton [1988] used word shaping to describe progressive schedules to accelerate learning. They found that progressive learning allowed their neural network to learn four times faster. This procedure was most helpful when learning the most difficult discriminations. Similar research confirmed these conclusions [Harston and Martinez, 1988].

6.2.4 Neural Mechanisms of Learning

There are two types of learning: operant and classical. This suggests that there may be two mechanisms of learning in the brain. However, only one mechanism of learning by the nervous system has been postulated. That is Hebb's rule [Hebb, 1949]. Details of Hebb's rule are presented in the chapter on the nervous system. If there is only one mechanism of learning in the brain, then there may be a single learning rule. Can the two types of learning be reduced to a single rule? Is there one learning rule which explains both operant and classical conditioning? Is one type of learning a variation of the other?

It may not be possible to resolve this theoretical argument. However, the implications are important for computer simulations of learning. If there is one mechanism for learning in the brain, then simulations must allow for all the variations of learning paradigms. On the other hand, if different types of learning suggest different mechanisms, additional learning algorithms are possible for simulation.

For now, there are two types of learning at the psychological level. Fittingly, there are two general approaches to computer simulated learning which appear to be related. Classical conditioning is an associative type. Direct association of input is found in BAM (i.e., Bidirectional Associative Memory) network [Kosko, 1989]. Operant conditioning is based on feedback. Neural network feedback exists in the form of the difference between the output and the correct response. This type of feedback is found in neural networks based on the Generalized Delta Rule [Werbos, 1974, 1989; Rumelhart, Hinton and Williams, 1987]. The feedback is used to make adjustments backwards through the layers of the network. These learning algorithms are discussed in more detail later in the book. Other general learning algorithms may appear, but for now this distinction is useful.

The neural network field includes other learning algorithms. The harmony theory and Boltzmann machines employ simulated annealing [McClelland and Rumelhart, 1986]. This approach borrows from the laws of physics. The relationship to traditional animal learning models or neurons may be analogous but not directly obvious. Unsupervised learning is called *competitive learning.* Here input patterns compete within interrelated neuron layers to determine a winner. Its similarity to animal learning is not obvious, however, similar data processing appears to be common in the nervous system.

6.3   LEARNED PERFORMANCE

Learned behavior does not occur in circumstances isolated to the learning environment. Learning and the performance of learned behaviors are embedded in a host of everyday situations. The dog may learn to sit in the training laboratory, but he also can perform the sitting trick at the dog show. The situation may be different but the dog can sit on the appropriate command. This phenomenon is called *generalization.* Generalization is also the ability to respond appropriately when the command or input is noisy. There may be extraneous input or noise, but the essential message comes through. The response is to the message not the irrelevant noise. Additionally, generalization enables the correct response to different input which have similarities to the training stimuli. Learning is related to a pattern within the total input presentation. However there must be some consistencies in the pattern which are constant and the essence of what is learned. A major neural network researcher, Stephen Grossberg, calls these consistencies the *invariants of the pattern.*

Programming a computer to generalize with traditional computer techniques has been difficult. Noise (i.e., extraneous input) has a devastating effect on a computer's performance. Background noise has a major detrimental effect on computerized voice recognition. Slight changes in the input data often keep the computer from making the desired response. Matching patterns or pictures is almost impossible to do unless the pictures are identical.

One of the most important features of neural network technology is generalization. Many neural networks are relatively insensitive to noisy input. Extraneous bits on an input pattern are not problematic. Appropriate results are often the outcome even with slightly different input patterns. As long as the general pattern is present, the network can function suitably. The ability to generalize makes neural network technology applicable to everyday noisy situations.

6.3.1   Discrimination

Pigeons can be taught to peck a key when a light is green and not peck the key when the light is red. This is discrimination. The ability to differentiate when a response is or is not appropriate is central to learning. Discrimination training is accomplished in associative (classical conditioning) learning by pairing different signals to those learning trials with reinforcements and without these stimuli. For example, one signal, a CS or green light, is presented with the UCS, meat powder, and another signal, a red light, is presented when the UCS is not presented. In this case the animal learns to respond with a CR, salivation, to the green light, but it also learns something in the case with the red light. Notice that the animal learns not to respond or salivate to stimuli (e.g., red light) not associated with the UCS (see Figure 6.3).

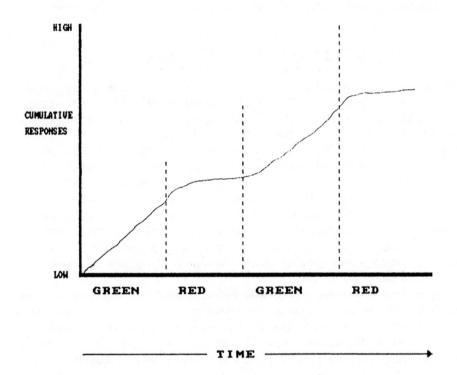

Figure 6.3   Discrimination between red and green stimuli. This discrimination results in high levels of behavior during the positive discriminative stimuli and little behavior during negative discriminative stimuli.

Similar results occur with operant learning. If reinforcement is available on trials with the green light signal, animals will learn to respond quickly. On the other hand, when reinforcement is not available with that signal, or with a different signal (e.g., red light) animals will respond slowly, if at all. The stimulus (i.e., green light) used in the reinforced trials is called the positive discriminative stimulus. The negative discriminative stimulus, red light, is associated with no reinforcements.

Discrimination training can be tricky. For example, a monkey can be trained to press a bar for food when a green light is on, and not press the bar when there is a red light. Afterward, we test the discrimination by using different lights of similar colors and watching the response of the monkey. If the training was successful, then the monkey will respond when green lights are on but not respond to red lights. Sometimes the animal does not respond as anticipated, but not because the monkey did not learn. It is possible that the monkey learned to respond to the brightest light, not to the green color. If the training lights were different brightness as well as color, then the monkey could have learned to respond to the difference in brightness, not color. By training with

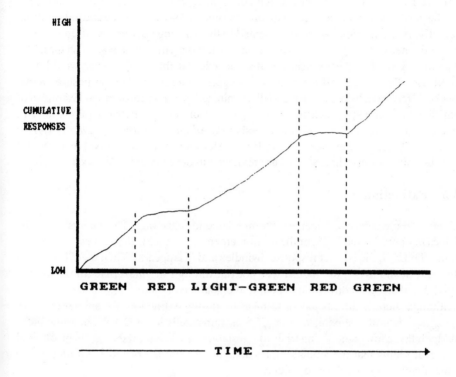

Figure 6.4   Generalization.

different lights which might vary in brightness and color the animal will respond on the basis of brightness, not color. In this case, the monkey might learn to generalize colors and discriminate between brightness.

Correct training to color, but not brightness, can be done by controlling the brightness as well as color of the training lights. If the red and green training lights are matched for brightness, then the animal can not learn to respond on the basis of brightness. The monkey has only one clue or difference to learn, and that is color. Another training trick would be to use red and green training lights with different intensities. This way, the monkey learns to respond to a green light of any intensity, but will not respond to red lights regardless of the brightness. The discrimination is to color and the generalization is to brightness (see Figure 6.4).

Training animals to learn the correct object or concept must be done carefully and with considerable control. This appears to be true for computerized neural networks as well. If a network must learn the difference between two lines on a graph, then training data must represent the material to be learned. Different lines on a graph may differ in a number of ways, such as shape and position. If shape was the desired concept to be learned by the network, then the training data should consist of different-shaped curves. Additionally, the position of the curves on the graph should be carefully controlled. There are two control techniques which could be used. The different training curves could be positioned at the same level to force the network to learn shape and not position. Alternatively, each shape could be presented to the learning network at different levels to force the network to generalize over position and discriminate between shapes.

Neural networks learn to discriminate, and without this ability, they would have limited use. This ability allows them to categorize and is the key purpose of many networks. Networks have been used to discriminate between sets of input. Examples of room discriminations, speech recognition, verbal or reading networks are found in Rumelhart and McClelland (1986). Grossberg's *Adaptive Resonance Theroy* (ART) network categorizes letters [Grossberg, 1988]. All these applications involve the neural networks ability to discriminate and are reviewed in detail later in this book.

6.3.2   Extinction

Extinction is the behavioral technical term related to forgetting. This occurs when the CS does not precede the UCS, or the reinforcement is withheld. Under these conditions without the UCS, the correct response dwindles and disappears. When there is no food at the end of the maze, the mouse will not run through the maze. He might wander around, but there is no direct movement to the goal (see Figure 6.5).

Although there is little behavior to measure during extinction, the animal does not forget. If the reinforcement, or the UCS is reinstated, learned behavior reappears quickly. It takes fewer retraining trials to re-establish the behavior than with the original training. It seems that the animal retains something from the original learning due to the fact that he can be retrained quickly.

Another type of forgetting is described when humans learn literature. Humans can forget a previously learned item if they must learn similar material just before or after its learning. This is called *proactive and retroactive inhibition.*

There are parallels to extinction and forgetting in neural network learning. When a back propagation neural network is taught too many things, it can forget simple problems previously learned. This seems similar to the retroactive inhibition from a human learning literature. Similar to relearning extinct behavior, forgotten problems are quickly retrained with the neural network. Often the retraining can be done in one trial (observations from the author's network research). These features are similar to the forgetting/extinction behavior of psychology.

6.4   MOTIVATION

Other aspects of animal behavior are important. Animals pay attention to special items in the environment, yet they ignore many things. Goal-oriented behavior is common place. What do these things have to do with computer simulations of the brain? Are they relevant to computer processing? These issues are briefly examined below.

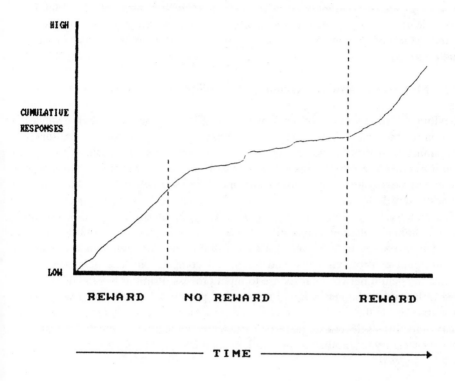

Figure 6.5   Extinction.

6.4.1   Attention/Orienting Behavior

A loud noise causes the cat to jump and look toward the noise. This is an orienting response common to all animals. If the noise is unimportant, the cat may go back to sleep. The cat may scramble and run away, if the noise came from a dog. The cat pays attention to stimuli in the environment such as dogs, birds and cat food. Yet the animal ignores things like flowers, music and breezes through the trees. Clearly the orienting response helps animals survive by ensuring a quick response to meaningful stimuli in the environment.

6.4.2   Habituation/Adaptation

Learning to ignore irrelevant things in the world is a large part of what humans and animals do. It may not seem like much, but think of all the things you do not pay attention to. For example, have you been thinking of the paint on the wall? What is the temperature of your feet? Can you feel the presence of your clothing? It seems that people are unaware of their surroundings. Not because they cannot pay attention to them, but they have learned not to pay attention. Many of these things are unimportant.

Learning to not pay attention is called *habituation* or *adaptation*. It is essential that animals learn not to pay attention to unimportant things. So insignificant items are ignored. Much of the lower brain activity is involved in controlling what we pay attention to.

6.4.3 Motivation — Goal-Oriented Behavior

Behavioral learning and performance is not an arbitrary process. Animal behavior is based on the biological system. The biological system functions to continue survival. Maintaining homeostasis and propagation are relevant. There are basic skills such as breathing and eating. Other skills are learned or developed. For example, the cat learns to hunt and the dog learns to do tricks. Advanced tricks or behaviors are built on prior abilities or development.

These motivational concepts are important for computerized simulations of brain activity. Brain simulations could utilize these attentional aspects to control computer performance. For example, Steven Grossberg (1988) uses four attentional mechanisms: priming, gain control, vigilance and internodal competition. The gain mechanism controls the neural network's response to input process. Priming is a top down sensitizing technique so the network will receive and pay attention to input. The vigilance factor autoadjusts the sensitivity to control the coarseness of the categorization. Internodal competition selects a single response to patters of stimuli. These mechanisms are deemed necessary to stabilize the network in complex input environments.

6.5 SUMMARY

This chapter examines learning techniques used for computer simulations of neural networks. In general, there is considerable similarity between the behavioral and computerized technologies. One can make comparisons with and draw from classic theories of learning: classical and operant conditioning. Each theory utilizes learning methods that are applicable to the development of neural network systems. The basics of behavioral technology have been reviewed and may prove useful in developing learning methods for artificial neural systems.

REFERENCES

Churchland, P.S. (1986). *Neurophilosophy Toward a Unified Science of the Mind-Brain*, MIT Press, Cambridge, MA.

Edelman, G.M. (1989). *The Remembered Present A Biological Theory of Consciousness*, New York, Basic Books.

Feigenbaum, McCordic, and Nii. (1990). *Rise of the Expert Company*, Vintage Books, New York, NY.

Grossberg, S. (Editor). (1988). *Neural Networks and Natural Intelligence*, MIT Press, Cambridge, MA.

Harston, C.T. and Martinez, O. (1988). "Shaping versus random training of simple and difficult problems on a pack propagation neural network," *Mid SE ACM FAll Conference*, Gatlinberg.

Hebb. D.O. (1949). *The organization of behavior*, Wiley, New York, NY.

Kimble, G. (1961). *Hilgard and Marquis's Conditioning and Learning*, Appleton Century, New York, NY.

Klopf, A.H. (1987). "A neuronal model of Classical Conditioning," *AFWAL-TR-87-1139*, Air Force Wright Aeronautical Laboratories.

Kosko, B. (1989). "Unsupervised learning in noise," *Proc. First Intl. Joint Conference on Neural Networks*, Washington, D.C., June 18-22, 1989, I. 7-17.

Simon, H.A. (1987). "The evolution of cognitive science," *Naval Research Reviews*, 23-30.

Skinner, B.F. (1971). *Beyond Freedom and Dignity*, Bantam/Vintage Books, New York, NY.

Rumelhart, D.E., McClelland, J.L. and the PDP Research Group. (Editors), (1987). *Parallel Distributed Processing*, MIT Press, Cambridge, MA, 318-362.

Rumelhart, D.E., Hinton, G.E. and Williams, R.J. (1987). "Learning internal representations by error propagation," *Parallel Distributed Processing*, Rumelhart, D.E., McClelland, J.L. and the PDP Research Group (Eds.), MIT Press, Cambridge, MA, 318-362.

Werbos, P. (1974). *Beyond Regression: New Tools for Prediction and Analysis in the Behavioral Sciences*, Ph.D. Dissertation, Harvard University, Cambridge, MA.

Werbos, P. (1989). "Maximizing long-term gas industry profits in two minutes in Lotus using neural network methods," *IEEE Trans. on Systems, Man, and Cybernetics*, (March-April), 315-333.

Wieland, A. and Leighton, R. (1988). "Shaping schedules as a method for accelerated learning," *Abstracts of the First Intl. Neural Networking Conference*, 231.

7

MULTILAYER FEEDFORWARD NEURAL NETWORKS I: DELTA RULE LEARNING

Alianna J. Maren

7.0 OVERVIEW

In this chapter and the following one, we will examine a group of neural networks which have a similar architecture and purpose. This is the group of feedforward networks, which are characterized by layered architectures and strictly feedforward connections between the neurons — no lateral, self-, or back-connections are allowed. These networks are all good pattern classifiers, and all undergo supervised learning. This means that they are taught to classify input into one of several *a priori* categories. This group includes the Perceptron, the ADALINE and MADALINE networks, the back-propagation network (also known as the back-propagating Perceptron), the Boltzmann machine, and the Cauchy Machine.

These networks differ primarily in the way in which they learn. The Perceptron, the ADALINE and MADALINE, and the back-propagating Perceptron all have somewhat similar learning laws. We consider them together in this chapter. The Boltzmann machine, and its related derivative, the Cauchy machine, have an entirely different historical antecedent, and a different basis for their learning laws. Their learning method draws on an optimization approach which comes from the statistical modelling of thermodynamic processes. We consider these two networks in the following chapter.

7.1 INTRODUCTION

There are many different types of neural networks, each of which has a different architecture, learning method, and performance capabilities. Often, the structure and dynamics of a network correspond readily to the network's function. In this chapter, we consider a class of multilayered, feedforward networks which are good for pattern recognition, signal filtering, data compression, and heteroassociative pattern matching. These include the Perceptron, the ADALINE and MADALINE, and the back-propagating Perceptron.

The Perceptron and the ADALINE and MADALINE networks are of substantial historical interest, and have paved the way for the development of other neural networks. The back-propagation network is probably the one most commonly used today, with the Boltzmann machine yielding similar performance and rivaling it for some applications.

7.2 THE PERCEPTRON NETWORK

The Perceptron, developed by F. Rosenblatt [Rosenblatt, 1958], was the first neural network to emerge. It substantiated the concept of the artificial neuron which is still used today, where each neuron computes a weighted sum of its inputs, and passes this sum into a non-linear thresholding function. The original Perceptron model was strictly feedforward and could be extended up through multiple levels. The learning algorithm for the Perceptron allowed it to distinguish between classes of input if they were linearly separable in terms of some decision space.

7.2.1 Multilayer Perceptron Architectures

Most of the networks which we will consider in this chapter have the same or very similar meso-structures. They differ in their micro-structures and in their learning laws.

The micro-structure for the Perceptron is the design of the neuron. The same neural design is used throughout. (This holds for most networks we will consider in this book; only a few have more than one different type of neuron per network.) In the case of the Perceptron, each neuron sums up the inputs times their weights, and subtracts a fixed threshold value. A binary (1, -1) transfer function is applied to this result. Figure 7.1 shows the architecture of a basic Perceptron.

At the meso-structural level, each Perceptron contains at least two layers, and could have many more. The Perceptron (and all other networks we will investigate in this chapter) needs at least two nodes on the first and following layers, and at least one node on its last layer.

7.2.2 The Perceptron Dynamics

The dynamics of the Perceptron set the stage for the dynamics of all other networks considered in this chapter, and also for most networks covered in this book. In the case of the Perceptron, the dynamics are quite simple. We begin by presenting a pattern. This means that if the input layer for the Perceptron is n units long, we input a vector of length n; one value for each neuron in the input layer.

Now, we move our attention to the next layer up. For each neuron in this layer, we calculate an input which is the weighted sum of all the activations from the first layer.

The Perceptron Network

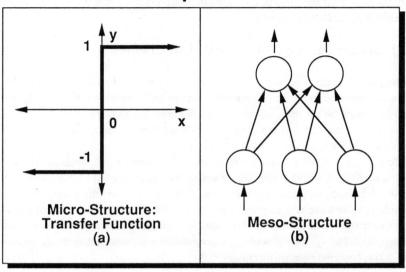

**Micro-Structure:
Transfer Function
(a)**

**Meso-Structure
(b)**

Figure 7.1   The basic Perceptron architecture. (a) At the micro-structural level, the neurons use the hard-limit transfer function. (b) At the meso-structural level, the typical Perceptron has two or more layers with feedforward connections.

The weighted sum is achieved by vector multiplying the activations in the first layer by a connection weight matrix. This is not the activation of this neuron, though. To obtain the activation, we subtract a threshold value (which is fixed for each neuron), and apply a transfer function. The transfer function is defined for each different network. In the case of the Perceptron, it is the binary, or threshold-logic transfer function shown in Figure 7.1.

Thus, the activation of the jth neuron in the second layer would be

$$y\_activ_j = f(\ w_{ij} \ input\_activ_i - \theta_j]$$ **Eq. 7.1**

Where y_activ_j is the activation of the jth neuron in the second layer

f is the transfer function, w_{ij} is the connection weight between the ith neuron in
the input layer and the jth neuron in the second layer
$input\_activ_i$ is the input to the ith neuron in the input layer
θ_j is the threshold of the jth neuron in the second layer

If more layers than one are used, this process is repeated, with each layer acting as the *input layer* to the layer just above it. As a result of differences among the values of the connection weights, different inputs can produce different outputs in the topmost layer.

Typically, in a Perceptron (and more especially for the other networks in this chapter), the connections move from a given layer to the one immediately above it.

There is nothing inherently wrong with skipping a layer, but our current learning laws do not make it easy for us to modify weights that skip layers. Thus, we stay with a simple layer-to-layer connectivity pattern.

7.2.3   Learning Through Adaptive Weight Changes

The Perceptron (along with all the other networks we will discuss in this chapter) undergoes supervised learning. This means that in order to train the network, a set of training data must be assembled, priorities established, and training patterns defined. We must be certain that our training data contains good *exemplars* of each of the categories which we wish our network to learn.

The Perceptron, as well as all other networks which undergo supervised learning, are taught in a separate training stage before actual use. Each time we train a network, we usually begin by assigning random initial values to all the connection weights. Usually, these random values are within some limits, such as -3 to +3. Then, we present each example from the training set, and use both the output that we want the network to produce (target_activ), along with the output the network actually does produce (output_activ) to generate a difference, Delta.

$$\text{Delta}_{(j)} = \text{target\_activ}_{(j)} - \text{output\_activ}_{(j)} \qquad \textbf{Eq. 7.2}$$

By applying the learning rule for our network type to this Delta, we change the connection weights in the network, and sometimes other modifiable parameters as well. Too great a change in any one weight is never made, because we want the weights to take on values that will allow the optimum overall response to all of the patterns in the training set. This means that no one pattern is allowed to influence a given weight too much. So each learning rule typically involves an equation for changing a connection weight by some scalar (constant or variable) α, times some function of the Delta.

There is one other factor involved in adjusting a connection weight. This relates to the Hebbian learning law discussed in Part I. The idea is that we want to make changes in the connection weight strengths which are proportional to the activation of the neuron making the connection. If the neuron is very weakly activated by an input pattern, we don't want it to cause a great change in the connection weights to the neurons in the next layer. On the other hand, a very strong input value should exert a very strong effect on the connection weight if the connection weight needs to be changed. (This argument is stronger in the case of the back-propagating Perceptron, which we will cover later in this chapter. All of the neurons in the back-propagation network can have a continuous range of activations, instead of just -1 or 1.)

In the case of the Perceptron, the learning rule is very simple. We change each connection weight by an amount equal to α times Delta times the output activation of the neuron i connecting the neurons.

$$w_{(ij, \text{ new})} = w_{(i,j \text{ old})} + \alpha \text{ input\_activ}_{(i)} \text{ Delta}_{(j)} \qquad \textbf{Eq. 7.3}$$

Linearly Separable Decision Spaces & Corresponding Perceptrons to Solve Them

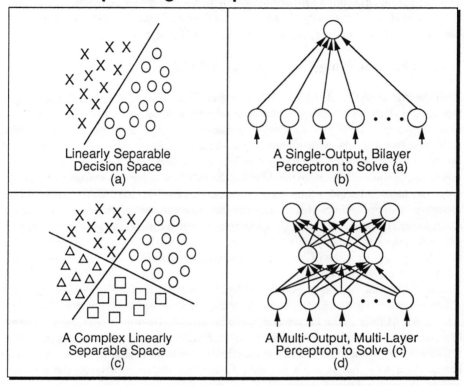

Linearly Separable
Decision Space
(a)

A Single-Output, Bilayer
Perceptron to Solve (a)
(b)

A Complex Linearly
Separable Space
(c)

A Multi-Output, Multi-Layer
Perceptron to Solve (c)
(d)

Figure 7.2   (a) A linearly separable decision space. (b) An architecture for a Perceptron which can solve this type of decision boundary problem. An "on" response in the output node represents one class, and an "off" response represents the other. (c) A more complex, yet still linearly separable decision space. (d) A Perceptron architecture which can learn to encode this space, but additional layers are required.

Note: if $Delta_{(j)}$ was 0 (no error in the output), the connection weight will not change. Otherwise, the connection weight will increase if the output was too low, and will decrease if the output was too high.

7.2.4   The Perceptron's Performance: Capabilities and Limitations

The basic capability of the Perceptron is that it can distinguish between linearly separable decision spaces, as shown in Figure 7.2. Each output corresponds to a binary decision about one of the linearly separable classes. The greatest weakness of the Perceptron is its limitation to this type of decision boundary. Later networks (such as the back-propagating Perceptron and the Boltzmann machine) used different learning rules to adjust their connection weights so that networks with three or more layers can solve for complex (non-linear) classification boundaries.

7.3   ADALINE AND MADALINE NEURAL NETWORKS

The ADALINE (*ADA*ptive *LIN*ear *E*lement) network was developed by Bernie Widrow at Stanford University shortly after Rosenblatt developed the Perceptron. The capabilities of this network are very similar to those of the Perceptron.

7.3.1   The Underlying Concept of the ADALINE

Both the ADALINE and the MADALINE (for *M*any *ADALINE*s) employ a slightly more sophisticated learning procedure than the Perceptron called the (Widrow-Hoff) Least-Mean-Squared (LMS) learning rule. This rule is also sometimes called the Delta Rule, because it works with minimizing a *delta* or *difference* between the observed output of the end neurons and the desired output of those same neurons.

The key distinction between the ADALINE and a two-layer Perceptron with a single output node is the basis for their learning laws. This basis is that the ADALINE learns using a LMS error reduction law to reduce the error between its actual output and the desired output, and a Perceptron uses a simple weighted difference. However, since the gradient of the LMS error yields the same equation as the Perceptron learning law, this is really just a difference of origination or conceptualization. The formulation of the learning laws is the same.

A real difference comes about in the way the learning laws are applied. The weight change for learning (the Delta) is calculated using different values in the Perceptron and the ADALINE. In the Perceptron, Delta is calculated using the real (binary) output of the node and the desired output. In the ADALINE, Delta is calculated using the neural activation before the binary transfer function is applied. This allows the weights to change in a way that is more sensitive to the real distance between the neuron's activation and the desired output.

7.3.2   The ADALINE and MADALINE Architectures

The ADALINE and MADALINE architectures are essentially the same as the Perceptron. The few differences are at the micro-structural level where they add a variable threshold instead of subtracting it (but since the thresholds can be positive or negative, this is just a difference in writing down the equations, not in any real performance). They both have binary transfer functions applied to the sum of the inputs times their weights plus the threshold value, yielding (1, -1) values for the nodes.

At the meso-structural level, the basic Perceptron and ADALINE/MADALINE architectures are identical. The ADALINE is limited to a single output node, and the MADALINE can have many. They are each two-layer networks. The ADALINE and MADALINE architectures are illustrated in Figure 7.3.

7.3.3   The Dynamics of ADALINE and MADALINE Networks

The dynamics of the ADALINE and MADALINE are the same as for the Perceptron.

The ADALINE & MADALINE Networks

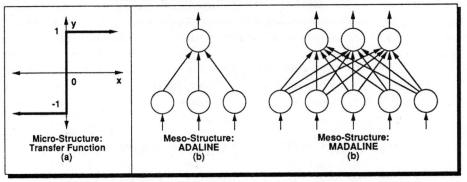

Figure 7.3   The ADALINE/MADALINE architecture. (a) At the micro-structural level, the neurons in this system also use the hard-limit transfer function. (b). The ADALINE has the same meso-structure as a Perceptron with a single output node. A MADALINE is a set of superimposed ADALINEs, where the same input nodes can feed to different outputs.

7.3.4   The Least-Mean-Squared Rule: The Learning Law for the ADALINE

Following the presentation of a pattern, and passing signals through the network in the usual manner, we can observe the actual output value at each node in the output layer and compare it with the targeted or desired value. The LMS error E is defined as:

$$E = \frac{1}{2} \sum_{j=i}^{N} (\text{target\_activ} - \text{output\_activ})^2$$

Eq. 7.4

Where Target_activ is the targeted activation at output node j,

Output_activ is the actual activation at output node j, and
n is the total number of nodes in the output layer.

In this case, E represents the total error observed over the entire layer of output nodes. It is the error which we wish to reduce, or minimize. To minimize this error, we take the derivative. This yields a scalar function of the same equation used for the Perceptron learning law, Eqn. 7.3. Again, note that the only difference is that we apply the learning law to a delta taken before the binary transfer function is applied in the case of the ADALINE/MADALINE networks, and after the transfer function has been applied in the case of the Perceptron.

7.3.5   Applications of the ADALINE and MADALINE

As adaptive filters, ADALINEs have been and are continuing to be used for multiple purposes. Applications include use as adaptive equalization filters in high-speed modems, and adaptive echo cancelers for filtering out echo in long distance telephone and satellite communications. Other applications include noise canceling and signal prediction. An example of a noise canceling application is to cancel maternal heartbeat in fetal electrocardiograph recordings. Combinations of ADALINEs into MADALINEs, using majority vote-takers, creates the possiblity of systems to descramble 2-D images. MADALINEs have some use for shift and rotation-invarient pattern recognition. See Widrow and Winter [1990], and Widrow, Winter and Baxter [1989] for a complete discussion of ADALINE and MADALINE operations and applications.

7.4   THE BACK-PROPAGATION NETWORK

The back-propagation network is probably the most well-known and widely used option among the current types of neural network systems available. This network is an outgrowth of earlier work on Perceptrons, with the addition of a *hidden layer* and use of the *Generalized Delta Rule* for learning. In essence, the back-propagation network is a Perceptron with a different transfer function in the artificial neuron and a more robust and capable learning rule.

We might note in passing that the term *back-propagation* refers to the training method by which the connection weights (and sometimes the node activation thresholds) of the network are adjusted. During operation, all information flow is feedforward, as with the other networks in this chapter.

Actually, two network models, the back-propagation network and the Boltzmann machine, came into prominence during 1986 [Rumelhart et al., 1986; Segnowski, 1986]. They share performance capabilities, and spearheaded the recent emergence of neural networks as an exciting new means of pattern recognition. Both networks typically require long periods of training in order to learn their pattern classes. The training sets for these networks have had to be presented many times (between 100–10,000 times) in order for the interconnection weights between the neurons to settle into a stable pattern which will correctly classify input patterns.

While both networks typically have high performance when given patterns similar to those they have learned, they do not have the ability to recognize new categories of patterns. (This is true of all networks which undergo supervised learning.) In order to be able to recognize and classify a new type of pattern, they need to be given examples of the new pattern type along with the corresponding category identification. Also, since training a new pattern involves changing already-stabilized interneuron connection weights, the network must also be retrained on previously known categories. This enables the new connection weights to yield correct decisions for all categories. The need for lengthy training (and retraining for new categories) is one of the major limitations of these two networks.

In the following subsections, we review and discuss the tasks at which the back-propagation method excels. We examine the micro-structure (especially the transfer function) and the meso-structure of this network. We describe the learning rule for back-propagation, and explore ways of adapting each of these components for either specialized applications or improved performance. We also identify the limitations of simple back-propagation, and applications for which it is not well-suited. In the remaining section of this chapter, we similarly address the Boltzmann machine. Later chapters discuss alternative networks which could be used when back-propagation or the Boltzmann Machine is not the preferred choice.

7.4.1   The Underlying Concept for the Back-Propagation Network

The key distinguishing characteristic of the back-propagation network is that it forms a mapping from a set of input stimuli to a set of output nodes using features extracted from the input pattern. This network can be designed and trained to accomplish a wide variety of mappings, some of which are very complex. This is because the nodes in the hidden layer(s) of the network learn to respond to features found in the input.

By features, we mean the correlation of activity among different input nodes. As examples, suppose that the input layer for a back-propagation network is a two-dimensional array of binary neurons. Features in this array could mean a pattern of horizontal or vertical node activations, or a corner junction between two line segments, or even pairs of activated neurons on opposite sides of the input array. These relationships between the activations of different input neurons provide a basis for a higher-level, more abstract representation of the input information in the second layer.

Because nodes in the back-propagation network learn to respond to features as the network is trained with different examples, the network develops the ability to generalize. For example, a back-propagation network can learn to distinguish between straight, concave, and convex curved lines — even if the lines to be tested occur in different locations in the input array than those used for training [Dietz et al., 1989]. In either of these last two cases, the network would probably respond correctly even if presented with a pattern which it has never seen before. The ability to make such complex distinctions, even when the presented pattern is different from those on which the network was trained, is due to the feature-detection and generalization abilities which are *trained* into the middle or hidden layer nodes.

In order for a back-propagation network to be successful for applications, the key issue is that the hidden layer nodes must be trained to recognize the right sets of features. (Since in practice there can be many right sets, we will say that they must learn to recognize an appropriate and sufficient set of features.) These features must be sufficiently general, so that the network can respond correctly, even when its input is different from those it has previously encountered.

7.4.2   The Architecture of the Back-Propagating Perceptron

The architecture of the back-propagation network is the same as that of the multilayer Perceptron. There are a few key distinctions that we make.

7.4.2.1   The Micro-Structure of the Back-Propagating Perceptron   The back-propagation learning law requires that the transfer function for each node be defined by a continuous function. This function should be asymptotic for both infinitely large positive and negative values of the independent variable (typically, the sum of the inputs). These conditions usually lead to a modified S (or sigmoid) shape for the transfer function.

The use of this kind of transfer function is one of the major differences between the back-propagation network and its predecessors, the Perceptron and the ADALINE. Each of these earlier networks used nodes with simpler transfer functions, and this limited their ability to be useful on the more complex pattern recognition problems.

The sigmoidal shape of the transfer function means that for most values of x (the independent variable, which here is sum of inputs), the value given by the transfer function is close to one of the asymptotic values. Usually the higher-valued asymptote is 1, and the lower one is 0 or -1. This allows the output value of the transfer function to usually be roughly grouped into one of two classes: high or low.

One important factor about the sigmoid function (or any other similarly shaped function) is that its derivative is always positive, and is close to 0 for either large positive or large negative values of x. Figure 7.4 illustrates both the sigmoid function and its derivative. The derivative has its maximum value when x is 0. This is important in helping the back-propagation learning law work effectively. This is because when we work out the equations of the learning law (given in Appendix A.1), we find that the changes made to the weights are proportional to the derivative of the activation. If this derivative is near 0, then the changes are small. This is desirable, because the derivative is near 0 when the activation value is near 0 or 1 — one of the two stable states. When the activation of the neuron is in the middle range, we want to change the neuron's output a good deal, driving it to produce a value near one of the (0 or 1-valued) stable states. The derivative is largest when the activation is in this middle range, and so the change in the weights (proportional to the derivative) is also fairly large. Thus, the transfer function not only gives us smooth and differentiable behavior, it actually helps the learning law work the way we want it to.

Note that as the activation of a neuron approaches either zero or one, the derivative approaches zero. (See Figure 7.4.) Because the learning or weight change is proportional to this derivative, a weight may change more slowly than desired for neurons which yield either a large or small activation. This can cause difficulties in training the network.

7.4.2.2   The Meso-Structure of the Back-Propagating Perceptron   The back-propagation network requires at least three layers of nodes, whereas the Perceptron and the ADALINE could each be instantiated with two layers. The layers for a back-propagation network are typically referred to as input, hidden or middle, and output. This architecture is shown in Figure 7.5. The hidden nodes are crucial in allowing the back-propagation network to extract features and generalize. The connections between the nodes are again like the Perceptron architecture: only feedforward connections are allowed. Further, all connections are typically between adjacent layers; the input layer nodes connect only to those of the hidden layer, and the hidden layer nodes connect only to the output layer. Usually, back-propagation networks are fully connected.

7.4.3   The Dynamics of the Back-Propagation Network

The back-propagation network, like all others in this chapter, has separate stages for learning and operation. Once the network has been trained, the learning process is stopped, and connection weights are fixed.

The "Sigmoidal" Transfer Function is Continuously Differentiable

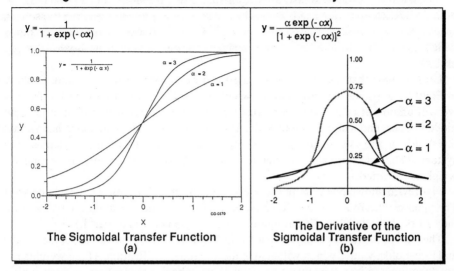

Figure 7.4   (a) The "sigmoid" transfer function, used by neurons in the back-propagating Perceptron (the back-propagation network), is continuous, monotonically increasing, and approaches asymptotes at both postive and negative infinity. (b) The derivative of this function yields a maximum when the function value crosses the 0 x-axis. It approaches 0 as the x-values approach either positive or negative infinity. This has important implications in creating an effective learning law.

In operation, the back-propagation works just like the Perceptron. No information is passed backward (or back-propagated) during actual network operation; the back-propagation refers strictly to the learning stage.

The Back-Propagation Network

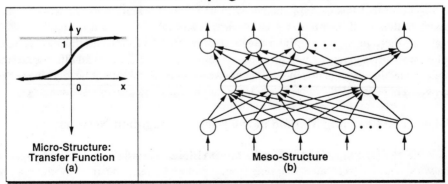

Figure 7.5   The architecture of the back-propagating Perceptron. (a) The "sigmoid" transfer function is used at the micro-structural (neuronal) level. (b) The meso-architecture of the back-propagation network is the same as the meso-architecture of the multilayerd Perceptron.

In Part I, we gave an illustration of the dynamics of the back-propagation network. This task was the X-OR (exclusive or) problem. Neither the simple Perceptron nor either of the ADALINE/MADALINE networks can solve this problem, because the solution space is not linearly separable (see Figure 7.2). However, one of the first demonstrations which brought attention to the back-propagation network was its ability to solve this problem.

Let's assume that we have selected a good architecture for this problem, as shown in Figure 7.6. (Later we will examine whether it is the best architecture, but for now, we will assume it is at least a good one.) This architecture has two input nodes, two hidden nodes, and a single output node. The architecture shown here also has variable thresholds on the two hidden and one output node, for a total of nine variables in the system. (The back-propagation network will also work when just the connection weights are varied.)

Figure 7.6 shows one possible set of weight and threshold values for this network. Try them on the four possible input combinations, and demonstrate for yourself that they work! (A detailed walk-through of one example is given in Part I.)

The goal of the back-propagation learning law is to find useful values for the weights and thresholds, values which will enable the desired classification. These weights may not be unique; in fact, they won't be. As with most back-propagation network applications, there are lots of different weight combinations which will work. However, finding a useful set of connection weight values — any useful set — is a challenge. The weight values could be very hard to come by unless some learning algorithm was used to drive the collection of weights towards a useful set of values over time.

Figure 7.7 illustrates an important aspect of the weight/threshold solutions found for a given problem. In this case, there are 32 possible combinations of signs for the variables in the top half of the network: the connection weights from the hidden to the output nodes and the thresholds in the hidden and output nodes. Of these 32 initial possibilities, we found that ten of them were produced as solutions when we trained the network using different random seed initial values, and were able to discern that another two combinations should be workable due to their symmetry with the solutions we found. Each symmetry class of weight and threshold values has potentially infinite possible values for weights and thresholds which will work.

The point that this example makes with us is that for almost any network we can envision, there will be multiple possible sets of similar weight values, and that within each set, there may be an infinite range of connection weights and thresholds. Each solution within a given set will probably be roughly proportional to all other solutions. This means that there is no single right answer, and there might not be any single best answer. There are likely to be a very large number of good or workable answers.

7.4.4   The Learning Method of the Back-Propagation Network

Like the Perceptron, ADALINE, and MADALINE networks, the back-propagation network is taught to create a mapping from the input pattern into an activation pattern in the output layer. Also, like these earlier networks, the back-propagation network undergoes supervised training.

Convergence to a Useful Set of Connection Weights &Thresholds for the "X-Or" Problem

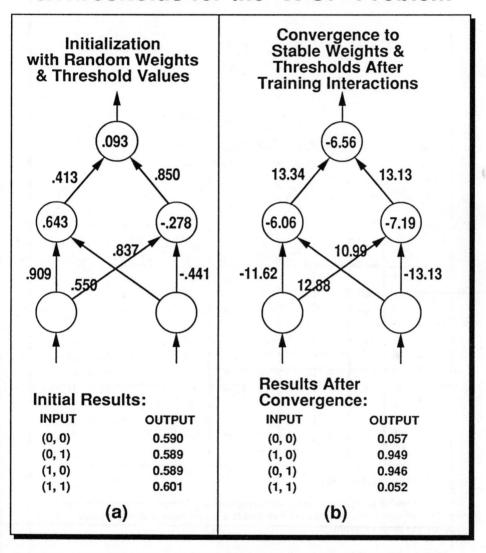

Figure 7.6   An example set of weights and thresholds for this X-OR configuration, shown as the "before training (initial randomly-assigned) weights" and "after training weights." There are several "sets" of similar weights and thresholds that will work, and infinitely many values within each set.

Signs of Weights and Thresholds for Solutions to "X-Or" Problem

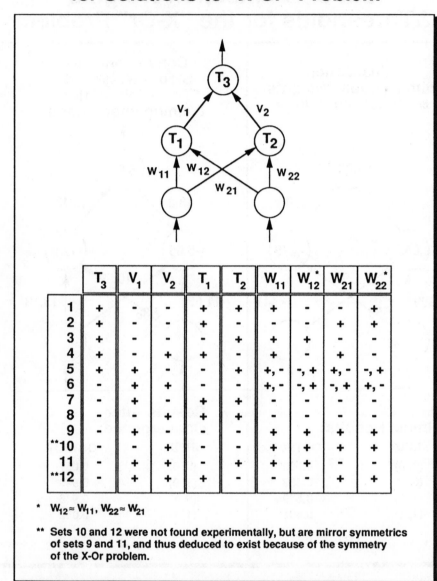

	T_3	V_1	V_2	T_1	T_2	W_{11}	W_{12}^*	W_{21}	W_{22}^*
1	+	-	-	+	+	+	-	-	+
2	+	-	-	+	-	-	-	+	+
3	+	-	-	-	+	+	+	-	-
4	+	-	+	+	-	+	-	+	-
5	+	+	-	-	+	+, -	-, +	+, -	-, +
6	-	+	+	-	-	+, -	-, +	-, +	+, -
7	-	+	-	+	+	-	-	-	-
8	-	-	+	+	+	-	-	-	-
9	-	+	-	-	-	+	+	+	+
**10	-	-	+	-	-	+	+	+	+
11	-	+	+	-	+	+	+	-	-
**12	-	+	+	+	-	-	-	+	+

* $W_{12} \approx W_{11}$, $W_{22} \approx W_{21}$

** Sets 10 and 12 were not found experimentally, but are mirror symmetrics of sets 9 and 11, and thus deduced to exist because of the symmetry of the X-Or problem.

Figure 7.7   An illustration of the different "sets" of possible connection weights and thresholds which yield solutions to the X-OR problem. (a) This simple architecture for the X-OR problem has nine adaptive variables; six connection weights and three thresholds. (b) Considering just the hidden-to-output layer connection weights and the three thresholds, there are 32 possible configurations, in terms of signs of these parameters. We have experimentally found twelve out of these 32 will allow solutions. Ms. Rosie Dupless produced the values shown in this figure.

The back-propagation network learns to distinguish among different pattern categories simultaneously. Each pattern will have a different type of influence on the change in the connection weights. Thus, it is important that the patterns in the training set presented in such a way that the changes in the connection weights move over time to values which optimize the network's response to all the pattern classes. We deal with this more completely in a later subsection on training and stability in the back-propagation network.

The major difference between the two networks is that the learning law for the back-propagation network is substantially more powerful than the learning law for its predecessors. The Perceptron used a straightforward method devised by Rosenblatt, and the ADALINE and MADALINE networks used the well-known LMS method, as applied by Widrow and Hoff.

The learning rule for the back-propagation networks is the Generalized Delta Rule, which is a generalized form of the LMS rule. The feature which most strongly distinguishes the Generalized Delta Rule from its predecessors is that the error is used to affect not just one set of weights (input-to-output, as with the ADALINE), but two sets of weights (input-to-hidden, and hidden-to-output). Some back-propagation networks also change the thresholds of the neurons in the hidden and output layers.

The Generalized Delta Rule uses the chain rule from differential calculus to calculate the way in which these weights (and thresholds) depend on each other. This is typically done in two stages. First, the connection weights between the hidden and output layers are adjusted, along with the activation thresholds in the output nodes. The difference between the actual value of the output node(s) and the desired value of the output node(s) is used to drive this stage, and it is analogous to the weight adjustment process in the Perceptron and the ADALINE.

In the second stage, the connection weights between the input and the hidden layers are adjusted, along with the activation thresholds in the hidden layer. The basis for this adjustment is different from the one used in the first stage, because neither the system nor the network designer knows what the target output of the hidden nodes should be. (Remember, in the first stage, the designer not only could, but had to specify what the appropriate output values should be.) This is where the back-propagation method developed by Werbos offers a unique and effective approach. The back-propagation method uses the adjustments and values to the hidden-to-output weights to help determine the changes made to the input-to-hidden weights.

The goal of the weight adjustments in the back-propagation method is to reduce the error in the output, or to reduce the difference between the actual output and the desired output. As before, this error is defined as LMS error, as given in Equation 7.4.

Even at this first stage, we have a significant difference between the back-propagation method of training and the previous use of the LMS training rule. Because we have specified that we wish to minimize error, we can use differential calculus to obtain a minimum — if the error equation operates on differentiable functions.

We can express the training rule in terms of a function, Delta. This function is applied to every connection weight w_{ij} after each presentation of a training pattern. Repeated use of the Delta function will usually lead the network to converge at a set of connection weights which minimizes the error for recognizing all patterns in the training set.

In order to understand how the Generalized Delta Rule works, we need to identify how the function Delta relates to both the error and to the values of the weights themselves. Then we can modify the connection weights according to the relationship:

$$w_{ij(new)} = w_{ij(old)} + \propto Delta(w_{ij, \, old}) - output\_activ \qquad \textbf{Eq. 7.5}$$

where w_{ij} represents the new and the old values of the connection weight between node i and node j,

α is a constant which controls the degree to which Delta affects the weights,

Delta is a function, and Delta ($w_{ij, \, old}$) is a value calculated for each $w_{ij, \, old}$ which determines the amount of change to be made in each weight, and output_active is the output activation of the jth neuron.

We set the Delta function to be proportional to the negative of the derivative of the error with respect to the connection weight.

$$Delta \, (w_{ij}) \propto \frac{-\delta E}{\delta w_{ij}} \qquad\qquad \textbf{Eq. 7.6}$$

To gain an interpretation of this equation, let us consider the relationship between the total error, E, and w_{ij}. The value for E is always calculated as a discrete value, following presentation of a pattern to the network. But it is reasonable to think of how E might change with the value of w_{ij}, and how that might be plotted. This is illustrated in Figure 7.8.

Suppose there is a value of w_{ij}, which we will call w_{ij} (min), for which the value of the error E is a minimum. Then our task is to find the value w_{ij} (min). (This is a simplified view of the problem, because Delta will have to be minimized for many different patterns with many different errors. It is because of these many different patterns and their differing targets for output activations that we cannot simply set the derivative of E with respect to w_{ij} to be 0, and solve for the resulting values. Instead, we must approach optimal values for w_{ij} incrementally.)

The derivative of E with respect to w_{ij} gives us the tangent of the error curve at any point defined by w_{ij} on the curve. From the graph shown in Figure 7.8, if the value of the error monotonically decreases as w_{ij} approaches w_{ij} (min) (e.g., as is shown on the left of w_{ij} (min)), then using this tangent gives a good method for changing w_{ij} so that it approaches w_{ij} (min). The steeper the curve is, the more rapidly the tangent will lead to the minimum value for energy. For shallow curves, the procedure still works, but convergence is slow. (Note that this method is very similar to Newton's method in numerical analysis for finding zero-crossings of curves.)

A problem can arise if the value of the error does not monotonically decrease as w_{ij} approaches w_{ij} (min). This is illustrated in the right-hand side of the graph shown in Figure 7.8, where there is a shallow minimum at w_{ij}(min*). If the method of using curve tangents described earlier is used in the vicinity of w_{ij}, then the process we have been describing will lead to convergence at w_{ij}(min*), and not at w_{ij}(min). However, there are way to overcome this difficulty, and we discuss them later in this section.

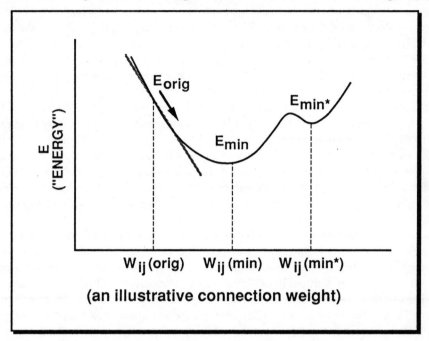

Illustration of how "Energy" of Network System Changes with Connection Weight(s)

(an illustrative connection weight)

Figure 7.8   Illustration of how the "energy" of the network, E, could vary with changes in connection weights, as diagrammed by variance against a single weight, w_{ij}. $w_{ij}(min)$ is the weight which will give the truest energy "minimum," but there are other weights, such as $w_{ij}(min*)$ which will yield "false minima." If the connection weights take on values which yield false energy minima, then the network will get "stuck" in a set of connection weights that do not yield a good solution to the problem.

Reverting now to our original discussion, we find that we need to have a way to calculate the derivative of the error with respect to the individual connection weights. The back-propagation method allows us to do this for all layers in the network. The effect of the error on the connection weights between the hidden nodes and the output nodes can be used to not only adjust those weights, but to back-propagate that influence through the net, affecting the weights connecting the input to the hidden nodes. This form of dynamic feedback is the most significant feature of back-propagation.

7.4.5   The Performance and Capabilities of the Back-Propagation Network

The performance of the back-propagation network is inextricably linked with its ability to generalize. This generalization and feature extraction is performed by the hidden layer neurons. Ideally, the knowledge which is stored in the hidden layer of a back-propagation network is abstracted from the information contained in the input patterns. This abstracted knowledge provides the basis for classifying the pattern into one of the categories available in the output layer. Typically, the hidden nodes can be trained to repsond to the presence of features in the input pattern. Then, the output layer will learn to respond to the presence or absence of features in the pattern, rather than to the exact spatial (or temporal) layer of the pattern itself.

In a fully-connected back-propagation network, a wide variety of features can be represented in the hidden layer. These features may be local, or they may describe different aspects of the overall pattern.

It is sometimes difficult to figure out exactly what features are being represented in the hidden layer. However, once a back-propagation network has been fully trained, it can be examined. A pattern can be given as input, and the network designer can identify which hidden nodes have become "activated" in response to that pattern. After repeating this process several times, with different patterns, the designer can usually identify the feature in the input pattern to which a given hidden node responds. This is an approach first used by Plaut, Nowlan, and Hinton [1986] and later by Gorman and Sejnowski [1988].

Sometimes, the features which become represented by hidden nodes seem very appropriate. For example, in a back-propagation network which is trained to distinquish among different block-letter alphanumeric characters, one hidden node may become trained to respond to vertical lines. Another may respond to horizontal lines, and yet another to a diagonal one. Sometimes, the feature to which a hidden node may learn to respond might be surprising to the designer. On examination, such features work well and have some rationale. For example, a back-propagation network which learns to recognize the letter "Z" may learn to respond to junctions of horizontal and diagonal lines in the upper right and lower left corners of the input array. Finally, sometimes it is hard to discern what some hidden nodes are doing when they respond to a given pattern.

Some work is beginning to accumulate that offer comparisions of various network performance capabilities. For example, an article by Pawlicki et al. [1988] compares the performance of linear auto-associative systems, threshold logic networks, back-propagation networks, Hopfield networks, and Boltzmann machines for recognizing hand-written characters. They find that the neural networks methods do not offer substantially improved performance over other techniques for this task, but acknowledge that combinations of neural network and symbolic pattern matching methods may be the most effective approach for such demanding problems. Also, their reported work does not employ any methods to optimize network performance, such as preprocessing or feature extraction. Because network performance depends on many variables, including the preprocessing of input data, selection of architectures and training sets, and selection of features or classes to be extracted, much more work needs to be done to determine how neural networks can be used for practical applications.

The ADALINE can perform linear class boundary separations, and sets of ADALINEs can be configured together for more complex decisions. ADALINEs and MADALINEs can operate very fast in real-time, as they have only simple multiplies, adds, and threshold non-linearities in the transfer function.

7.4.6   Applications of the Back-Propagation Network

The back-propagation network has been successfully applied to a variety of practical pattern recognition tasks, including phoneme identification [Waibel & Hampshire, 1989; Rossen and Anderson, 1989], text reading [Sejnowski & Rosenberg, 1987], air combat maneuver selection [McMahon, 1990], handwritten character recognition [Burr, 1987], image compression [Cottrell et al., 1987], and medical diagnosis [Yoon et al., 1989; Mulsant & Serban-Schreiber, 1988].

Back-propagation networks are also useful for signal filtering [Klimasauskas, 1989; Anderson and Montgomery, 1990] and detecting segment boundaries or discontinuities in time-varying signals [Vierhinthan and Wagner, 1990]. Using a simple back-propagation network, Malferrari et al. [1990] have been able to detect discontinuities in sigma-log readings from oil well drilling. These discontinuities are difficult to detect using conventional moving average window methods when the discontinuities occur in multiple. It is also difficult to discern real discontinuities from spurious ones. The results of the back-propagation network were on the order of 99% accuracy, a few percentage points above the accuracy of moving average window detection of discontinuities.[1]

REFERENCES

Anderson, B., and Montgomery, D. (1990). "A method for noise filtering with feedforward neural networks: Analysis and comparison with low-pass and optimal filtering," *Proc. of Third Intl. Joint Conf. on Neural Networks* (San Diego, CA; June 17-21, 1990), I-209-I-214.

Anderson, K., Cook, G.E., Karsal, B., and Ramaswamy, K. (1990). "Artificial neural networks applied to arc welding process modeling and control." *IEEE Trans. on Industry Applications, 26* (Sept.-Oct.), 824-830.

Burr, D.J. (1987). "Experiments with a connectionist text reader," *Proc. First Int'l. Conf. Neural Networks* (San Diego, CA; June 21-24, 1987), IV-717 — IV-724.

Cottrell, G.W., Munro, P., & Zipser, D. (1987). "Image compression by back-propagation: An example of extensional programming," *ICS Report 8702*, University of California, San Diego, CA.

Dietz, W., Kirch, E., and Ali, M. (1989). "Jet and rocket engine fault diagnosis in real time," *J. Neural Network Computing* (Summer), 5-18.

Gorman, R.P., and Sejnowski, T.J. (1988). "Analysis of hidden units in a layered network trained to classify sonar targets," *Neural Networks, 1*, 75-89.

Klimasauskas, C. (1990). "Neural nets and noise filtering (with a case study by S. Melnikof)," *Dr. Dobb's Journal* (January), 32-38, 42, 44, 47, 48, 96-100.

Malferrari, L., Serra, R., & Valastro, G. (1990). "An application of neural networks to oil well drilling," *Proc. Int'l. Neural Networks Conf.* (Paris, France, July 9-13, 1990), 127-130.

---

1   Simulations were performed by Ms. Rosie Dupless, as part of a project in a class taught by the author. She also produced the values shown in Figure 7.7.

McMahon, D.C. (1990). "A neural network trained to select aircraft maneuvers during air combat: A comparison of network and rule based performance," *Proc. of Third Intl. Joint Conf. on Neural Networks* (San Diego, CA, June 17-21, 1990), I-107–I-112. v1, pp 107-112.

Minsky, M. L., and Papert S. (1967). "Linearly unrecognizable patterns," *Proc. Symp. on Appl. Math. 19, Amer. Math. Soc*, 176-217.

Minsky, M.L., and Papert, S. (1987). *Perceptrons* MIT Press, Cambridge, MA, and reprinted in J. Anderson and E. Rosenfeld (Eds.), *Neurocomputing,* (1988). MIT Press, Cambridge, MA, 161-170.

Mulsant, B.H., & Servan-Shreiber, E.A. (1988), "Connectionist approach to the diagnosis of dementia," *Proc. of the Twelfth Annual Symposium on Computer Applications in Medical Care,* 245-250.

Pawlicki, T.F., Lee, D.-S., Hull, J.J., & Srihari, S.N. (1988). "Neural network models and their application to handwritten digit recognition," *Proc. United States Postal Service Advanced Technology Conference* (May 3-5, 1988), II-751–II-767.

Plaut, D.C., Nowlan, S.T., and Hinton, G.E. (1986). *Experiments in Learning by Back-Propagation*, Carnegie Mellon University Tech. Report CMU-CS-86-126.

Rosenblatt, F. (1958). "The Perceptron: A probabilistic model for information storage and organization in the brain," *Psych. Review, 65,* 386-408, and reprinted in J. Anderson and E. Rosenfeld (Eds.) *Neurocomputing,* (1988). MIT Press, Cambridge, MA, 92-114.

Rossen, M.L., and Anderson, J.A. (1989). "Representational issues in a neural network model of syllable recognition," *Proc. First Intl. Joint Conf. Neural Networks* (Washington, D.C., June 18-22, 1989), I-19–I-25.

Rumelhart, D.E., Hinton, G.E., and Williams, R.J. (1986). "Learning internal representations by error propagation." In D.E. Rumelhart, J.L. McClelland, and the PDP Research Group (Eds.), *Parallel Distributed Processing, Vol. I.*, MIT Press, Cambridge, MA.

Sejnowski, T.J., & Rosenberg, C.R. (1987). "Parallel networks that learn to pronounce English text," *Complex Systems, 1,* 145 — 168.

Veerhinathan, J., and Wagner, D. (1990). "A neural network approach to first break picking." *Proc. Third Intl. Joint Conf. Neural Networks* (San Diego, CA, June 17-21, 1990), I-235–I-240.

Waibel, A., and Hampshire, J. (1989). "Building blocks for speech," *Byte,* (August) 235-242.

Werbos, P. (1974). *Beyond Regression: New Tools for Prediction and Analysis in the Behavioral Sciences.* Ph.D. dissertation, Harvard University, Cambridge, MA.

Werbos, P. (1988). "Generalization of back-propagation with application to a recurrent gas market model," *Neural Networks, 1,* 339-356.

Werbos, P. (1989). "Back-propagation and neurocontrol: A review and prospectus," *Proc. Int'l. Joint. Conf. on Neural Networks* (Washington, D.C., June 18-22, 1989), I-209 — I-216.

Werbos, P., (March-April, 1989). "Maximizing long-term gas industry profits in two minutes in Lotus using neural network methods," *IEEE Trans. on Systems, Man, and Cybernetics, SMC-19,* 315-333.

Widrow, B., and Hoff, M.E. (1960). "Adaptive switching circuits," *Institute of Radio Engineers, Western Electronic Show and Convention, Convention Record*, Part 4, 96-104, and reprinted in *Neurocomputing* J. Anderson and E. Rosenfeld (Eds.), MIT Press, Cambridge, MA, 126-134.

Widrow, B. (1988). "ADALINE and MADALINE — -1963," *Proc. of First IEEE Intl. Conference Neural Networks*, (San Diego, CA, June 21-24, 1988), I-145 — I-157.

Widrow, B. and Winter, R.G. (1990). "Neural Networks for adaptive filtering and adaptive pattern recognition," in S.F. Zornetzer, J.L. Davis, and C. Lau (Eds.), *An Introduction to Neural and Electronic Networks*, Academic Press, San Diego, 149-269.

Widrow, B., Winter, R.G., and Baxter, R.A. (1988). "Layered neural networks for pattern recognition," *IEEE Trans. Acoustics, Speech, and Signal Proc.*, *36*, 1109-1118.

Yoon, YO., Brobst, R.W., Bergstresser, P.R., Peterson, L.L. (1989). "A desktop neural network for dermatology diagnosis," *Journal of Neural Network Computing, 1* (Summer), 43-52.

8

MULTILAYER FEEDFORWARD NEURAL NETWORKS II: OPTIMIZING LEARNING METHODS

Harold H. Szu and Alianna J. Maren

8.0 OVERVIEW

The Boltzmann machine and its refinement, the Cauchy machine, are two networks which are structurally and dynamically similar to the back-propagating Perceptron, and which can perform similar tasks. They differ from the Perceptron-like networks in that they learn using an energy state optimization method derived from statistical considerations rather than a Delta rule. This approach to learning allows them to perform optimization tasks as well as pattern recognition.

8.1 THE BOLTZMANN MACHINE

The Boltzmann machine has both conceptual and performance similarities to the back-propagation network. Both have hidden nodes, and both need to be trained to match input patterns to previously determined categories. Both networks were developed at about the same time, and were applied to the same types of problems. Often, the same researchers were investigating both the back-propagation and the Boltzmann machine networks.

There are several reasons why the Boltzmann machine never achieved the popularity of the back-propagation network. First, the performance of the two networks was very similar, so there was no need to favor one over the other because of performance or capability considerations. Thus, the main criteria for selecting a network became ease of learning (on the part of the researcher), and ease of actually using it for a particular application. The back-propagation network was easier to both learn and to use.

The back-propagation concepts and learning rule come out of a fairly direct approach to minimizing an energy (or distance function) using differential calculus. Anyone who can follow a chain rule argument can understand the derivation for the learning rule. It takes less than a page to code the algorithm, and the learning time (while large) is accomplished within a single phase of learning.

In contrast, the Boltzmann machine concepts come out of statistical mechanics, expressed in either information theoretic and/or statistical thermodynamic formalisms. Not many people in the neural networks field have the background to understand either of these approaches. Its learning rule takes much longer to write down than the back-propagation learning rule, is more complex, and takes more time when training a network. As a result, we have seen an explosion of applications of the back-propagation method, while interest in the Boltzmann machine has dwindled.

Despite these factors, there is some real value in studying the Boltzmann machine. The lines of reasoning used by the researchers who developed this approach evidence an attack on some basic problems, and the way that they approached these problems sheds light on some fundamental issues in neural networks today. Thus, even if we chose not to use the Boltzmann machine for an application, it is useful to know something about it which might spur some creative thinking that can help solve the problems which come up when we take on the more interesting applications.

8.1.1 The Concepts Underlying the Boltzmann Machine

A major distinction to make at the outset is that even though the Boltzmann machine looks like the back-propagation network (and thus like a Perceptron), the origination point for the Boltzmann machine was not the Perceptron. It was the Hopfield network. (We will cover this network in the next chapter.)

Also, the question that the inventors of the Boltzmann machine (Geoffrey Hinton, Terrence Sejnowski, and David Ackley) were not asking was how to make a network which could classify non-linearly separable decision spaces; rather, they were asking directly about generalizabilty. This is a very subtle, almost minute distinction, but it led them on a very different path. They began with a Hopfield-like network (which is an autoassociator), but were looking at it from a heteroassociative perspective, so that one field of the neurons should represent or respond to some sort of high-level order among the pattern presented in another field of neurons. The simple pairwise relationships which form the basis for the Hopfield network would not facilitate high-level pattern recognition, so they needed to do something special. They did this by introducing the concept of hidden neurons; neurons which connected to input and output neurons, but which were invisible to normal view.

So the starting place for the Boltzmann machine was something like a long vector of neurons, one or more fields of which represented some input pattern(s), and a final field which represented the desired output. Instead of connecting the neurons directly (as is done in the Hopfield network and several other networks we'll consider), each of the input neurons connected to several hidden neurons, and each of the hidden neurons connected to each of the output neurons. (Rearrange the way we view this, and we have a Perceptron architecture. But they were not interested in Perceptrons at the time.)

Their main concern after this architectural modification was to get some sort of learning rule that would adjust the weights so that when an input pattern was presented (or clamped, to use their phrase) to the input field of neurons, the appropriate output pattern should appear in the output field.

Now, the context for the Hopfield network is that of an energy space defined by a Lyapunov function. The solutions (stable states) of the Hopfield network are the minima in this energy space [Hopfield, 1982]. Hopfield developed the concept of using the properties of sets of neurons as a model for cooperative computation. He considered only bi-state artificial neurons, and showed how the function for their probabilistic distribution among the two states might be found. In 1984, Hopfield extended his work to include neurons with continuous activation values.

Given this initial context, the researchers developing the Boltzmann machine looked for a way to distribute the probabilities of the hidden layers among two possible states (0 and 1, just like the Hopfield network). The difference was that these neurons were accessed indirectly via their connections to the input and output neurons, so that a different type of energy function had to be used. Also, the Hopfield approach to setting up the connection weight values was very simple. They needed a more robust method which would learn the generalizable features of the input patterns and represent them as connection weights.

Hinton, Sejnowski, and Ackley found inspiration for their learning method from the work done by Kirkpatrick, Gellatt, and Vecchi [1983]. Kirkpatrick et al. developed a *simulated annealing* approach to finding optimal values for state assignments in large systems. To do this, they used a statistical mechanical approach applied to combinatorial optimization. In short, they were looking at ways to find optimal probability distributions for large ensembles of bistate (0 and 1) units. This article forms the basis for learning in the Boltzmann machine.

By putting together the simulated annealing method (for faster, more optimal learning) with Hopfield's ideas on bi-state systems of neurons, they came up with the Boltzmann machine [Hinton, Sejnowski, and Ackley, 1984; Hinton and Sejnowski, 1986]. Their early work presents the Boltzmann machine in the context of a network which solves constraint optimization problems.

8.1.2   The Architecture of the Boltzmann Machine

The architecture of the Boltzmann machine is the same as that of the back-propagation network, except that the neurons have bistate values (0 or 1). (See Figure 8.1.)

8.1.3   The Dynamics of the Boltzmann Machine

Once trained, a Boltzmann machine operates just like a back-propagation network.

8.1.4   Simulated Annealing: The Learning Law for the Boltzmann Machine

The learning method for Boltzmann machines is called simulated annealing. The only similarity between it and the back-propagation learning law (the Generalized Delta

The Boltzmann Machine Network

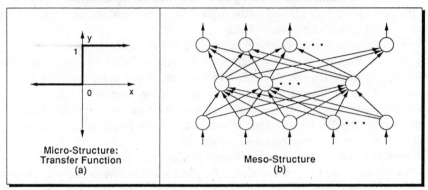

Figure 8.1   The architecture of the Boltzmann Machine.

Rule) is that both are methods for supervised learning. Networks take a long time to learn with either rule, and need to be retrained extensively if they will be taught new material. The simulated annealing approach to learning is sometimes called stochastic or statistical, because it relies on generating random events and evaluating their effect in terms of desired goals and probability distributions.

Each of the three networks we have studied up to now uses a learning rule which ideally takes the network's connection weights towards a state which minimizes the difference between the desired and the real outputs of the network. The problem with these approaches (even with the back-propagation network) is that sometimes the connection weights take on values which trap the network in a local minimum, as described in the previous subsection. The goal of using the simulated annealing method is to avoid such local minima trapping.

The essence of the simulated annealing learning law is that we make an analogy between the energy state of the entire network and the energy state of a physical solid which is slowly being cooled. We will pretend that each individual unit in the solid (atoms, molecules, etc.) can take on one of two possible states; a high-energy state (1) or a low-energy state (0). The free energy (a term from thermodynamics) of the solid is a combination of two factors; the combined energies of the individual units and the negative of the entropy (the disorder among the units) times the temperature of the solid.

The key to this analogy is that whether a solid is in a high-energy or low-energy state, it is always in equilibrium. Equilibrium is found at the lowest point on the free energy curve. (We can think of the energy curve shown in Figure 8.2 as a free energy curve.) If we lower the temperature slowly, the solid will have an opportunity to find this lowest point, even if there are many other shallow minima. We want to accomplish the same thing with the Boltzmann machine network. We want to create some sort of artificial "temperature" so that as we slowly reduce this "temperature," the connection weights take on values that put the network at the global minimum for the energy curve, and don't get trapped in one of the local shallow minima.

Let's envision a physical analogy to this energy surface. Suppose that we hold in our hands a large tray of firm plastic that has some pockets or indentations in it. This would

Illustration of how "Energy" of
Network System Changes with Connection Weight(s)

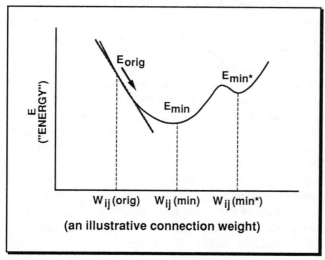

(an illustrative connection weight)

Figure 8.2   Illustration of how the "energy" of the network, E, could vary with changes in connection weights, as diagrammed by variance against a single weight, w_{ij}. $w_{ij}(min)$ is the weight which will give the truest energy "minimum," but there are other weights, such as $w_{ij}(min*)$ which will yield "false minima." The Boltzmann machine learning method allows weights to take on values corresponding to the deepest minima of the energy surface.

be like the bottom half of one of those cartons of eggs that we buy at the grocery store. Now, some of those indentations are very shallow, and some are very deep. This represents the energy "surface" (which is really multidimensional). Let's suppose that we have a single egg-sized ball (or egg, if you prefer) in this tray. Although this is just a single one, we want to think of it as corresponding to the entire collection of weights that we want to optimize. Our goal is to shake the tray so that the ball goes into the deepest of the indentations. How will we do this?

We could shake the tray continuously, and hope that just by keeping this up, the ball will fall into the hole. But how will we know when the ball has reached the deepest indentation? This can be tricky, especially if we don't know in advance which of the indentations is deepest. So we adopt a strategy. We shake the tray, pretty hard at first. Shake it some more. Shake some more, a bit more gently. Then more gently yet. Finally, just the barest of tremors. Now, let's look into the tray. Chances are, the ball is in the deepest pocket. That's because as we shook the tray more and more gently, the ball could be shaken out of shallow indentations, but not the deepest ones. There's no absolute guarantee that it would have fallen into the very deepest one in the tray, but it is very likely in one of the deepest.

We want to take this idea and apply it to finding optimal connection weights in a neural network. To do this, we return to our original analogy, that of a solid which will undergo simulated annealing. We need to see how the "shaking the tray" strategy fits in with annealing a solid, because this leads directly to the neural network.

Recall that our hypothetical solid was composed of identical units, each of which could be in one of two possible energy states, high (1) or low (0). The proportional number of units in each state is a function of temperature at high temperatures, there are more high-energy units than there are at low temperatures. This proportion can be expressed as a probability distribution. This temperature-dependent probability distribution will be a key factor in the simulated annealing process.

"Shaking the tray" corresponds to setting the temperature of the solid. "Shaking the tray hard" corresponds to a high temperature, with high kinetic energy for the ball in the tray. "Shaking the tray gently" corresponds to low temperature. The ball has low kinetic energy, and can't easily move out of a deep pocket. The pockets in the tray correspond to energy minima in the free energy surface. There are numerous local minima, but relatively few deep minima. These minima represent equilibrium states for a solid, and correspond to optimal connection weight values in the neural network.

We've already set up an analogy between the neural network and the solid. Now, let's see what would happen if we had some sort of artificial temperature which governed the energy state of the neural network, just as it governs the energy of the solid. As we slowly reduce the temperature from a high value to a low one, we lower the proportion of neurons that are in a high energy state.

If this were a physical solid with multiple energy states, the probability of finding a unit in any given energy state would be:

$$P(E) = A \exp(-E/kT), \qquad \text{Eq. 8.1}$$

where:

E is the energy of the state,
P(E) is the probability of finding a unit in that state,
A is a normalizing constant,
T is the temperature (called Boltzmann's equation), and
k is called Boltzmann's constant.

When we have just two states, we can rewrite Eq. 8.1 as a function of the difference in energy between the two states. This becomes:

$$P(0) = \frac{1}{1 + \exp(-\Delta E/T)} \qquad \text{Eq. 8.2}$$

where P(0) is the probability that a given neuron is in state 0, the low-energy state,

ΔE is the difference in energy between states 1 and 0, and
T is the artificial temperature.

This equation appears quite often in the Boltzmann machine literature, and it may cause some confusion to those who come from a back-propagation perspective. Note that the form of this equation is the same as that of the sigmoid function shown in Figure 7.4. The apparent similarity is purely coincidental, the equation for the sigmoid function defines a transfer function which scales the input into a neuron between the range of 0 and 1. Equation 8.2 represents a probability distribution between two binary states.

Let's interpret this equation. We'll assume that ΔE is a positive value, since state 1 is higher in energy than state 0. We want to consider the effect on the probability distribution as we decrease T. When T is large, delta E/T is small, and exp(- ΔE/T) is approximately exp(0), which is 1. This means that P(0) is about 1/2, or half of the neurons will be in a high energy state. (You might wonder why only half, not all. The answer resides in the entropy term which contributes to the free energy. For a derivation, see Maren [1987].) When T is small relative to E, ΔE/T is large, and exp(-ΔE/T is very small. This means that P(0) is about 1, or that most of the neurons are in the low energy state.

We use the probability distribution, generally expressed in the form of Eqn. 8.1, in an algorithm to adjust the weights in the Boltzmann machine. We start by giving the weights random initial values, and then cycle through the following process:

1. Select a time-dependent temperature function, such as

$$\text{Eq. 8.3}$$

    where T_0 is the initial value for T. This function should decrease T with increasing time. Start with a large initial value for T_0.
2. Select an input pattern from the training set. Present it to the network, calculate the Delta between the output produced by the network and the desired output, as shown in Eqn. 7.5. This is the objective function, or (LMS) error, which we will minimize. (There is some similarity to the ADALINE and the back-propagation methods in selecting this as the term to be minimized, but the way we go about it for the Boltzmann machine is different.) Call this objective function value Error_1.
3. Randomly pick one of the connection weights. Change it a little. Find the new objective function or error, using the same input pattern as above. Call this Error_2.
4. If Error_2 is less than Error_1, we have found a "better" value for that connection weight, and we keep it.
5. If Error_2 is greater than Error_1, we have found a "worse" value connection weight. Normally, we wouldn't keep it. But to break out of local minima, we keep these "not-so-good" weight values some of the time. Specifically, we do the following:
    a. Set Delta_Error = Error_1 - Error_2.
    b. Compute the value P, where

$$P = \exp(-c/T), \qquad \text{Eq. 8.4}$$

    Where c is a constant incorporating the energy difference as shown in the Boltzmann distribution (Eq. 8.1).
    c. Randomly generate a (0,1) value according to the probability distribution specified by Eqn. 8.3.
    d. If the value generated is 1, keep the change. If the value is 0, discard it.

When we keep a non-optimal weight change, we are doing the neural network equivalent of allowing a unit to be in a higher energy state than its neighbors, which is permissible because of the overall high energy of the system.

We continue the weight adjustment by beginning again with step (3), until we have worked through all of the weights. Then, select a new pattern at step 2 until all the patterns have been exercised. Then, we increase the time count, which leads to a lower temperature (step 2). We repeat the entire process (steps 2-5) with increasingly low values of temperature until the difference in errors is zero. Then, we input a new pattern (step 1), and repeat the entire process all over. This process may have to be repeated numerous times, with multiple presentations of each input pattern.

The entire process discussed above, while complex, is actually a simplification of the more complete Boltzmann learning method. For a more detailed description, see Simpson [1990] and Hinton and Sejnowski [1986].

8.1.5 Performance Characteristics of Boltzmann Machine Networks

While the simulated annealing method tends to find a solution (and not get trapped in local minima, it can take one or two orders of magnitude more than even a *vanilla* back-propagation network to converge. Since a fair body of work has been done in finding ways to both speed up the back-propagation method and to guarantee convergence (getting out of local minima traps), there is not a strong impetus to use the Boltzmann machine approach unless there is some specific reason to use an optimization or constraint satisfaction approach to problem solving.

Sejnowski and others rapidly applied both the Boltzmann machine and the back-propagation method to a number of pattern recognition problems. With Charlie Rosenberg (then a Princeton graduate student), Sejnowski developed NETtalk, a Boltzmann machine-based system which learned to produce speech from text [Sejnowski and Kosenberg, 1987]. NETtalk exhibits some similarities to human performance: a power law for learning, generalization after learning, graceful degradation with partial damage, and rapid learning after partial damage.

Working with Kienker and Hinton, Sejnowski also showed that the Boltzmann machine can recognize mirror, rotational, and translational symmetries in a pixel-type (2-D) input layer. In their 1986 paper, they explored not only the abilities of the Boltzmann machine to recognize such abstract pattern characteristics, but also presented some very interesting work on knowledge representation in hidden units.

8.2 THE CAUCHY MACHINE: A REFINEMENT OF THE BOLTZMANN MACHINE

Recent developments in Boltzmann machine include a deterministic Boltzmann learning procedure proposed by [Hinton, 1989], and the "Cauchy Machine," invented by Szu [1987a,b], which uses both random walks and flights to generate new states. Both approaches offer a faster learning procedure than the original Boltzmann machine [Hinton et al., 1984]. In a very interesting paper, Pasberry and Schnitger [1989] augment the "classical" Boltzmann machine model, and show that in some cases Boltzmann machines may not be much more powerful than combinatorial circuits built from Boolean threshold gates. They make a number of useful comments about the practical implementation of Boltzmann machines. An electronic chip implementation of Boltzmann machine has been developed by Alspector et al. at Bellcore [1987], and its optical version by Farhat at the University of Pennsylvania [1987]. Similarly, an electronic Cauchy machine has been designed by Takefuji and Szu [1989], and its optical version by Scheff and Szu [1987]. Recently, a Gaussian machine based on both the minimization of Helmholtz's free energy and the maximization of entropy has been studied and implemented in a chip by Akiyama et al. at Keio University [1990]. Since the major ingredient in Boltzmann-like machines is the simulated annealing algorithm, we shall compare Boltzmann machine and Cauchy machine in terms of different algorithms: Boltzmann Annealing (BA) and Cauchy Annealing (CA). Then, we will review two applications: (1) in finding the global minimum solution of the Traveling Salesman Problem (TSP) and (2) in search for the mini-max feature in image processing.

We shall discuss the sequential algorithms used in the above parallel machine implementations as follows. In BA, Gaussian random process is used to generate new states in the sequential algorithm. Geman & Geman [1984] has proved that the cooling schedule T(t) must be inversely proportional to the logarithm of time t, in order to guarantee convergence to the global minimum. This relatively slow convergence is due to the bounded variance of the Gaussian process which constrains the neighborhood of successive samples. This bounded variance random walk is called a local search strategy. If, on the other hand, one uses an infinite variance Cauchy random process, a faster cooling schedule that is inversely proportional to time t has been deduced by Szu [1987a,b] in one dimension and Szu & Hartley [1987a,b] in arbitrary higher dimensions in order to solve the bearing-fix problem with multiple sensors and multiple targets. We call this new class of algorithms a semi-local search strategy because they permit occasionally long steps (the so-called L_vy-Doob diffusion) far from the neighborhood of the previous sample. These random flights are indicative of the divergence of the second moment of the Cauchy probability distribution.

8.3.1   Simulated Annealing: The Concept Underlying Both the Boltzmann and Cauchy Machine

In a convex optimization problem, one can start at any point in the function space, measure the local gradient, and take a step in any direction which is lower in altitude than the current position. Repetition of this process will assure asymptotic convergence to the minimum (i.e., optimum) solution. In a non-convex problem, the optimization function has multiple local minima, each with different depths, for which the optimum is defined to be the global minimum. The application of local gradient techniques to non-convex optimization creates a problem where one gets caught in a local minimum with no way of determining if the local minimum is also the desired global minimum. One solution to this dilemma is to permit steps whose magnitude and direction are dependent on the local gradient and to add random noise in an annealing process [Wasserman, 1989].

Further, for the algorithm to converge, the magnitude of the random component of the step size must decrease in a statistically monotonic fashion. In the physical annealing process these steps can be equated with Brownian motion of a particle, traveling at statistical velocity V, over an inter-sample time Δ+. The expectation of V^2 is linearly related to the temperature of the particle. The simulated annealing community [Kirkpatric et al.,1987] therefore refers to the "temperature" of the random process and uses the term "cooling schedule" to refer to the algorithm for monotonically reducing the temperature. In a biological system kept at constant thermal temperature, the "temperature" must mean any external neurochemical control substance that can change and gradually reduce the mean square fluctuation of the system agitations, e.g., adrenal, caffeine, to name a few.

An annealing methodology requires three major steps: (1) The generation of a new search state by means of a random process covering all phase space without the barrier of an energy landscape; (2) The acceptance criterion of the new state, based on the energy landscape property at the new and the old states; (3) The cooling schedule, for quenching the random noise used to generate a new state together with an appropriate change in the new-state acceptance criterion.

8.2.1.1 State-Generating Probability Density The Boltzmann machine uses a Gaussian probability density to generate the incremental displacement X between the old state x and the new state x'as follows:

$$G_T(x'|x'=x+ X) = (\tfrac{1}{2} \pi \sqrt{T}) \exp(- X^2/T) \qquad (8.5)$$

Based on the Central Limiting Theorem (CLT), any random variable with a bounded variance approaches the Gaussian distribution in the large sampling limit.

The Cauchy state generating probability density is:

$$G_T(x'|x'=x+ X) = [T/\pi(T^2 + |X|^2)] \qquad (8.6)$$

Both distributions can be expanded in Taylor series and become identically quadratic for small displacements. This means that locally they are both identical to random walks. When the second moment is taken, however, the Cauchy density produces an infinite divergence while the Gaussian density gives the value of the temperature. This shows that the Cauchy distribution will generate random flights in long steps (Leavy flights), and that the CLT does not apply.

The random displacement X can be easily generated by a uniform angle distribution between $\pm\pi/2$ by a light beam deflected from a suspended mirror on a flat screen as demonstrated previously for an optical Cauchy machine [Scheff & Szu, 1987]. The displacement X is measured from the center

$$X = T \tan (\theta) \qquad (8.7)$$

[Proof : Using d $\tan(\theta)/d\theta = 1/(1+\tan(\theta)^2)$, we replace $\tan(\theta)$ with X/T giving Eq. 8.7]

8.2.1.2 Local and Distributed Acceptance Criteria The primary difference between sequential simulations and parallel implementations of simulated annealing is that the former relies on a centralized acceptance criterion (an *uphill* energy concept) while parallel versions require a distributed criterion (an against peer-pressure concept). The total system energy is convenient for a top-down design, but is not suited for parallel implementations. Any criterion based on the total system energy requires a central processor to tally the contribution from all distributed processors. If each processor is waiting for a centralized decision, the speed of parallel execution will be slowed down.

A natural choice for a distributed acceptance criterion is one based on the interaction forces carried by local communication links. These interactions can be related to the entire energy landscape. For example, the natural phenomenon occurring in a water-ice phase transition is a parallel and collective computing without any central processor control, and yet a slow cooling/annealing schedule ensures a low energy crystalline state of ice. In other words, what really happens during the occasional up-hill climbing of the energy landscape, to de-trapping a meta-stable crystalline state, is an occasional thermal fluctuation against-peer-pressure. This fluctuation manifests itself via the interacting Coulomb forces which communicate instantaneously among all processors/molecules, rather than through the posterior energy landscape, which is a concept that only facilitates our understanding. The simulated annealing approach to neural computation is similar to the liquid-solid phase transition which promises the minimum energy crystal state if it is cooled down properly. If the energy change $\Delta E = E_{new} - E_{old}$ is less than zero, one accepts the new state. On the other hand, if the energy change ΔE greater than zero, then the following value is computed,

$$P_T = 1/[1 + \exp(-\Delta E/T)] \qquad\qquad \text{Eq. 8.8}$$

which is larger than a random number generated uniformly, then the uphill state is accepted, otherwise rejected. Such an energy landscape formula can be thought of two state normalized transition probability: $\exp(-E_{new})/[\exp(-E_{new}) + \exp(-E_{old})]$ and therefore works well on a conventional serial machine for one neuronic decision at one time. For a Gaussian noise model, the appropriate Metropolis acceptance criterion cannot be integrated to an elementary quadrature, which yields, by the most steep descent approximation, the energy landscape concept ΔE incurred by neuronic decisions. Hinton and Sejnowski have later interpreted the acceptance criterion Eq. 8.8 as the energy change for each neuron, ΔE_i which can be used to derive a specific hidden layer probability weight learning, in order to derive a local acceptance criterion (c.f Appendix of [Hinton & Sejnowski, 1986]).

A one-dimensional optical neural network utilizing CA has been developed as Cauchy Machine[Scheff & Szu, 1987]. However, a local distributed VLSI design could not be completed until a distributed acceptance criterion was derived for Cauchy density [Takefuji & Szu, 1989]. If the total input u_i to the McCulloch-Pitts model of a binary neuron is defined as follows,

$$u_i = S_j T_{ij} v_j$$

then, consistent with the Metropolis acceptance [1953], the output v_i is locally set to be one only if random numbers generated within [0,1] are less than the acceptance function which is integrated exactly for each total input as follows:

$$(1/\pi T) \int_0^\infty dx/ [1 + ((x - u_j) / T)^2 = (1/2) + \arctan(u_j/T(t)) /\pi$$

Eq. 8.9

In the case of annealing, the inverse of the cooling schedule is defined as the piece-wise constant gain coefficient G_n at a positive integer time point t_n:

$$G(t_n) = 1/T (t_n) = G_n$$

Eq. 8.10

Then, the output v_i also fluctuates within a finite bound described as both firing rate transfer functions:

$$v_i = \sigma_{1n}(u_i) = 1/(1 + \exp(-u_i(G_N)), \text{ or}$$

$$v_i = \sigma_{1n}(u_i) = (1/2) + \arctan(u_i G_n)/p$$

Eq. 8.11

Note that Eq 8.11 is almost identical to the standard sigmoidal/logistic function: $1/[1 + \exp(-u_i G_n)]$, except that the arctangent function becomes slightly rounded near the central region. In the case of the sigmoidal function, the slope σ'_n is proportional to the gain coefficient G_n

$$\sigma'_n = dv_i/du_i = G_n v_i(1-v_i)$$

Eq. 8.12

When $T = 0$, the infinite gain G implies an infinite slope. In this limit, both firing rate transfer functions become a binary step function $v_i = \text{step} (u_i)$ describing a binary neuron model. Thus, the annealing process gradually changes a sigmoidal neuron toward a binary neuron.

8.2.1.3 Annealing Cooling Schedules   The cooling schedule is critical to the performance of the algorithm. For a given random process, cooling at too fast a rate will likely "freeze" the system in a non-global minimum. Cooling at too slow a rate, while reaching the desired minimum, is a waste of computational resources. The technical problem then is to derive the fastest cooling schedule that will guarantee convergence to the global minimum. With this understanding, the term "cooling schedule" is synonymous with "fastest permissible cooling schedule" during which the complete phase space is guaranteed to be available for the search at all time.

Without any knowledge of energy landscapes, one can only hope to derive an appropriate cooling schedule for a specific random process. The necessary condition is that at any temperature the phase space is always accessible infinitely often in time (i.o.t.). In other words, an inappropriately fast cooling schedule may quench the i.o.t. availability of some remote states, and hence, not possible find global minimum. The specific energy landscape and an appropriate acceptance criterion must be taken into consideration to determine whether the minimum will be actually be found.

For a Gaussian random process, Geman and Geman [1984] has proved the simulated annealing cooling schedule of the temperature T(t), which must be decreased from a given sufficiently high enough temperature T_0 down to a zero degree temperature, according to the inverse logarithmic formula

$$T = T_0/ \log (1+t) \qquad\qquad \text{Eq. 8.13}$$

Thus, in the interest of speeding up the annealing process and yet maintaining the capability of finding the global minimum, Szu et al. applied the Cauchy colored noise to the problem, instead of the Gaussian random process. The resultant cooling schedule for an arbitrary initial temperature is derived as:

$$T = T_0/ (1+t) \qquad\qquad \text{Eq. 8.14}$$

This is known as the Cauchy Annealing (CA), as opposed to the Gaussian white noise annealing known as the Boltzmann Annealing (BA). The mathematical truth in both proofs is based on the fact that the infinite series of the inverse time steps to be divergent from arbitrary initial time point t_0

$$\sum_{t=t_0}^{\infty} \frac{1}{t} = \infty$$

where either the exponential of the logarithmic in time in BA or the inverse square of linear time in CA can produce the identical series.

$$\text{Isu}(t=\text{tsdo5}(0),\text{ßß}, f(1,t)) = \text{ßß}. \qquad\qquad \text{Eq. 8.15}$$

CA is $t/\log(t)$ faster than the Gaussian white noise simulated annealing algorithm which, in turn, is superior to the conventional Monte Carlo method in which the temperature is held at a constant.

8.2.1.4   Applications   We will consider two typical applications as follows: **constraint satisfactions:** Travelling Salesman Problems (TSP),which attempt to find the shortest possible routes through given number of cities, can be stochastically solved by [Szu,1990] generating noise via the fat-tail Cauchy probability density, $T/p \ (T^2 + X^2)$. The noise must be quenched with the inversely linear cooling schedule:$T = T_0/(1+t)$ described above. Moreover, the schedule must be followed consistently for every time step, both in generating new states and in visiting some of the states whenever the acceptance criterion is met.

The performance of CA is calibrated by comparing with the results obtained by an exhaustive search through all possible TSP solutions. This is possible due to a new factorial number representation for each TSP configuration by an integer n described as follows.

We need one dimensional coding scheme for TSP search space that is 1-1 unique. Because the combinatorial nature of the TSP, a factorial number base system is adopted for coding as follows: (A) The real line x is sampled by the set of real integers x, using the function: Int( ); (B) Then, integers are made periodically in the module base set of

(N-1)!, using the function: Mod(,); and (C) Such an integer number can represent a state of a valid tour since a factorial base set is related to the tour order permutations. Thus, one represents the integer in term of the factorial number base system by calculating the most significant numbers denoted by index (, , ,...).

$$X_{new} = S_n \, index_n \times n! \qquad\qquad \textbf{Eq. 8.16a}$$

$$X_{new} \_ (index_{N-1}, index_{N-2}, ..., index_1, index_0) \qquad\qquad \textbf{Eq. 8.16b}$$

sequentially for all n beginning with N-1 down to 0. To produce the set of index, one considers for example N=5 for five cities denoted by city: #1, #2, #3, #4, #5. Given $X_{old}= 0 = $ (#1,#2,#3,#4,#5) as a reference (the diagonal matrix element of Hopfield-Tank), where the arbitrary tour order is that city #1 is visited first, etc. One finds

$$X_{new}= 15 = 0x0! + 1x1! + 1x2! + 2 \times 3! + 0x4! \_ (\#1,\#4,\#3,\#5,\#2)$$

Where the representation index=(0, 1, 1, 2, 0) is obtained with respect to the base set (0! ,1! , 2! , 3!, 4!).

Thus, the energy corresponding to each of the possible round-trip routes through n cities, 4 ßß n ßß 10 cities has been reported [Szu, 1990]. While the exhaustive search through a 'hundred thousand possible cases had used several hours of CPU time on a Mac II (e.g., five hours for ten cities implies 50 hours for 11 cities), CA took about ten minutes or less to find global minima. The shortest tours agreed with those found by CA. CA is superior because the search required a sampling of less than 1% of the states, with another 2% sampling to verify the stability. Thus, traditional Monte Carlo random sampling should be replaced with the CA algorithm.

8.3.2   Image Processing and Pattern Recognition

Geman and Geman [1984] have applied Boltzmann annealing to the problem of noisy image restoration. Smith et al. [1983] have also applied BA to radiology image reconstructions. Szu and Scheff [1990] have shown that CA can be also useful in pattern recognition. In particular, they have used a mini-max cost function to investigate the self-extraction of unkown features, previously accomplished using self-reference matched filters [Szu et al. 1980; Szu & Blodgett, 1982; Szu & Messner, 1986]. Let the critical feature of the template class-c be denoted as $f_c(x,y)$. Then, a space-filling curve, the Peano N-curve, is employed to replace the traditional line-by-line scanning sampling, in order to preserve the neighborhood proximity relationship.

The performance criterion seeks to minimize the distance between the image template I_c of the c-class (c=1,2), to minimize the inner product between classes $<f_c|f_c'>$, and to maximize the distance $|f_c - f_c'|2$ between two feature vectors. Thus, the mini-max energy for the determination of the global minimum associated with the unknown feature f_c is:

$$E(f_c) = \alpha \sum_{c \neq c'} (< f_c \mid f_{c'} >) + b \sum_{c = 1,2,...} \mid f_c - I_c \mid^2 + \sum_{c \neq c'} d/\mid f_c - f_{c'} \mid^2 \qquad \text{Eq. 8.17}$$

The Lagrangian multipliers a = 10 and d=10 are set higher than b =1 to reflect the less important fact that feature f_c should resemble image I_c.

8.3.3   Proofs of Both Cooling Schedules

There are a number of similarities in the proofs of the cooling schedules for the CA and BA algorithms in D-dimensional vector spaces. For the convenience of comparison, the proofs will be demonstrated in parallel. In locating the minimum, one must start at some position or state in a D-dimensional space, evaluate the function at that state, and generate the next state vector. The CA and BA algorithms are different in that CA uses a Cauchy distribution and BA uses a Gaussian distribution in their respective state-generating functions. Both the BA and CA algorithms will use as their next state either the current state vector or the next state vector provided its incremental cost increase is less than the time dependent noise bound which is temperature (and therefore time) dependent.

The CA algorithm requires that state-generating be infinitely often in time (i.o.t.) (in the sense of accumulation in time defined by the negation below) whereas the BA requires the state-visiting be i.o.t.. At some cooling temperature $T_c(t)$ at time t, let the state-generating probability of being within a specific neighborhood be lower bounded by g_t. Then the probability of not generating a state in that neighborhood is upper bounded by $(1 - g_t)$. To ensure a globally optimum solution for all temperatures, a state in an arbitrary neighborhood must be able to be generated i.o.t., which however does not imply ergoticity, the latter requiring actual visits i.o.t. To prove that a specific cooling schedule maintains the state-generation i.o.t., it is easier to prove the negation of the converse, namely the impossibility of never generating a state in the neighborhood after an arbitrary time t_0.

Mathematically, this is equivalent to stating that the infinite product of $\mid 1 - g(t) \mid$ terms is zero. Taking the Taylor series expansion of the logarithm of the product, one can alternatively prove that the sum of the $g_o(t)$ terms in infinite. One can now verify cooling schedules in a D-dimensional neighborhood $bbc \mid \Delta X_0 \mid$ and arbitrary time t_0. Among the various Levy-Doob distributions (including Cauchy, Holtzmach, and Gaussian) there are two different classes, local (as in CA) and semi-local (as in CA). There exists an initial temperature T_0 and for t > 0, such that

BA
$$T_a(t) = T_0/\log(t) \qquad \text{Eq. 8.17a}$$

CA
$$T_c(t) = T_0(t) \qquad \text{Eq. 8.17b}$$

$$g_t = \exp\left(\frac{|\Delta x_0|^2}{T_a(t)}\right) T_a(t)^{D/2}$$

Eq. 8.18a

$$g_t = \frac{T_c(t)}{(T_c^2 - |\Delta x_0|^2)^{(D+1)/2}} = \frac{T_0}{t\,|\Delta x_0|^{D+1}}$$

Eq. 8.18b

$$\sum_{t=t_0}^{\infty} g_t \geq \exp(\log(t)) = \sum_{t=t_0}^{\infty} \frac{1}{t} = \infty$$

Eq. 8.19a

$$\sum_{t=t_0}^{\infty} g_t \geq \frac{T_0}{|\Delta s_0|^{D+1}} \sum_{t=t_0}^{\infty} \frac{1}{t} = \infty$$

Eq. 8.19b

8.3 SUMMARY

The Boltzmann machine and its refinement, the Cauchy machine, are two networks which are structurally and dynamically similar to the back-propagating Perceptron, and which can perform similar tasks. They learn using an energy state optimization method derived from statistical considerations. This approach to learning allows them to perform optimization tasks as well as pattern recognition.

REFERENCES

Ackley, D.H., Hinton, G.E., and Sejnowski, T.J. (1985). "A Learning algorithm for Boltzmann machine," *Cognitive Science, Vol. 9,* pp. 147-169.

Akiyama, Y., Anzai, U., and Aiso, H. (1990). "The Gaussian machine: A stochastic, continuous Neural Network Model," *J. Neural Network Computing,* Summer.

Alspector, J., Guputa, B., and Allen, R. (1989). "Performance of Stochastic Learning Microchip," In: "Advances in Neural Information Processing Systems I," Touretzky, D. (Ed.), Morgen Kaufmann.

Farhat, N.H., and Psaltis, D. (1987). "Optical Implementation of Associative Memory Based on Models of Neural Networks," In: "Optical Signal Processing," J.L. Horner (Ed.) (Academic, New York).

Geman, S., and Geman, D. (1984). "Stochastic relaxation, Gibbs distributions, and the Bayesian restoration of images," *IEEE Trans. on Patt. Anal. Mach. Int., Vol. PAMI-6,* pp. 614-634, November.

Hinton, G. (1989). "Deterministic Boltzmann learning performs steepest descent in weight-space," *Neural Computation, 1,* 143-150.

Hinton, G., Sejnowski, T., and Ackley, D. (1984). *Boltzmann Machines: Constraint Satisfaction Networks That Learn.* Carnegie Mellon University Technical Report CMU-CS-84-119.

Hinton, G.E., and Sejnowski, T.J. (1986). "Learning and relearning in Boltzmann machines." In D.E. Rumelhart, J.L. McClelland, and the PDP Research Group (Eds.), *Parallel Distributed Parallel, Vol. 1,* MIT Press, Cambridge, MA.

Hopfield, J. J. (April, 1982). "Neural networks and physical systems with emergent collective computational abilities," *Proc. Nat'l. Acad. Sci. USA, 79,* 2554-2558.

Hopfield, J. J. (May, 1984). "Neurons with graded response have collective computational properties like those of two-state neurons," *Proc. Nat'l. Acad. Sci. USA, 81,* 3088-3092.

Kirkpatrick, S., Gellatt, C. D. and Vecchi, M. D. (1983). "Optimization by simulated annealing," *Science, 220,* 671-680.

Maren, A.J. (1987). *A Tutorial on the Boltzmann Distribution and Energy Minimization for Neural Network Ensembles,* University of Tennessee Space Institute, Advanced Human Capabilities Engineering Laboratory Tech. Report AHCEL-87-1.

Metropolis, N., Rosenbluth, A.W., Rosenbluth, M.N., and Teller, A.H. (1953). "Equations of state calculations for fast computing machines," *J. Chem. Phys. Vol. 21,* pp. 1087-1091, June.

Parberry, I., and Schnitger, G. (1989). "Relating Boltzmann machines to conventional models of computation," *Neural Networks, 2,* 59-67.

Scheff, K., and Szu, H. (June 21-24, 1987). "1-D optical Cauchy machine infinite film spectrum search," *Proc. IEEE First ICNN* San Diego, CA, III-673–III-679.

Sejnowski, T.J., Kienker, P.K., and Hinton, G.E. (1986). "Learning symmetry groups with hidden units: Beyond the Perceptron," *Physica, 22D,* 260-275.

Sejnowski, T.J., and Rosenberg, C.R. (1987). "Parallel networks that learn to pronounce English text," *Complex Systems, 1,* 145-168.

Simpson, P.K. (1990). *Artificial Neural Systems: Foundations, Paradigms, Applications, and Implementations,* Pergamon, New York.

Smith, W.E., Barrett, H.H., and Paxman, R.G. (1983). "Reconstruction of objects from coded images by simulated annealing," *Optics Letters, Vol. 8,* pp. 199-201, April.

Szu, H., Blodgett, J., and Sica, L. (1980). "Local Instances of Good Seeing," *Optical Comm., Vol. 35,* pp. 317-322.

Szu, H., and Blodgett, J. (1982). "Self-reference Spatiotemporal Image-Restoration Technizue," *J. Opt. Soc. Am., Vol. 72,* pp. 1666-1669.

Szu, H., and Messner, R. (1986). "Adaptive Invariant Novelty Filter," *Proc. IEEE, V. 24,* p. 519.

Szu, H. (1987a). "Fast Simulated Annealing," In: "Neural Networks for Computing," *AIP Conf. Vol. 15,* pp. 420-425, J. Denker (Ed.), Snow Bird, UT.

Szu, H. (1987b). "Nonconvex Optimization," *SPIE Vol. 968,* pp. 59-65.

Szu, H., and Harley, R. (1987a). "Fast Simulated Annealing," *Phys. Lett. A, Vol. 122,* pp. 157-162, June.

Szu, H., and Harley, R. (1987b). ""Nonconvex Optimization by Fast Simulated Annealing," *Proc. IEEE, Vol. 75,* pp. 1538-1540, November.

Szu, H. (1989). "Reconfigurable Neural Nets by Energy Convergence Learning Principle based on extended McCulloch-Pitts Neurons and Synapses," *Intl. Joint Conf. Neural Networks*, p. I-485–I-496, Washington, D.C., June 18-22.

Szu, H., and Scheff, K. (1989). "Gram-Schmidt Orthogonalization Neural Nets for Optical Character Recognition," *Intl. Joint Conf. Neural Networks, Vol I* p. 547-555, Washington, D.C., June 18-22.

Szu, H., and Foo, S. (1989). "Space-Scanning Curves for Spatiotemporal Representations, useful for large scale neural network computing, *Intl. Joint Conf. Neural Networks*, Washington, D.C., Jan. 15-18.

Takefuji, Y., and Szu, H. (1989). "Parallel Distributed Cauchy Machine," *Intl. Joint Conf. Neural Networks*, p. I-529, Washington, D.C., June 18-22.

Szu, H., and Scheff, K. (1990). "Simulated Annealing Feature Extraction from Occluded and Cluttered Objects," *Intl. Joint Conf. Neural Networks*, Washington, D.C., Jan 15-18.

Szu, H. (1990). "Colored noise annealing benchmark by exhaustive solutions of TSP," *Proc. Second Intl. Joint Conf. Neural Networks*, p. I-317–I-320, Washington, D.C., Jan 15-18.

Wasserman, P.D. (1989b). "A Combined Back-Propagation/Cauchy Machine Network," *J. Neural Network Comp.*, pp. 34-40, Winter.

Wasserman, P.D. (1990). *Neural Computing: Theory and Practice," Van Nostrand Reinhold, New York.*

9

LATERALLY-CONNECTED, AUTOASSOCIATIVE NETWORKS

Alianna J. Maren

9.0 OVERVIEW

Whereas multilayer networks can extract features and thus perform good pattern classification, single-layer fully laterally connected networks can only store one pattern at a time. This enables them to perform autoassociation, by constantly refining the pattern until it is dynamically stable during network operations. This dynamic stability arises when the network's units take on values which correspond on an energy minimum in an energy space. Dynamic stability is an important issue in all networks which have recurrent processing through their units, such as the ones we will consider in this chapter.

This chapter covers three types of autoassociators. Two of these networks, the Hopfield/Tank network and Anderson's Brain-State-in-a-Box network (BSB), are well-known. A new autoassociative technique called Sparse Distributed Memory (SDM), developed by Kanerva, is also included. Taken together, they form a *core set* of autoassociative network methods.

The Hopfield/Tank network (or Hopfield network) and the Brain-State-in-a-Box network are both well-known examples of single-layer, laterally connected networks. They both perform autoassociation, and the Hopfield network is also an optimizer. They have somewhat different structures, and very different learning rules. Their dynamics of operation are very similar. The performance level of the Brain-State-in-a-Box is somewhat higher than that of the Hopfield network.

A new network, the Sparse Distributed Memory, has an interesting and novel approach to representing patterns in a truly distributed manner. In this, it is conceptually very different from any of the other autoassociative networks. It is best used for very large, sparsely-coded patterns.

9.1 INTRODUCTION TO ASSOCIATION NETWORKS

In the previous two chapters, we examined multilayer feedforward networks that work as pattern classifiers. In this chapter, we begin to examine a different type of network — those which act as pattern associators. Now we could split many a hair on making a distinction between these two concepts. However, we are trying to make some distinctions in a realm where distinctions are difficult, even strained. We are trying to get a coherent clustering of most related neural network types together, so that they can be better understood. It is in this context that we want to shift our attention from the process of classification to the process of association.

There are two basic types of *association networks*. Autoassociative networks correlate a pattern with itself. Heteroassociative networks associate one pattern with another one. Resonating networks are a special class of heteroassociative networks. Each of these network types has different structures and dynamics.

9.2 AUTOASSOCIATIVE NETWORKS

Autoassociative networks can regenerate a noise-free, complete pattern from one that is noisy or incomplete. As these networks were among the first associative networks to be developed, they often had simple architectures, generally single-layered. Each different type of autoassociative network has a different way of learning patterns, and can hold only a certain number of patterns in its memory. The Hopfield/Tank (or Hopfield) network and the Brain-State-in-a-Box network exemplify this type of single-layer network.

There are several important uses for autoassociative networks. The most obvious use is to create clean patterns. This is often done as part of a larger process. For example, the input to a classification network may be put through an autoassociative stage first to filter the data (described more fully in the last chapter of Part II). Alternatively, it is possible to have a classification scheme or heteroassociation process which outputs a distributed pattern instead of simply activating a single node. This output pattern might be passed through an autoassociative processor to clarify a pattern which might have been ambiguous.

There are a number of other possibilities for autoassociative processes which have not yet been widely tapped in the current research. One would be to use an autoassociative network such as Kohonen's Learning Vector Quantization network (which we discuss in the next chapter) or Kanerva's Sparse Distributed Memory network (which we discuss at the end of this chapter). Either of these yields neighborhood relationships in their mapping, so that similar classes of input are topological neighbors in their representation in the network. This type of property can be useful for exploring relationships between similar patterns. This can be used to aid common sense reasoning (a still largely unachieved goal in the artificial intelligence community), to facilitate database mining, to establish relationships between fuzzy classes, and other tasks.

Most current autoassociative methods use spatially-occurring patterns as input, and generate a spatially-occurring output which is temporally stable and unchanging. A few auto- and heteroassociators take in temporally-varying input and generate corresponding temporal output. A few methods allow associative networks to work with both spatially and temporally varying input and output. We briefly touch on these methods in this and the following chapter. We give more specific methods in Part IV.

9.2.1   The History of Autoassociative Networks

In 1967, Shun-Ichi Amari published a paper which set a mathematical foundation for a learning theory (error-correction method) for adaptive pattern classification. In 1972, he expanded on his work to show how a self-organizing network could form a representative pattern from a set of stimulus patterns, and fix this new pattern as a stable state of the network. He considered this self-organizing network to form a model for associative memory. Amari's rigorous mathematical treatments do not make for easy reading, but form part of the classical basis for the field of neural network computing.

In 1972, one of the first books on neural networks appeared. *Parallel Models of Associative Memory*, edited by James Anderson and Geoffrey Hinton, helped establish the context for an emerging field that dealt with neurally-based computation as a means of mimicking intelligent processes.

Steve Grossberg also developed some early autoassociative networks: the Additive and Shunting Grossberg networks. His early work [Grossberg, 1968 & 1973] became the precursors to his more robust and general Adaptive Resonance Theory network (ART). We will consider the work of Gail Carpenter and Steve Grossberg on ART in substantial detail in a later chapter on heteroassociative networks, along with Bart Kosko's Bidirectional Associative Memory systems.

9.3   THE HOPFIELD/TANK NETWORK

Of all the autoassociative networks, the Hopfield (or Hopfield/Tank) network is the most widely known today. Developed in 1986, it is a fairly recent network. This network has achieved prominence because it is relatively easy to implement in VLSI chips. Also, it illustrated a clear application of the physics of energy surface minimization to finding stable solutions in the network's pattern of activation. This application contributed to a sense of some underlying mathematical formalisms which could be used to establish a basis for renewed work in neural networks.

The Hopfield/Tank network is useful both as an autoassociator and for optimization tasks. In this chapter, we will primarily consider it as an autoassociator, since that is the way in which all the networks presented here operate. We discuss the optimization capabilities of the network briefly in the section of the capabilities and limitations of the network, and again in Part IV.

9.3.1   The Concepts Underlying the Hopfield Network

The major concept underlying both the Hopfield network and Anderson's Brain-State-in-a-Box is the same. This concept is that a single network of interconnected, binary-value neurons can store multiple stable states.

Suppose we create a network of binary-valued neurons, where each neuron is connected to the other, but not back to itself. (There is no direct feedback in this network.) Assume all the connection strengths are symmetric. This means that for any two neurons, i and j, then $a_{ij} = a_{ji}$, where a is the strength of the connection between these two neurons.

This network can have a set of stable states. For each stable state, each binary-valued neuron takes on a value (0 or 1) so that when it acts on its neighbors (via the connection strengths described in the previous paragraph), the values of each neuron do not change. If we give this network an input pattern close to one of these states, and let the neurons affect each other, then the network can converge to that nearby state.

The implication of this capability is that we can use the network as an autoassociator. First, we create a network which stores a number of stable states. We will create a matrix which stores the connection weights between the neurons. Then, we can present a pattern to the network. This gives each neuron a value. We let the neurons affect each other via their connection weights. If is at all close to one of the patterns or stable states which the network has learned, then the neuron values will change to be those of that learned pattern. This gives us the potential ability to input a noisy or partial pattern into the network, and regenerate the closest learned pattern. The concepts underlying the Hopfield network were published by Hopfield [1982, 1984], and by Hopfield and Tank [Hopfield & Tank; 1986, Tank & Hopfield; 1987]. These papers discuss how the convergence of a presented pattern follows a path of energy minimization for the total energy of the network.

9.3.2   The Hopfield Network Architecture

This network can be implemented as a single layer of binary-valued (0,1) neurons. (Extensions to continuously valued neurons are also possible.) Each neuron is symmetrically connected to all other neurons in the network, but not back to itself. This architecture is shown in Figure 9.1.

9.3.3   Dynamics of the Hopfield Network

The input pattern should be read directly into the single network layer. Once the input pattern is read in, a neuron is chosen at random. It receives input from each of the other neurons, where each input is multiplied by its connection weight. It sums up these inputs, and passes the value through a bi-state transfer function. Thus, if the summed value is greater than or to a threshold (e.g., 0), then the neuron is "on" (1), if less than or equal to the threshold, it is turned "off" (0). This is the single processing step in the Hopfield/Tank network. This process continues until all neuron values are stable; that is, none of them change value when they go through this process. An illustration of this process was given in Part I. Somewhat more complex dynamics are possible for continuously valued Hopfield networks [Hopfield, 1984, Hopfield and Tank, 1986].

After the neurons interact with each other, the output will be the value of each neuron in the single-layer network. This should correspond to one of the known (stored) patterns.

The Discrete and Continuous Hopfield Network

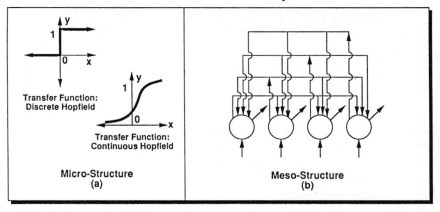

Figure 9.1   The Hopfield network is one of the simplest autoassociative networks. (a) The micro-structure for the "discrete" Hopfield nework specifies that all neurons are simple bi-state neurons. The transfer function for the "continuous" Hopfield network is continuous. (b) The meso-structure is a single layer of laterally-connected neurons without recurrent (self) connections.

9.3.4   Learning in Hopfield Networks

The Hopfield network must learn all of its patterns in a single learning step in which the values for the connection matrix are defined. This means that we have to have a complete set of all the patterns which we want the network to learn at the very outset. These patterns should be orthogonal, or as nearly orthogonal as possible.

The connection matrix is created via an algorithm that presumes a Hebbian-type of learning of connection weight values between neuron synapses. It is basically a sum of the pattern vectors, each multiplied by their transpose. Thus, if A is the connection matrix, and we have a set of n pattern vectors X_i, then

$$A = \sum_i^{n_i} X_i^T X_i$$

Eq. 9.1

Figure 9.2 illustrates the process of creating a Hopfield network matrix for a simple network storing two four-element orthogonal patterns.

9.3.5   The Performance of Hopfield Networks

Difficulties with Hopfield nets may arise if the exemplar patterns selected for storage do not form stable states in the interactive convergence procedure. If the Hopfield network does converge (during operation) to a stable state, then that state will hold for the remainder of its existence. However, the corollary property of this network is that it is not adaptive; it cannot learn in real time. Cohen and Grossberg [1983] have written

Creating a Connection Weight
Matrix for a Discrete Hopfield Network

Use Equations: $\overline{Y}_s = 2\overline{V}_s - \overline{1}$

$$\overline{W} = \sum_{s=1}^{N} (\overline{y}_s\,\overline{y}_s^{T} - \overline{I})$$

Where:

$\overline{Y}_s$ is the 5th vector in an set of N vectors
$\overline{Y}_s^{T}$ is the transformed vector
$\overline{I}$ is the identity matrix, $\overline{1}$ is a vector of 1's
$\overline{W}$ is the connection weight matrix

For an orthogonal two-vector set

$\overline{V}_1 = (1\ 1\ 0\ 0)$, $\overline{V}_2 = (0\ 0\ 1\ 1)$, $\overline{Y}_1 = (1\ 1\ -1\ -1)$, & $\overline{Y}_2 = (-1\ -1\ 1\ 1)$.

$\overline{W}$ is found as:

$$\begin{bmatrix} 1 \\ 1 \\ -1 \\ -1 \end{bmatrix} [1\ 1\ -1\ -1] - \overline{I} = \begin{bmatrix} 1 & 1 & -1 & -1 \\ 1 & 1 & -1 & -1 \\ -1 & -1 & 1 & 1 \\ -1 & -1 & 1 & 1 \end{bmatrix} - I = \begin{bmatrix} 0 & 1 & -1 & -1 \\ 1 & 0 & -1 & -1 \\ -1 & -1 & 0 & 1 \\ -1 & -1 & 1 & 0 \end{bmatrix}$$

$$\begin{bmatrix} -1 \\ -1 \\ 1 \\ 1 \end{bmatrix} [-1\ -1\ 1\ 1] - \overline{I} = \begin{bmatrix} 1 & 1 & -1 & -1 \\ 1 & 1 & -1 & -1 \\ -1 & -1 & 1 & 1 \\ -1 & -1 & 1 & 1 \end{bmatrix} - I = \begin{bmatrix} 0 & 1 & -1 & -1 \\ 1 & 0 & -1 & -1 \\ -1 & -1 & 0 & 1 \\ -1 & -1 & 1 & 0 \end{bmatrix}$$

$$\overline{W} = \begin{bmatrix} 0 & 1 & -1 & -1 \\ 1 & 0 & -1 & -1 \\ -1 & -1 & 0 & 1 \\ -1 & -1 & 1 & 0 \end{bmatrix} + \begin{bmatrix} 0 & 1 & -1 & -1 \\ 1 & 0 & -1 & -1 \\ -1 & -1 & 0 & 1 \\ -1 & -1 & 1 & 0 \end{bmatrix} = \begin{bmatrix} 0 & 2 & -2 & -2 \\ 2 & 0 & -2 & -2 \\ -2 & -2 & 0 & 2 \\ -2 & -2 & 2 & 0 \end{bmatrix}$$

Figure 9.2   Illustration of the process of operating a Hopfield network connector weight matrix.

a significant paper which defines stability criterion for associative networks, including the Hopfield network and others.

There are several problems inherent in the use of Hopfield networks. First, the maximal number of patterns which can be stored is only about .15 times the number of nodes in the network. Also, there is no guarantee that even that limited number of patterns can be stored. If the patterns are close to each other (i.e., very non-orthogonal), the network will not converge to a stable state which distinctly represents each pattern. Finally, it is possible that a noisy pattern presented to the network will not produce convergence to its exemplar.

There is yet another problem with the original Hopfield/Tank algorithm. The Hopfield network can be used as an optimizer as well as an autoassociator. In this context, Wilson and Pawley [1988] assessed the stability of the Hopfield network for an example optimization problem, the "Traveling Salesman Problem." They found a severe difficulty in scaling up the Hopfield network. The problem was that once a neuron's activation had been downgraded from a 1 to a 0 (or if it originally started at a value of 0), the Hopfield/Tank algorithm made it very difficult to return to a 1 value. This makes it difficult for a pattern with lots of 0's to associate with the right exemplar.

Many researchers have proposed solutions to this problem. One of the most effective seems to have been proposed by Harold Szu [1988]. He modified the original Hopfield/Tank equations, added the additional constraint of local excitation (to be used when a neuron went to a 0 state), and suggested the use of binary neurons for fast but accurate computation.

Kennedy and Chua [1987] identified another error in the Hopfield/Tank network equations. They analyzed this network using circuit theory to show that the circuit (network) minimized the total content function, but that a bounded solution was not guaranteed. They proposed a modification to the nonlinearity function used to implement constraints in the problem, and were thus able to guarantee the existence of a bounded solution.

Bruck and Sanz [1988] have also reviewed the known properties of the Hopfield network. They show that there is a large class of mappings which cannot be performed by neural networks, explore the concept of memory capacity of a network, show how the network can perform local search, and demonstrate some limitations on using this network as a pattern recognizer.

To address the problem of low pattern storage, a group of researchers [Chen et al, 1986; Simpson, 1990] developed a higher-order method for associative networks. They showed that by using higher-order correlations between elements and between patterns within sets of patterns, they could achieve dramatic increases (a hundredfold) in the numbers of patterns which could be effectively stored. It is also worth noting that Venkatesh [1986] has written a classic article on the maximal storage capacity of a neural network. (The results reported by Chen et al. seem to exceed the predicted limits reported by Venkatesh; this is due to incorporation of the richer data storage brought about by using higher-order correlations between neurons.)

The primary reason that the Hopfield network has attracted so much attention is that several researchers have shown that it is a relatively easy network to build in hardware using currently available VLSI technology. (As mentioned earlier, the weights are non-adaptive.) However, given the strong impetus by DoD to design and build neuromorphic chips with adaptive weights, the emphasis currently placed on the Hopfield network could wane in the next few years.

An example of the type of network which could be more effective than a simple Hopfield network for autoassociation is the *Hamming network*, developed by Richard Lippmann [Lippman, 1987; Lippman et al., 1987]. This network is covered in the last chapter of this part of the book, which deals with hybrid networks. Also, the *Brain-State-in-a-Box* network (presented in the next section) appears to be more robust than the Hopfield/Tank network.

9.4   THE BRAIN-STATE-IN-A-BOX NETWORK

Anderson's *Brain-State-in-a-Box* (BSB) network [Anderson et al., 1977; Anderson, 1988] has some conceptual similarities to the Hopfield network, and is actually a precursor to the simpler Hopfield network. The BSB network is also more robust than the Hopfield network and has a better inherent memory capacity. Like the Hopfield network, the BSB performs autoassociative recall of binary-valued pattern vectors. The BSB pattern vectors are encoded as patterns of -1 and 1 values, whereas the Hopfield binary patterns are in terms of 0 and 1 values. To be effective, the stored BSB pattern vectors should be orthogonal or nearly orthogonal. The memory capacity of the BSB network is equal to the dimension of the network, which is the same as the number of orthogonal pattern vectors which can be stored.

9.4.1   The Concepts Underlying Anderson's Brain-State-in-a-Box

In developing the BSB, Anderson used as his major concept the fact that a single network of interconnected, binary-value neurons could have multiple stable states. Anderson correlates the two possible extreme values of the neurons in his model (-1 and 1) with minimal and maximal firing rates of a neuron. This assumes an implicit correlation of -1 with a minimal but positive (or null) level of neural activity. His model also assumes that the neurons perform identically; the maximal firing rates are the same, and are correlated to a +1 neural value in his model.

9.4.2   Brain-State-in-a-Box Architecture

In the BSB, neural activity is bounded by these -1 and +1 values, and can take on any value between these two limits. Thus, while the inputs to this network are binary-valued, the intermediate and output neural activities are continuous-valued. This leads to a model which is partially analog and partially binary, as shown in Figure 9.3 (a).

Anderson further assumes that the single layer of neurons in his model are fully connected via a symmetric weight matrix (like the Hopfield network). Unlike the Hopfield network, the neurons in the BSB also feed back directly to themselves, as shown in Figure 9.3 (b). This allows a neuron to directly affect its own value as the network converges towards a stable state.

The connection strengths between two neurons increase via a Hebbian-type of learning. In order to put bounds on the possible unchecked growth of activity of each neuron, Anderson put (-1,1) limits on the possible values of neural activity (or firing rates).

The Brain State-in-a-Box Network

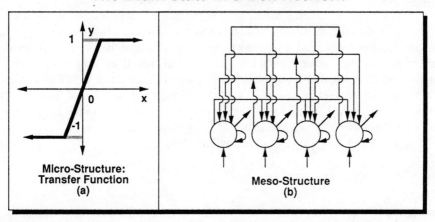

Figure 9.3   The architecture of the Brain-State-in-a-Box (BSB) network is very similar to that of the Hopfield/Tank network. (a). At the micro-structural level, the neurons may take on a range of activations from -1 to 1. (b). At the meso-structural level, the BSB network is a single layer of laterally connected nodes with recurrent (self) connections.

9.4.3   Dynamics of the Brain-State-in-a-Box

After an input pattern has been presented, convergence can generally be achieved within eight to twenty steps. Each step of the convergence process produces a modification of the input pattern vector, where the vector values move towards the corners of the BSB hypercube. For each step, with an original pattern X,

$$X(t+1) = f[X(t) + AX(t)], \qquad \text{Eq. 9.2}$$

Where f is the hard-limit transfer function shown in Figure 9.2 (a)

A is the connection matrix

The convergence method used in the BSB usually leads to one of the known, stable states appearing as output. This will generally be the pattern closest to the presented input pattern. Cohen and Grossberg [1983] demonstrated that the BSB network is a subclass of a type of globally stable networks.

9.4.4   Learning in the Brain-State-in-a-Box Network

Learning in the BSB is a separate activity from pattern association. However, patterns can be learned throughout the operation of the network. The learning rate is governed by a single parameter, α.

Each pattern, once learned, is temporally stable in the BSB. Learning takes place by modifying the connection weight matrix A, which is initially formed in a manner similar to the Hopfield network connection matrix. For this to occur, one should have an starting set of exemplar vectors which will be used to produce the initial connection matrix A.

The BSB network is usually thought of as an autoassociation network, but can function as a heteroassociation network as well. Let's think of the exemplar patterns as a set of patterns {h}, where h_i is the ith pattern in the set. If we want to consider the heteroassociative possibilities of the BSB, then we will decompose each pattern vector h_i into two pattern vectors, f_i and g_i, so that h_i is the concatenation of f_i and g_i. The meso-structure in the heteroassociative case would use two layers or fields of neurons, with connections from neurons in one field going to those in another.

Thus, the initial correlation matrix A will be given by:

$$A = h_i h_i^T \text{ (for the autoassociative case)} \qquad \text{Eq. 9.3a}$$

$$A = g_i f_i^T \text{ (for the heteroassociative case)} \qquad \text{Eq. 9.3b}$$

It is possible to build up A over time, to reinforce an exemplar, or to extract a pattern exemplar from a category if enough patterns within that category exist (and are similar enough) so that an exemplar can be established. This is done by incrementing A by an amount Δ A with each pattern presentation:

$$A(\text{new}) = A(\text{old}) + \Delta A, \qquad \text{Eq. 9.4}$$

$$\text{where } \Delta A = \alpha h_i h_i^T \text{ (for the autoassociative case)} \qquad \text{Eq. 9.5a}$$

$$\text{and } \Delta A = \alpha g_i f_i^T \text{ (for the heteroassociative case).} \qquad \text{Eq. 9.5b}$$

Now, unlike the discrete Hopfield network (but similar to the continuous Hopfield), the BSB network can continue to refine its weight vectors to optimally extract the exemplar patterns even when the input data is noisy or incomplete. This is done by continuous learning, which is applied after every pattern presentation to the network. This type of learning is error-based learning, or a modification of the Widrow-Hoff approach (first introduced in the first chapter of this part of the book as the learning method for the ADALINE network). This learning requires that there be a supervisor, or some knowledge of what the desired output should be. The amount of increase when the modified Widrow-Hoff learning is applied is:

$$\Delta A = \alpha(h'_i - Ah_i)h_i^T \text{ (for the autoassociative case),} \qquad \text{Eq. 9.6a}$$

where h'_i is the desired output vector, or exemplar, when h_i is used as input

$$\Delta A = \alpha(g_i - Af_i)f_i^T \text{ (for the heteroassociative case),} \qquad \text{Eq. 9.6b}$$

where g_i is the desired output associated with the input vector f_i.

The BSB system has been shown to be stable when there is no error in the input/output encoding, because there is no learning in that situation.

9.4.5   The Performance of the Brain-State-in-a-Box Network

The memory capacity of the BSB network is equal to the dimension of the network, which translates to the number of orthogonal pattern vectors which can be stored. Thus, if the network contains N neurons, then the memory capacity is N patterns.

Unlike the Hopfield network, spurious states are rare. The only occasion in which a spurious state might arise would be when a pattern element gets stuck at a value of 0, and does not move towards either of the -1 or +1 extreme. Even when this occurs, the remaining pattern elements will correspond to those of the closest stored pattern.

9.4.6   Applications of the Brain-State-in-a-Box Network

Penz et al. [1989] has shown how a BSB network modified for analog representation can be applied to anti-radiation homing missiles.

Rossen and Anderson [1989] have used the BSB as a filter for the results of a classification network. This illustrates the type of fusion of networks into systems which we discuss in more detail in Chapter 10.

9.5   KANERVA'S SPARSE DISTRIBUTED MEMORY NETWORK

Recently, Pentti Kanerva has introduced a new model for associative memory. This model is interesting, but its usefulness awaits determination. In order to use this model, a person has to have a lot of memory space available in a computer. The input patterns should be large (on the order of a 100-10,000 elements) and binary. Each specific feature of a pattern is represented by a binary on/off value.

9.5.1   The Concept Underlying the Sparse Distributed Memory Network

The basic concept underlying the Sparse Distributed Memory (SDM) network lies in the fact that a pattern of vector length n is stored in a distributed manner about its location in n-dimensional space. When a pattern is stored, it is not just the original pattern which is stored but a distributed representation of it: a large number of patterns which are within a certain metric distance of the original pattern. A representation of this storage would look somewhat like a probability density function of an electron around an atomic nucleus.

A specific and useful aspect of sparse distributed memories is that similar patterns are stored in similar locations. That is, if a binary pattern n is stored in a location x, then reading from x retrieves n. What is more important is that reading from a nearby location to x, say x', retrieves a pattern similar to n. This quality also exists in the Learning Vector Quantization network (which will be discussed in the next chapter), but that

**Storage in the
Sparse Distributed Memory**

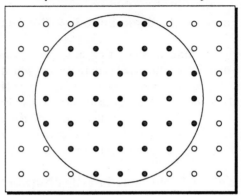

Figure 9.4 In a Sparse Distributed Memory (SDM) network, a binary pattern of high dimensionality is stored by writing a copy of that pattern into each of the locations within a critical radius around the pattern's location. The pattern's value is conceptually the same as its location in memory.

network is ideally suited for patterns of low to moderate dimension. The sparse distributed memory approach is well suited for patterns of large dimension.

A brief introduction to Kanerva's SDM approach can be found in [Kanerva, 1986]. For those who are willing to go into more depth, he has written a delightful book, *Sparse Distributed Memory* [Kanerva, 1988].

9.5.2 The Architecture of the SDM Network

A sparse distributed memory can be constructed by using a memory space with large dimension. The patterns stored can be multivalued, but the simplest type will use binary patterns. In this approach to autoassociative memory, the patterns stored and their memory locations will be indistinguishable. For data retrieval, a presented pattern acts as a pointer to retrieve a datum from the storage location given by that pattern. This retrieved datum, consisting of a pattern of the same dimension as the first, can be a new address for data retrieval.

A particularly noteworthy aspect of the SDM is that a pattern is stored not just in one location, as one pattern, but by activating all patterns within a certain metric distance of the original key pattern. This means that a single pattern is redundantly stored. This is a major difference between the SDM and all other neural networks. While other networks have distributed storage by storing patterns instead of single-valued symbolic or numeric data, the SDM goes one better. Not only is the original data encrypted and stored as a pattern, the pattern itself is distributed and multiply stored. This concept is illustrated in Figure 9.4.

To implement this, each memory word (or memory location) is broken up into bit locations. Each bit location will require three lines or connections; an address decoder for selecting the location, an address encoder which allows its contents to be read, and a third connection which will allow the contents of that location to be updated. Activity on the encoder-decoder line(s) can activate the location for a read or write operation, but the data input line is necessary to change the value stored in that location.

9.5.3   The Dynamics of the SDM Network

When a pattern must be read out, the key pattern is used to find the nearest pattern stored in memory. This triggered pattern is used as a new key. This cycle is repeated until a trigger pattern regenerates itself on readout. Pattern readout is stable if the key pattern is within a certain radial distance of the stored, *attractor* pattern. If the key pattern is too distant (in terms of Hamming metric similarity, which is the number of different bits, and corresponds to the physical distance) from the stored pattern, it will either generate a pattern activation sequence which will fall into the attractive basin of another pattern, or it will meander and be stopped only by a control process which limits the number of iterative read steps.

Because there is a possibility of divergence, and associative recall sequences wandering throughout the storage space, the SDM uses an internal check (measuring the distance between successive words recalled) to determine whether the recall sequence is converging or diverging. Generally, this can be established within 10 to 20 iterations. Divergent sequences are stopped after they have been determined to be divergent. There is no global stability as has been demonstrated for the Hopfield network.

The original pattern may not be stored exactly. (Recall that the storage space needed for implementation of this memory is substantially less than the storage space that would be implied by taking two to the power of the pattern dimension.) Thus, accurate pattern readout is guaranteed by averaging all of the patterns generated during the read-out process.

9.5.4   Learning in the SDM Network

When a word (pattern vector) is stored in an SDM, a copy of it is written into the location corresponding to the pattern vector itself. However, in keeping with the distributed memory concept, a copy of this word is written into each of the other locations within a critical radius of the target location. When multiple words are written into a storage location, the word which actually becomes represented there is an average of the words which have been stored in that spot. Decoding is possible because when a key pattern vector (location) is presented to the system, it generates a large number of words from within the local area. These are averaged to produce the original target word.

9.5.5   The Performance of the SDM Network

A major consideration with all neural networks is their pattern memory. In the case of the SDM, the number of patterns is sparsely encoded relative to the number of possible orthogonal patterns given the network's high dimension. Thus, memory considerations in the SDM more appropriately relate to the storage necessary to create and handle the network.

The SDM network does not require as much storage space as it would initially appear. At first glance, the SDM approach would suggest that if pattern vectors are n units long, that 2 (to the power of) n 2^n memory locations would be needed. However, this is not the case. Only a small fraction of these memory locations are needed for operation when selected randomly from among the total theoretically possible locations. This still does not make the SDM method the preferred choice for those of us working on PC's. However, for those of us who want to work large-scale, the SDM method can be just fine. (Also, we need to be dealing with applications which require long pattern vectors for pattern specification. Not all applications have that need.) Recently, Surkan [1989] has described a modified SDM which requires substantially less storage space and processing time.

9.6 SUMMARY

Single-layer, laterally connected, recurrent processing networks can act as autoassociators. The two well-known networks of this sort are the Hopfield/Tank network and the Brain-State-in-a-Box (BSB) network. These networks both have symmetric connectivity. The BSB metwork has self-connections, the Hopfield network does not. The dynamic stability of the Hopfield network allows us to have confidence that the network processes will converge to a stable solution. This network can act as an optimizer as well as an autoassociator. This network can be readily implemented in existing VLSI technology. Unfortunately, the basic Hopfield network has many problems, including low memory capacity and a tendency to produce spurious results. The BSB network has somewhat better performance, but still has relatively low memory capacity. It is also dynamically stable. A new network, the Sparse Distributed Memory network, may offer useful performance on very large, sparse pattern vectors.

REFERENCES

Amari, S. (1967). "A theory of adaptive pattern classifiers," *IEEE Trans. on Electronic Computers, EC-16*, 299-307.

Amari, S. (1972). "Learning patterns and pattern sequences by self-organizing nets of threshold elements," *IEEE Trans. Computers, C-21*, 1197-1206.

Amari, S., & Maginu, K. (1988). "Statistical neurodynamics of associative memory," *Neural Networks, 1*, 63-74.

Anderson, J.A., Silverstein, J.W., Ritz, S.A., & Jones, R.S. (1977). "Distinctive features, categorical perception, and probability learning: some applications of a neural model," *Psych. Review, 84*, 413-451, and reprinted in J.A. Anderson and E. Rosenfeld (Eds.) (1988). *Neurocomputing* MIT Press, Cambridge, MA.

Anderson, J. A. (1983). "Cognitive and psychological computation with neural models," *IEEE Trans. on Systems, Man, & Cybernetics, Vol. SMC-13*, 799-814.

Bruck, J., & Sanz, J. (1988). "A study on neural networks," *Int'l. Journal of Intelligent Systems, 3*, 59-75.

Chen, H.H., Lee, Y.C., Sun, G.Z., Lee, H.Y., Maxwell, T., & Giles, L.C. (1986). "High order correlation model for associative memory," in J.S. Denker (Ed.), *AIP Conference Proc. 151, Neural Networks for Computing*, American Institute of Physics, New York, NY, 86-99.

Cohen, M., & Grossberg, S. (1983). "Absolute stability of global pattern formation and parallel memory storage by competitive neural networks," *IEEE Trans. on Systems, Man, & Cybernetics, SMC-13*, 815-825.

Grossberg, S. (1968). "Some nonlinear networks capable of learning a spatial pattern of arbitrary complexity," *Proc. Nat'l. Acad. Sciences, 59* (1968), 368-372.

Grossberg, S. (1973). "Contour enhancement, short-term memory, and constancies in reverberating networks," *Studies in Applied Mathematics, 52*, 217-257.

Hinton, G.E., & Anderson, J.A. (Eds.) (1989; revised ed.). *Parallel Models of Associative Memory* Lawrence Erlbaum, Hillsdale, N.J.

Hopfield, J.J. (1982). "Neural networks and physical systems with emergent collective computational abilities," *Proc. National Academy Science, USA, 79*, (April), 2554-2558, and reprinted in J.A. Anderson and E. Rosenfeld (Eds.) (1988). *Neurocomputing* MIT Press, Cambridge, MA, 460-464.

Hopfield, J.J. (1984). "Neurons with graded response have collective computational properties like those of two-state neurons," *Proc. Nat'l. Acad. Sciences USA, 81*, 3088-3092, and reprinted in J.A. Anderson and E. Rosenfeld (Eds), *Neurocomputing* MIT Press, Cambridge, MA, 579-584.

Hopfield, J.J., & Tank, D. (1986). "Computing with neural circuits: A model," *Science, 233*, 625-633.

Kanerva, P. (1986). "Parallel structures in human and computer memory," in J.S. Denker (Ed.), *AIP Conference 151, Neural Networks for Computing*, New York, NY, American Institute of Physics, 247-258.

Kanerva, P. (1988). *Sparse Distributed Memory* MIT Press, Cambridge, MA.

Keeler, J.D. (1988). "Comparision between Kanerva's SDM and Hopfield-type neural networks," *Cognitive Science, 12*, 299-329.

Kennedy, M.P., & Chua, L.O. (1987). "Unifying the Tank and Hopfield linear programming circuit and the canonical nonlinear programming circuit of Chua and Lin," *IEEE Trans. Circuits and Systems, CAS-34*, 210-214.

Lippmann, R.P., Gold, B., & Malpass, M.L. (May 12, 1987). "A comparison of Hamming and Hopfield neural nets for pattern classification." *MIT Lincoln Lab. Tech. Report 769* Lexington, MA.

Lippmann, R.P. (April, 1987). "An introduction to computing with neural nets," *IEEE ASSP Magazine*, 4-22.

Penz, P.A., Katz, A., Gately, M.T., Collins, D.R., & Anderson, J.A. (June 18-22, 1989). "Analog capabilities of the BSB model as applied to the anit-radiation homing missile problem," *Proc. Int'l. Joint Conf. on Neural Networks* Washington, D.C., II-7–II-11.

Rossen, M.L., & Anderson, J.A. (1989). "Representational issues in a neural network model of syllable recognition," *Proc. First Int'l. Joint Conf. on Neural Networks* Washington, D.C.; June 18-22, 1989, I-19–I-25.

Simpson, P.K., (1990). *Artificial Neural Systems*, Pergamon, New York.

Surkan, A.J. (1989). "Fast trainable classifier by a modification of Kanerva's SDM model," *Proc. Int'l. Joint Conf. on Neural Networks* Washington, D.C.; June 18-22, I-347–I-350.

Szu, H. (1988). "Fast TSP algorithm based on binary neuron output and analog neuron input using the zero-diagonal interconnect matrix and necessary and sufficient constraints of the permutation matrix," *Proc. Second Int'l. Conf. on Neural Network* San Diego, CA; July 24-27, II-259–II-266.

Tank, D.W., & Hopfield, J.J. (1987). "Collective compuatation in neuronlike circuits," *Scientific American, 257,* 104-114, 158.

Venkatesh, S.S. (1986). "Epsilon capacity of neural networks," in J.S. Denker (Ed.), *AIP Conference Proc. 151, Neural Networks for Computing* American Institute of Physics, New York, NY., 440-445.

Wilson, G.V., & Pawley, G.S. (1988). "On the stability of the travelling salesman problem algorithm of Hopfield and Tank," *Bio-Cybernetics, 58,* 63-70.

10

VECTOR-MATCHING
NETWORKS

Alianna J. Maren

10.0 OVERVIEW

The Learning Vector Quantization network and the self-organizing Topology-Preserving Map are two leading networks which operate on a vector matching principle, rather than the usual network operation of summing connection-weighted inputs from afferent neurons. The Topology-Preserving Map is a refinement of the earlier Learning Vector Quantization network. The Learning Vector Quantization network is useful for autoassociation, data compression, and as a preprocessor for pattern classification. The Topology-Preserving Map is useful for (reduced dimensional space) mappings, and for optimization and control applications.

10.1 INTRODUCTION

The two types of networks which we discuss in this chapter, the Learning Vector Quantization (LVQ) network and the self-organizing Topology-Preserving Map (TPM) network, are very different from all the networks considered up to this point. We consider them here because they can be thought of as single-layer networks, and function as autoassociators and optimizers, thus giving them an applications potential similar to that of the networks considered in the previous chapter.

The first important distinction to make is that these networks have a unique structure. At the micro-structural level, none of the units attempt to emulate biological neurons. They do not take in afferent signals and output a correspondingly transformed value. At the meso-structural level, we note that the networks do not have connections, in the way that the networks described in all the previous chapters had. However, each *neurode* (or pattern vector) can be related to similarly-valued pattern vectors in a topological sense.

The learning rules and the dynamics are also markedly different from those for other neural networks. Further, the resulting data representation is very different; the Topology-Preserving Map, in particular, allows for an unusual form of data compression.

The type of network we will describe here is so radically different from the others that our previous experience with (neural) networks may be usefully (if temporarily) erased from one's mind. If we can explore this new type of network without any preconceived notions of what a network should be, it will be much easier to comprehend the principles of operation for the networks described in this chapter.

10.2 THE KOHONEN LEARNING VECTOR QUANTIZATION NETWORK

The Kohonen self-organizing *Learning Vector Quantization* (LVQ) network [Kohonen, 1988a, 1989a] is unique among those discussed so far in that it is the first one to organize its own representation of categories among the input data. All other networks have either had supervised training methods in which the network was taught to recognize an exemplar pattern via adaptive weight-changing algorithm, or had fixed weights and were unable to learn.

10.2.1 The Concept Underlying the LVQ Network

The basic concept underlying the Kohonen network is that we can distribute a set of vectors across a space so that the way they *span* the space mimics the probability distribution of a set of training data. This is an efficient data compression scheme which can be used for codebook accesses and similar tasks.

As an example, suppose that we have 100 training examples, and we want to represent them with a set of 10 vectors. Let's further assume that these vectors are two-dimensional. (Note that this means that we will be using a non-orthogonal set, as we could have at most two orthogonal vectors in a two-dimensional space. This is another significant difference between the Kohonen network and those we have just considered.)

An additional implication of selecting a set of classification vectors is that we can predefine the number of classes which will be used, but that we won't know exactly what each class will represent, so there are no predefined *exemplar* patterns. We will use the Learning Vector Quantization training method to redefine the values of the *classification* vectors. Their values at the end of this process will be the exemplar vectors for each of the classes we will define.

For this to take place, we have to redefine our concept of a neuron for the Learning Vector Quantization network. Obviously, the dimensionality of the classification vectors must be the same as the dimensionality of the training vectors. Let's keep this in mind, because this is a drastic modification of the way we have conceptualized neurons prior to this.

When a pattern is presented, only those neurons in the vicinity of the new pattern can adjust their vector element values to move towards the new pattern. As learning progresses, the distribution of neurons takes on the same distribution of vector orientations and lengths as that found in the training set. This makes the LVQ network particularly useful for modeling sets of data in which the distribution of data vectors is initially unknown.

10.2.2   The Architecture of the LVQ Network

At the micro-structural level, there is a key distinction between the neurons in the Kohonen network and the neurons of the other networks we have studied. In the Hopfield and BSB networks, each neuron moved toward one of a pair of binary values, and the restored pattern was distributed across all the neurons in the network. In the case of the Kohonen network, we have a much more localized representation. Each neuron will become an exemplar for a class. When an output pattern is selected, a single neuron is activated. This is the neuron which the input pattern most closely resembles.

The issue is not just that a single neuron is selected (instead of having a pattern spread across a set of neurons), it is also that the dimensionality of this neuron must correspond to the dimensionality of the pattern vectors. Thus, if the pattern vectors are two-dimensional, each neuron must have two response elements, one to represent a value in each of the pattern dimensions.

Let's return to our earlier example of using a set of ten neurons to encode the spatial distribution of 100 two-dimensional vectors. In this case, we would be training ten neurons consisting of two *values* each, one for each dimension. This would be a total of 20 elements (ten sets of two each) which will be trained. Because this concept of a neuron is so different from that used in all other networks (and has no neurophysiological basis), we might call each set of elements a *neurode* instead of an artificial *neuron* for this network.

The Kohonen network neurodes can take on a wide range of values in order to mimic the values expressed by input pattern vectors. They may be normalized in some way (e.g., the activations of the neurons may be scaled so that their sum across all neurons for a pattern representation would equal 1), but this does not detract from the fact that this network facilitates discriminating between non-orthogonal patterns.

At the meso-structural level, there are some useful similarities between the meso-structure of the Kohonen network and other autoassociation networks. Like the Hopfield and BSB networks, the Kohonen network can be thought of as a single layer of neurodes. They are not explicitly interconnected; that is, one neurode does not act directly on another. However, the topological distance between the values of the neurodes will influence how they respond in adapting to a presented input pattern. This is shown in Figure 10.1. The *distance* which defines the neighborhood within which a set of neurodes may respond is fixed at the outset by the network designer, and does not change during the course of learning. (This is in contrast to the Kohonen Self-Organizing Topology-Preserving Map, which is discussed in the next section.)

10.2.3   The Dynamics of the LVQ Network

The goal of the dynamics of this network (as with all other autoassociation networks) is to be able to take as input a pattern vector, and to output the *closest* pattern vector that defines one of the stored classes. In this case, the output will be one of the vectors

The Learning Vector Quantization Network

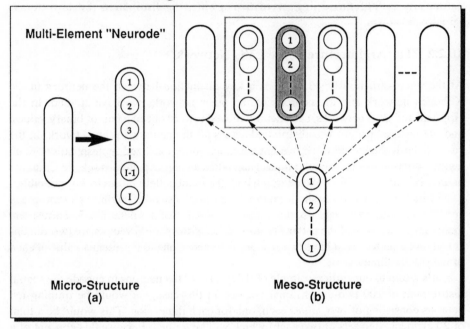

Figure 10.1   A Kohonen Learning Vector Quantization network. (a). At the micro-structural level, neurodes consist of multiple elements, each of which may be continuously valued. The dimensionality of each neurode, I, is the same as the dimensionality of the input patterns. (b). At the meso-structural level, the neurodes are arranged in a single layer. Instead of working with connection weights, the Learning Vector Quantization learning algorithm adjusts the values stored in each element of the neurode.

which we used to create a *category*. This vector will have taken on a value which best represents, or is an exemplar, of that category.

Let's consider the dynamics of the LVQ network using an example such as the one shown in Figure 10.2, which is adapted from one used by Caudill Butler [1989]. Figure 10.2 (a) shows how the original distribution of neurodes might appear if they are given an initial random distribution between 0 and 1 for both the x and y components of the two-dimensional vectors. Suppose that the training set has a distribution over three quadrants of the circle with the greatest distribution in the upper right quadrant, as is graphically shown in Figure 10.2 (b). Repeated training on vectors in this set, when presented with this probability distribution, will result in dispersal of the exemplar vectors somewhat like that shown in Figure 10.2 (b). After training, when a new input pattern is presented (shown as a dotted arrow in Figure 10.2(c)), the *winning* neurode (shown as a bold arrow) is activated. In this example, the neurodes are each two-dimensional. Much higher dimensionalities have been used in some applications.

To find the closest neurode in the exemplar set to a presented pattern, we compute the total distance (using the Euclidian metric) between each neurode in the exemplar set and the input pattern. The winning neurode is the one with minimum distance. Only

The Learning Vector Quantization Network: Training a Distribution of Pattern Vectors

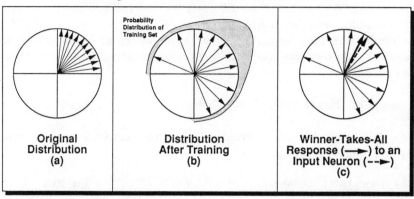

Original
Distribution
(a)

Distribution
After Training
(b)

Winner-Takes-All
Response (——►) to an
Input Neuron (--►)
(c)

Figure 10.2 A Kohonen Learning Vector Quantization network consisting of ten trained "exemplar" neurodes of two dimensions each. They "span" a two-dimensional space with the same distribution as their training set. (a). The neurodes in an original random distribution between values of 0 and 1 for both x and y values. (b). The distribution of neurodes after training by presentation of patterns with the distribution over three quadrants, as shown. (c). An input vector (dotted arrow) finds its closest match (bold arrow) among the neurodes in the set of exemplars.

this neurode is activated in response to the input pattern, which makes this a winner-take-all type of network. When the network is in operation, activation of the winner neurode is the final step. When the network is learning, finding the winner neurode is the first step in the learning process. When the network is in regular operation, the winning neurode represents the output of the network.

10.2.4   Learning in the LVQ Network

There are separate learning and response phases for the Kohonen LVQ network. Learning may require thousands of presentations of patterns from the training set.

The first steps in training the LVQ network are to select a training set which spans the likely space for input patterns. The LVQ network exemplar neurodes will take on values whose distribution reflects the probability distribution across the training set, so obtaining a comprehensive and well-balanced training set is important.

We assign initial random values to the members of the set of exemplar neurons. Even these initial values should be within the range of the training set data. If there is any a priori knowledge about the distribution, it should be used in assigning the initial values to the neurodes.

Next, we train the neurodes by presenting, individually, different patterns from the training set. We calculate the distance between the input vector x and each of the j exemplar neurodes v as follows:

$$\text{dist}_j = (x_i - v_{j,i})^2 \qquad\qquad \text{Eq. 10.1}$$

Where i runs from 1 to I, the dimensionality of both the input and the exemplar vectors. We select the exemplar vector v_j whose distance $dist_j$ is a minimum. This is the *winning* neurode. So far, there is no difference between this step and the process of exemplar selection discussed in the previous section.

The learning takes place by modifying both the exemplar vector v_j and its *neighbors* so that both it and the neighbors take on values closer to those of the input pattern. We do this by setting

$$v_j' \text{ (new)} = v_j' \text{ (old)} + alpha(t)[x - v_j'(old)] \qquad \text{Eq. 10.2}$$

Where alpha(t) is a scalar-valued function of t, which decreases with increase in t. A typical formulation of alpha(t) is:

$$alpha(t) = alpha1[1 - t/alpha2] \qquad \text{Eq. 10.3}$$

Where alpha1 might have a value of .1, and alpha1 might be assigned the value of the number of expected presentations which will be made during network teaching. Alpha(t) might be held constant during the first 500 or so pattern presentations. This vector adaptation process would be applied to all vectors j' in the neighborhood of the winning vector j. This neighborhood is defined as all exemplar vectors within a certain distance d^* of v_j, using the distance metric just defined. This distance is typically fixed for the entire learning process.

10.2.5   The Performance and Capabilities of the LVQ Network

There is no clearcut a priori basis for selecting the number of neurodes for use in an LVQ network. Simple common sense and caution, combined with experimentation in each example, will probably yield the best results.

Also, there is the possibility that the set of neurodes may not take on a distribution which effectively represents the training set. This is especially true if the training set of data includes examples which are so far from the original distribution of neurodes that it is difficult for the neighborhood modifications in neurode activations to take on the values of the training data. This approach works best if the original distribution of the neurodes is at least some approximation of the real distribution in the training data. And, of course, the network does not respond favorably when presented with an input pattern that is far removed from the values taken on by the network's neurodes.

10.3   THE SELF-ORGANIZING
TOPOLOGY-PRESERVING MAP

Self-organizing Topology-Preserving Maps (TPM), developed by Teuvo Kohonen, have been used for speaker-independent phoneme recognition. Topology-Preserving Maps are the "network of choice" for applications involving mapping distributed sensory information (e.g., tactile information) into a two-dimensional or three-dimensional representation. They may also be useful for robotic arm control and for optimization applications.

A Topology-Preserving Mapping is said to exist in the network if the network's response to the input patterns has a similar topological relationship as that which exists among the input patterns. Cooperative excitation of nearest neighbors leads to projection mappings of input vectors into spaces which may even have a different dimensionality. These mappings preserve topographic relations (even hierarchical structures) which may have existed among input data. This process can be used for reducing the dimensionality of complex input data, and for pattern recognition.

10.4.1   The Concept Underlying the Topology-Preserving Map

We have previously discussed the basis for Topology-Preserving Maps earlier in this chapter, when we examined the LVQ network. In the LVQ network, a mapping vector would win a competition with its neighbors to respond to an input vector. The TPM network is similar to the LVQ network in that a winning neurode responds to an input signal that influences its nearest neighbors respond also. The TPM network is different from the LVQ in that the activity of the winning neurode in the LVQ had only an (implicit) excitatory effect on its nearest neighbors. A winning neurode (or a neurode which responds at all to the input stimulus) in the TPM network has both an excitatory and an inhibitory effect on its neighboring neurode.

This type of lateral cooperative-competitive interaction affects the network so that the location of a neurode in the network where the response to an input pattern is obtained becomes specific to a certain characteristic feature of the input pattern [Kohonen, 1982, 1989]. If there is some topological order among the input patterns, the same order will be found in the laterally-connected cooperative-competitive network. For example, a network which responds to audio signals will become tonotopic; that is, the neurodes will arrange themselves spatially so that those which respond to low tones are at one end of the network, those which respond to middle tones are in the middle, and those which respond to the highest tones are found at the other far end of the laminar network. This self-organization results from the lateral cooperative-competitive interaction among the neurodes, and is similar to earlier work done by von der Malsburg [1973].

10.4.2   The Architecture of the TPM

The neurodes in a TPM are hard-limit neurodes. The activation of a neurode i at time t is given by:

$$y_i(t) = \Sigma \, (x_i(t)) \qquad\qquad\qquad \text{Eq. 10.4}$$

Where $x_i(t)$ is the sum of all input into the neurode at time t Σ is the transfer function applied to x.

An example of such a transfer function might be:

$\Sigma(x) = 0$ for x,
$\Sigma(x) = x$ for 0xz, where z is some positive constant,

and

$$\Sigma(x) = z \text{ for } x = z.$$ Eq. 10.5

A TPM operates over several clock cycles (in contrast to the LVQ, which completes its response in a single cycle), and so the input for a neurode at any time t is a function of both the current external input stimulus and the time-dependent input it is receiving from its laterally-connected neighbors. An example equation showing how the sum of all inputs to neurode i at any time t might be calculated is given by:

$$x_i(t) = s_i(t) + \sum_{j=-M}^{M} \delta_j \times x_{(i+J)(t-1)}$$

Eq. 10.6

In this case, $s_i(t)$ is the time-dependent external input coming into neuron i at time t. Commonly, the external input is assumed to last for only a single cycle of operation. δ_j specifies the strength of the lateral connection, which is a function of the distance j of a neighboring neurod from neurode i. M specifies the extent of the neighborhood over which lateral connections exist.

The local neighborhood connections mimic a "Mexican hat" type function which centers around the local maximum of input excitation. This function, which can be created by superimposing three Gaussian distributions, mimics the on-center, off-surround pattern of the receptive field of a retinal neuron. Figure 10.3 (a) shows how this function appears, and Figure 10.3 (b) shows how a simple interpretation of this function can be used for creating a TPM.

A simple TPM uses only a single layer of neurodes, often arranged in a two-dimensional matrix form. Each neurode is connected to its neighbors within a neighborhood of radius M, according to a function such as that shown in Figure 10.3(b).

10.4.3   The Dynamics of the TPM

In developing a TPM, Kohonen [1989] describes how *activity bubbles* corresponding to activation of an area around the best matching mapping vector are created using both cooperative and competitive connections. Figure 10.4 shows how an activity bubble is formed over time. A single sinusoidal input is presented to the network. The pattern which forms in response sharpens over time, due to the lateral connections described above.

Figure 10.5 illustrates how the vector "neurodes" in a TPM can be ordered relative to each other, leading to a two-dimensional (in this illustration) mapping of the topographic relationship among a set of higher-dimensional data.

10.4.4   Learning in the TPM

Connection weights are typically fixed in a topology-preserving map, so no change or learning of these weights takes place during operation, nor is there a learning phase. Some recent enhancements to the basic TPM by DeSieno [1988] and Hodges and Wu [1990] yield faster convergence. The *learning with a conscience* method of DeSieno also provides a better mapping.

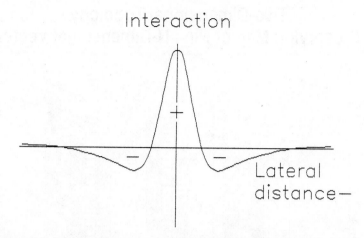

Figure 10.3a The "Mexican-hat function" can be used either to define a pattern of lateral excitation, inhibition, and excitation, or to indicate connection strengths from cells in one layer of a network to a neuron in a higher layer. Figure taken from those used by Teuvo Kohonen in *Self-Organization and Associative Memory*, 3rd. Ed. (Berlin: Springer-Verlag, 1989). Reproduced with permission of Springer-Verlag, Heidelberg.

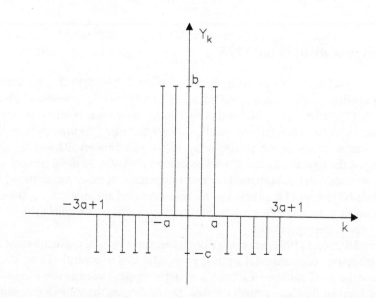

Figure 10.3b A simplified version of this function can be used to define values for lateral excitatory and inhibitory connections among neurons. Figure taken from those used by Teuvo Kohonen in *Self-Organization and Associative Memory*, 3rd. Ed. (Berlin: Springer-Verlag, 1989). Reproduced with permission of Springer-Verlag, Heidelberg.

Two-Dimensional Topology - Preserving Map of Nine N-Dimensional Vectors

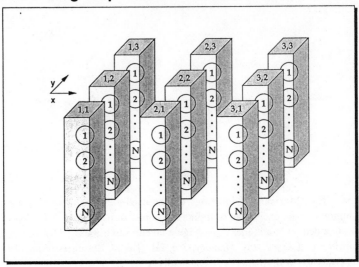

Figure 10.4   The two-dimensional ordering reflects the relative distances between different vector values.

10.4.5   Applications of the TPM

There are several possible interesting applications of TPMs. Speech interpretation is one such application [Kohonen, 1988], leading to a "phonetic typewriter" prototype application. (The languages used for this application have been Finnish and Japanese, each of which has words which are readily interpreted based on their pronunciation.) Transcription accuracy of the prototype system varies between 92 and 97 percent, depending on the speaker. Isolated-word recognition is in the 96 to 98 percent range.

A similar approach has been used for text-independent speaker recognition [Naylor et al., 1988; Naylor and Li, 1988]. The Kohonen method automatically generates the desired number of templates necessary, and compares well with results achieved using the K-Means clustering method.

Ritter and Schulten [1988] have described three major types of applications of TPMs: sensory mapping, combinatorial optimization, and motor control. They illustrated [1986] how the TPM could be used as a somatopic map between tactile receptors on a simulated hand surface and a model cortex. Depending on the vectors describing the input data, this can lead to dimensionality reduction as well as topographically organized mapping.

Angeniol et al. [1988] have applied the self-organizing feature map approach to the travelling salesman problem. Their results compare favorably with those obtained by researchers using other methods, giving near-optimal solutions within a reasonable time. Factors which may favor the self-organizing approach to this type of problem are

Clustering of Activity in a One-Dimensional Array

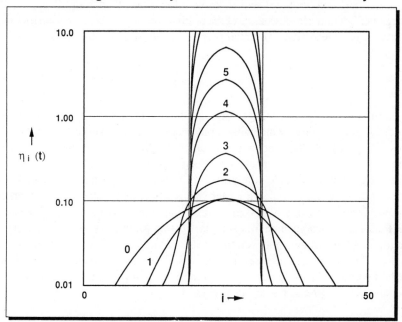

Figure 10.5   (a) A sinusoidal (smoothly varying) stimulus is presented to a network with lateral connections, where connection strengths are defined by the "Mexican-hat function" shown in Figure 10.3(b). (b) The pattern is presented once, but the operation of the network over a series of time steps leads to a sharpening of neural activation into a well-defined "activity bubble." Figure taken from T. Kohonen, *Self-Organization and Associative Memory*. Copyright Springer-Verlag, Heidelberg. Reproduced by permission.

that the total number of nodes and connections is proportional to the number of "cities" (in the TSP problem), which leads to good scalability. Hardware based on analog technology may be suitable for implementing this approach, and only a single parameter (governing the rate by which the activity bubble is reduced) needs to be tuned. Heuter [1988] has used the Kohonen model in learning to adapt a ring of nodes to solve the same TSP problem. He has also obtained good results, although his method requires more parameters which need to be tuned.

Ritter et al. [1989] have used TPMs to simulate visuo-motor-coordination. A key issue which they address in their simulations is using the isolated quality of the system to an advantage, by having the neural system observe and learn from its own reactions. They have recently extended their work to guiding a simulated robot arm in three-dimensional space [Martinetz et al., 1989].

Self-organizing feature maps have also been used to design codebooks for signal processing [Kohonen, 1988 (a), Naylor and Li, 1988]. When applied to image compression [Nasrabadi and Feng, 1988; Ahalt et al, 1989], the results compare favorably with those of the well-known Linde-Buzo-Gray algorithm.

Most recently, TPMs have been used to encode semantic information [Ritter and Kohonen, 1989] and temporal sequences of information [Tolat and Peterson, 1989]. Although neither of these extensions of the original network have been used for practical applications yet, they offer substantial potential for useful development.

10.4 SUMMARY

The Learning Vector Quantization (LVQ) and the self-organizing Topology-Preserving Map (TPM) both operate by matching input vectors against a set of (trained) exemplar vectors. This network does not have explicit connections between neurodes (vectors). The number of exemplar vectors must be specified in advance, but during training, they take on a distribution to match the probability distribution of the training data set of vectors. In the Topology-Preserving Map, the arrangement of exemplar vectors forms a mapping from a possibly high dimensional input space into a lower dimensionality space. Further, the arrangement of exemplar vectors (or trained neurodes) corresponds to the topological distance between the vector values comprising the elements of the neurodes. These networks are useful for autoassociation and codebook data compression (the Learning Vector Quantization), and for mappings, data compression, optimization, and control (the Topology-Preserving Map).

REFERENCES

Ahalt, S.C., Chen, P., Krishnamurthy, A.K. (1989). "Performance analysis of two image vector quantization techniques," *Proc. First Int'l. Joint Conf. Neural Networks*, Washington, D.C.; June 18-22, I-169–I-175.

Angeniol, B., de la Croix Vaubois, G., and le Texier, J.-Y. (1988). "Self-organizing feature maps and the travelling salesman problem," *Neural Networks, 1,* 289-293.

Caudill, M., and Butler, C. (1989). *Naturally Intelligent Systems,* MIT Press, Cambridge, MA.

DeSieno, D. (1988). "Adding a conscience to competitive learning," *Proc. Second IEEE Conf. on Neural Networks,* San Diego, CA; July 24-27, 1988, I-117–I-124.

Hodges, R.E., and Wu, C.-H. (1990). "A method to establish an autonomous self-organizing feature map," *Proc. Second Int'l. Joint Conf. Neural Networks*, Washington, D.C.; January 15-19, I-517–I-520.

Heuter, G.J. (1988). "Solution of the traveling salsesman problem with an adaptive ring," *Proc. Second Int'l. Conf. Neural Networks,* San Diego, CA; July 24-27, I-85–I-92.

Kohonen, T. (1989). *Self-Organization and Associative Memory,* Third Edition, Springer-Verlag, Berlin.

Kohonen, T. (1989). "A self-learning grammar, or "Associative memory of the second kind," *Proc. Int'l. Joint Conf. on Neural Networks,* Washington, D.C., I-1–I-5.

Kohonen, T. (1988). "The "neural" phonetic typewriter," *Computer, 21,* 11-22.

Kohonen, T. (1988 b). "Learning Vector Quantization," *Abstracts of the First Annual INNS Meeting,* Boston, MA, 303.

Kohonen, T. (1987). "Self-learning inference rules by dynamically expanding context", *Proc. IEEE First Int'l. Conf. on Neural Networks*, San Diego, CA; June 21-24, II-3-II-9.

Kohonen, T. (1982). "Clustering, taxonomy, and topological maps of patterns," *Proc. 6th Int'l. Conf. Pattern Recognition*, October, IEEE Computer Society Press, Silver Spring, MD, 114-128.

von der Malsburg, C. (1973). "Self-organizing of orientation sensitive cells in the striate cortex," *Kybernetik, 14,* 85 -100.

Martinetz, T.M., Ritter, H.J., and Schulten, K.J. (1989). "3D-neural-net for learning visuomotor-coordination of a robot arm," *Proc. First Int'l. Joint Conf. Neural Networks*, Washington, D.C.; June 18-22, II-351-II-356.

Nasrabadi, N.M., and Feng, Y. (1988). "Vector quantization of images based upon the Kohonen self-organizing feature maps," *Proc. Second IEEE Int'l. Conf. Neural Networks*, San Diego, CA; July 24-27, I-101-I-105.

Naylor, J., and Li, K.P. (1988). "Analysis of a neural network algorithm for vector quantization of speech parameters," *Abstracts of the First Annual INNS Meeting*, Boston, MA, 310.

Naylor, J., Higgins, A., Li, K.P., and Schmoldt, D. (1988). "Speaker recognition using Kohonen's self-organizing feature map algorithm," *Abstracts of the First Annual INNS Meeting*, Boston, MA, 311.

Ritter, H., and Kohonen, T. (1989). "Self-organizing semantic maps," *Biol. Cybernetics, 61,* 241-254.

Ritter, H.J., Martinetz, T.M., and Schulten, K.J. (1989). "Topology-Preserving Maps for learning visuo-motor-coordination," *Neural Networks, 2,* 159-168.

Ritter, H., and Schulten, K. (1986). "On the stationary state of Kohonen's self-organizing sensory mapping," *Bio. Cybernetics, 54 ,* 99-106.

Ritter, H., and Schulten, K. (1988). "Kohonen's self-organizing maps: Exploring their computational capabilities," *Proc. Second Int'l. Conf. Neural Networks*, San Diego, CA; July 24-27, I-109-I-116.

Tolat, V.V., and Peterson, A.M. (1989). "A self-organizing neural network for classifying sequences," *Proc. First Int'l. Joint Conf. Neural Networks*, Washington, D.C.; June 15-19, II-561-II-568.

11

FEEDFORWARD/FEEDBACK (RESONATING) HETEROASSOCIATIVE NETWORKS

Alianna J. Maren

11.0 CHAPTER OVERVIEW

In this chapter, we turn our attention to heteroassociative networks — networks which can associate one pattern with another. These networks have a distinctive architecture. More importantly, the two networks which are the focus of this network — Adaptive Resonance Theory (ART) networks and Bidirectional Associative Memory (BAM) networks — have a property which makes them unlike any of the networks we have covered up to now. They can learn a new pattern, or pattern class, at any point of their operation; this means that they have the ability of responding to novel input by recognizing it as such, and creating a new pattern class to represent this input. This capability makes these networks uniquely interesting and useful for many potential applications.

11.1 INTRODUCTION

Let's begin by discussing the comparison of heteroassociative networks with the two other functional types of networks we've examined: feedforward networks for pattern recognition, and autoassociative networks. The basis for our comparison is threefold:

- Function (what do they do)
- Form (how are they constructed)
- Dynamics (including both training and operation)

We've already mentioned that there is a similarity between the functions of pattern recognition and heteroassociation. Let's generalize and say that (simple) pattern recognition networks should select one category from a group of known categories when given an input pattern, and heteroassociative networks should regenerate an entire pattern. (The pattern may stand for a category, or vice versa, but let's not split hairs.)

Distinguishing the function of autoassociative networks from those of heteroassociative networks is easy; an autoassociative network associates a pattern with the exemplar pattern which the input was supposed to represent. This would be the best match from a set of known, stored patterns. Heteroassociative networks associate an input pattern with a different known, stored output pattern.

This provides a clear guideline to distinguish between the architectures of autoassociative and heteroassociative networks. Autoassociative networks can get by with one layer of neurons. This layer will start off holding the initial, presented pattern, and end up representing the autoassociated pattern. We might add additional layers to read data in or out, but the functionality may be accomplished in a single layer.

In contrast, heteroassociative networks need at least two layers. One holds the input pattern, the other holds the output, or heteroassociated pattern. We can't overwrite the initial pattern with the associated one. This is because if the network's initial try at matching doesn't succeed, it will need access to the original pattern again. Thus, the initial pattern must be stored in an input layer where it can be re-accessed.

Neural connectivity is an important architectural (form) issue in comparing these different types of networks. In feedforward networks, the connections are strictly feedforward. The algorithms that change the values of the connection weights may use information about the error between a desired output and the actual output to influence the connection weights. We have to use a lot of imagination to see how this is biologically plausible, but at least the networks work. The point is that there is no direct transfer of information from the upper levels to the lower levels of the network. There are no lateral connections, either.

The autoassociative networks have lateral communications. Briefly reviewing what we covered earlier, in the Hopfield network, each neuron connects to all others except itself. In the Brain-State-in-a-Box, each neuron connects to all others and back to itself as well. In the Learning Vector Quantization Network, the idea of local connection is implicitly simulated when a winner neuron and it's nearest neighbors all learn together. (Kanerva's network is a little unusual; let's say that it also implicitly simulates local connectivity.)

Heteroassociative networks are different. First, the recent ones (both ART and BAM types) have both feedforward and feedback connections between the input and output layers. Second, they may or may not have lateral communications, depending on their specific design. The important characteristic is the feedforward/feedback connectivity, and lateral connectivity usually plays a subordinate role.

The heteroassociative networks which we will investigate here have more subtle dynamics than those previously discussed. In retrospect, the networks we have covered previously were quite simple. Specifically, supervised learning typically led to stable (read static) systems; the weights were fixed after the training cycle ended. The Hopfield network also had fixed weights, and so was stable both in terms of its dynamics and its learning. The Learning Vector Quantization and Brain-State-in-a-Box networks had distinct training phases in which they could learn from a training set of patterns. After training was over, their weights did not change any more. These systems were also dynamically stable.

In contrast, one of the major concerns that we have about the heteroassociative networks will be both their stability of association through time and their stability of dynamic processing. This is because the two systems on which we will focus — the ART and the adaptive BAM-type networks — undergo unsupervised learning. Their weights change over time as new patterns are presented to the system. There is no distinction between the training phase and the operation phase, nor is there typically a distinct set of *training data* and *test data*. Even the number of categories to which these networks associate can change (increase) over time. This makes these systems dynamically varying.

When systems change over time, their stability becomes a major issue. For that reason, the leading researchers in heteroassociative networks (Carpenter, Grossberg, and Kosko) have all devoted substantial attention to exploring dynamic behavior and stability issues. These explorations are typically in the form of mathematical proofs.

The first heteroassociative networks were devised by Shun-Ichi Amari, N. Nakano, Teuvo Kohonen, James Anderson, and Steve Grossberg. Amari, Nakano, and Kohonen each took a mathematical approach: Anderson was interested in replicating certain forms of biological and psychological associative behaviors, and Grossberg was deliberately trying to make neural networks that would model the classical conditioned-response association that we can observe in animals and in ourselves. For some time now, Gail Carpenter and Steve Grossberg have worked together developing the Adaptive Resonance Theory networks based on biologically plausible models. More recently, Bart Kosko has developed the Bidirectional Associative Memory (BAM), which is somewhat simpler architecturally but has elegant mathematical foundations. Both the ART and BAM networks have been developed into families of network topologies.

In the following sections, we explore the individual differences among networks which allow these capabilities. Before we do that, we'll briefly address some of the historically important concepts in neural heteroassociation. This work lays a foundation for the ART and BAM-types of networks.

In 1972, Tuevo Kohonen and James Anderson independently developed a neurally based model for a simple heteroassociative network [Anderson, 1972; Kohonen, 1972; Anderson & Rosenfeld, 1988]. As expected, Kohonen (an electrical engineer) gave a mathematical treatment of the network. Anderson (a psychologist with an interest in mathematical neurophysiology) addressed the necessary mathematics, but spent most of his time demonstrating how the network corresponded to neurophysiological systems.

The network which both researchers presented can be viewed as a two-layer network with a single feedforward connecting matrix. The neurons are all of the linear summation type with no limit to their *firing frequencies* (expressed as activations). The values of elements in the connection matrix change in accord with a simple Hebbian law for synaptic learning. The network performs well only when orthogonal input vectors are used.

This network is not only conceptually similar to the Hopfield and Brain-State-in-a-Box networks (for autoassociation), but forms a logical precursor to the resonating networks of Carpenter and Grossberg, and of Kosko. Like the Hopfield network, it learns all of its associations at once. It cannot learn new associations once trained. This ability to learn new associations is an important characteristic of the ART and BAM-type networks, which we will address in the next two sections.

11.2 THE CARPENTER/GROSSBERG ADAPTIVE RESONANCE THEORY NETWORK

Adaptive Resonance Theory (ART) networks are again radically different from anything we have encountered up until now. If we were biologists, we could think of this as looking at a whole different phylum. This comes about as a result of Steve Grossberg's work (which was the basis for ART) which was built up over a twenty-year period. During this time, Grossberg and his colleagues were actually investigating very different issues from those who developed the well-known feedforward networks.

ART networks are most useful as pattern recognizers, although with some modification, they can be pattern heteroassociators as well. These networks can work on binary or real-valued input, in any desired manner of representation. (1 and 2-dimensional arrays are common.) Their ability to generalize is limited because ART networks lack the middle layer of neurons which act as feature-recognizers in Perceptron-type networks. However, the ability of an ART network to create a new pattern classification when it observes a new type of pattern makes it highly attractive for applications such as automatic target recognition, seismic signal processing, and other spatial and spatio-temporal pattern recognition tasks.

11.2.1   The Concepts Underlying the ART System

Let's briefly outline the major distinctions between ART networks and the type of feedforward network discussed earlier. It is important to make these distinctions because functionally, ART systems and feedforward networks can be said to do a similar task — pattern classification. Thus, we need the underlying development of such different architectures.

Grossberg, Carpenter, and their colleagues went beyond using Hebbian law, which is favored still by many researchers (especially for associative networks). They used more detailed knowledge of conditioned and unconditioned (Pavlovian) associations as a guide, and also deliberately invoked more subtle and descriptive models of actual neural activities to obtain the learning rules. This was because they wanted a single network that could do two things:

- Learn new patterns of pattern categories (i.e., learning plasticity), and at the same time
- Retain knowledge of previously learned patterns or pattern categories (i.e., learning stability).

Formulating a network that could do one or the other would have been easy. Designing a network that could do both was not. They postulated this as the stability/plasticity dilemma.

One of the key features in achieving learning plasticity is the use of *pattern resonance*. Both the ART systems and Kosko's BAM systems use resonance of a pattern in the output layer with a pattern in the input layer establishing a good heteroassociative pattern match. This idea of resonance is still unique to just these two types of neural network systems.

A *resonating network* has two main layers. The first layer receives and holds the input patterns. The second layer responds with a pattern classification or association to the input pattern, verifying that its association with the input is correct by sending a response or return pattern back to the first layer. If this return pattern is correct, then there is a match. If the return pattern is substantially different from the input pattern, then the two networks will communicate back and forth; that is, they will *resonate* until there is a match.

These resonating networks are by no means simple. In fact, the underlying conceptual and theoretical basis for their designs is more complex than that of the feedforward networks discussed earlier. However, they are well worth investigating because they possess two useful properties: real-time learning and self-organization.

The real-time learning of the ART network is of two types. First, the system can identify a novel pattern and learn it. Second, every time the system recognizes a pattern, a little learning occurs. This amounts to fine-tuning the weights for that pattern category. The learning is non-Hebbian, and can be described in terms of differential equations.

The ART and BAM systems are self-organizing; they create new categories when given novel patterns. In practical design terms, this self-organizing capability is both a blessing and a curse. It can be useful when recognizing novel patterns and classifying them as such. However, in both ART and BAM, the categories evolve over time. We can influence what the categories represent by introducing certain types of training data in certain sequences, but we do not have absolute control of (or certain knowledge of) over category representation that we have when we ourselves select the categories. This is perhaps the most distinguishing feature between a feedforward classifier and the ART and BAM heteroassociative networks.

11.2.2   The Architecture of the ART System

To achieve the goal of making a network which had both learning plasticity and learning stability, Carpenter and Grossberg created a complex architecture composed of three interacting subsystems: the attentional subsystem, the orienting subsystem, and the attentional gain control. All ART systems have these subsystems. Since several variations of ART have been developed, we concentrate here on ART 1, which recognizes binary patterns [Carpenter and Grossberg, 1987a]. ART 1 architectures and processes are the basis for all other ART systems. For simplicity. we may refer to ART 1 as ART. Art 2 recognizes analog patterns [Carpenter and Grossberg, 1987b].

The attentional subsystem is used to recognize and classify previously learned patterns. The values encoded for the connection weights in this subsystem (its Long-Term Memory, or **LTM**) lend the ART system its stability in terms of retaining learned categorizations.

The orienting subsystem is invoked when ART encounters a new pattern. Under these circumstances, the orienting subsystem shuts down any attempt to continue matching with old pattern categories, and sets up a new category to respond to the new type of pattern. Thus, the attentional subsystem performs the actual pattern recognition, and the orienting subsystem acts on the attentional subsystem to enable it to respond to novel patterns.

The attentional gain control stabilizes system operation. It also allows ART to be *primed* to receive *expected* patterns. It acts to increase sensitivity or attention in response to *expectations*. Figure 11.1 shows the overall interaction between the attentional subsystem, the orienting subsystem, and the attentional gain control in ART 1.

The architecture of the attentional subsystems is shown in Figure 11.2. It consists of two layers. The lower one is consistently referred to by Grossberg and his colleagues as F_1, and the upper one as F_2. To be consistent with their work (and to make your life easier, should you decide to read their papers), we will use the same notation as they do for all aspects of the ART system. Table 11.1 shows the variable naming conventions used for the ART system.

The nodes in the first layer of the attentional subsystem, F_1, are more complex than usually encountered. They have to keep track of two values. In addition, the input vector must be stored and kept accessible; it will be used again if the first attempts at matching

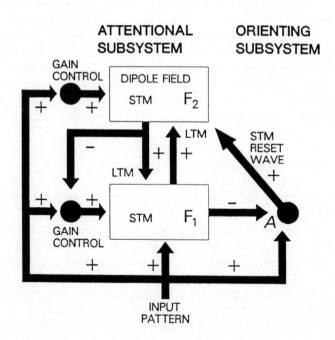

Figure 11.1   The Adaptive Resonance Theory (ART) network has three major subsystems; the attentional subsystem, the orienting subsystem, and the attentional gain control. Figure taken from *Neural Networks, 1,* S. Grossberg, "Non-linear neural networks: Principles, mechanisms, and architecutres." © 1988, Pergamon Press, Inc. Reproduced with permission of Pergamon Press, Inc. and S. Grossberg.

Architecture of the Attentional Subsystem of the Carpenter/Grossberg Adaptive Resonance Theory Network

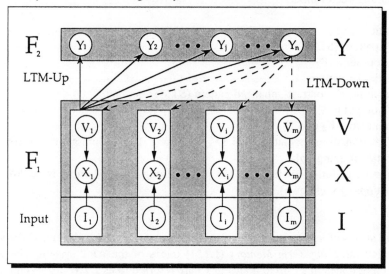

Figure 11.2   The attentional subsystem of an ART network has two layers; F_1 and F_2. F_1 contains three related arrays of values, I, B, and X, which are input, top-down matching from F_2, and buffer (short-term memory), respectively. The input vector I has the same length as B and X. F_2 contains the pattern classes. There are two sets of connection weights; one going up (LTM_up), the other down (LTM_down).

do not work. The vectors **X** and **B** in F_1 and **I** (the input) will all have the same length. Each node in F_1 will store:

- The value which is stored in that node as short term memory, or X_i, and
- The value which will come back from F_2 as a top-down template attempting to match at node **i**. This will be B_i.
- Each node **i** in F_1 will correspond to a node in the input vector, I_i.

Generally, the value which is stored as short term memory in X_i is the same value received as input I_i. It would be possible to code an ART system without making two separate vectors. However, it is also possible to code a more advanced architecture in which some process operates on **I** before it is stored as short term memory **X**. (For example, the values of the pattern elements could be normalized). As mentioned earlier, we are staying strictly with the Carpenter/Grossberg notation. It is more complex than is necessary for building a simple ART system, but makes it easier to understand their work.

Thus far, we have described F_1, which has *complex nodes,* each of which will hold two values: X_i and U_i. For each node, there is also a corresponding I_i.

F_2 can be represented as a vector **Y**, where each node is *simple* and holds a single value Y_j.

Table 11.1   Variables for Art Network

(Assuming vector input)

I = input vector
X = F_1 pattern (STM)
S = F_1 output to F_2
T = initial F_2 pattern
Y = resulting F_2 contrast-enhanced pattern (STM)
U = F_2 output to F_1
B = input from F_2 into F_1;
S - T = transition mediated by LTM_up [Long Term Memory traces (or weight connections) going up]
U - B = transition mediated by LTM_down [Long-Term Memory traces (or weight connections) going down].

Two arrays of interconnection weights connect the F_1 and F_2 layers. These weights are called the *Long Term Memory* (LTM) traces of the ART system. The weights connecting nodes in F_1 up to nodes in F_2 are referred to as **LTM_up**. Weights going down are called **LTM_down**. Both directions are fully connected.

The main component of the orienting subsystem is an activity monitor called A which interacts with both F_1 and F_2, and with the input itself, or with I. A can be both excited and inhibited in its interactions with these three components. It is the complex combination of excitation, inhibition, and disinhibition which turns A on and off, allowing it to affect the ART network or not, as the need may be. The strength of the operation of A is governed by a single parameter, the *vigilance* parameter.

The attentional gain control component of the orienting subsystem is affected by F_2 and operates on F_1. Its purpose is to stabilize system operation, so that the signals coming from F_2 down to F_1 do not themselves excite further activity in F_2. Another function of the attentional gain control is that it allows ART to be primed to receive expected patterns. It acts to increase sensitivity or attention in response to expectations.

Table 11.2 summarizes what we have covered thus far in terms of the ART 1 configuration. We have just reviewed the components, and the next subsection will deal with ART processes.

11.2.3   The Dynamics of the ART System

The dynamics of the ART system are a result of interactions between different ongoing, interacting processes. Each is designed to contribute to the *stability* or retention of previously learned material, while endowing the network with enough *plasticity* to enable its learning new material. First, we will consider the basic dynamics of the ART system. We will illustrate the ART dynamics with an example of how ART handles a known pattern. This will focus our attention on the attentional subsystem. Then we will consider how ART handles an unknown pattern. This will involve greater activity from the orienting subsystem, and will invoke the major ART learning processes.

Table 11.2   Art 1 Configuration Summary:

- There are two components to the ART system; the attentional subsystem and the orienting subsystem.
- The attentional subsystem has two components or layers:
  - The bottom layer, denoted F_1, and
  - The top layer, denoted F_2,
- The orienting subsystem also has two components; the main orienting subsystem (A), and gain control. Each interacts with the attentional subsystem at both levels (F_1 and F_2).

11.2.3.1   Basic ART Dynamics — the Attentional Subsystem   When ART receives an input pattern, that pattern is stored as the vector I (of length n). (Remember, each node i in F_1 stores two different values, X_i, and B_i. There are a total of m nodes in F_1 to match the input vector.) Right away, I is copied over into X. From now on, I will not change — it stores the original input pattern. However, X may change. It can be affected by information coming down later from F_2. At the beginning, X is set equal to I, and B may be set to 0.

The first major processing stage is called *bottom-up adaptive filtering* and *contrast enhancement*. These processes create a new vector, S, from X. In simple ART systems, these steps are not elaborated, and S is set equal to X.

The next step is to pass the information in layer F_1 up to F_2. All that is needed is to multiply the values in X (of length n) by the upwards long term memory matrix (**LTM_up**, which has dimensionality nxp), and store the resulting values in F_2 in vector T, which has length p.

These values of T_j in F_2 undergo a competitive process (similar to the competitive process in the Learning Vector Quantization network). Only one value of T_j will win. Suppose the **j**th element in F_2 wins. Then the winning value is stored as the **j**th element in vector Y, which has the same dimension as vector T. This winning node Y_j is what the ART system thinks is its best match for the input pattern.

Next, the system does what is called *top-down template matching*. First, we copy vector Y over into vector U. (This is basically a *hook* in the ART system which allows insertion of sophisticated cooperative processes. Since we are describing a very simple ART, we will stay with Carpenter/Grossberg notation, and just copy vectors over instead of changing them.)

There is already a set of downward-connecting weights, stored in the matrix **LTM_down**, which Carpenter and Grossberg refer to as **LTM**$_{ji}$. These values are different from the values in **LTM_up**, or **LTM**$_{ij}$. We multiply vector U (of length p) by **LTM_down** (of dimension pxn), to obtain vector B (of length n). Vector B is stored in F_1. B is the top-down template which is produced by F_2. We will compare it with I to see how accurate the match is between the known pattern generated from F_2 as B and the input pattern I.

11.2.3.2   Basic ART Dynamics — Attentional Gain Control   There are now three vectors of length m stored as input or in F_1. These are I, X, and B. I is the input. B is the best match produced by layer F_2 to this input. **X** is a buffer. It was originally the same as **I**, but now we will change it so that the values stored in **X** reflect how well **B** matches **I**. This is an important step in the ART system, and is different from anything we have seen thus far in other neural networks.

The way in which Carpenter, Grossberg, and their colleagues have handled this is to define a *flag* called *attentional gain control*. When the first input arrives to the system, the flag is set on. When F_2 sends down a pattern **B** to F_1, the flag is set off. At each step of the forward and backward process of sending information from F_1 to F_2, and from F_2 to F_1, the flag setting influences how values are stored in the buffer vector **X**. This rule is very simple, and is called the *2/3 Rule*. This means that for a value to be stored in the ith location of **X**, two out of three things must be true. These three true possibilities are:

- Flag on
- X_i on
- B_i on

To summarize how this works, either there must be an on value in the corresponding ith location in **I** and the flag must be on, or the corresponding ith locations in both **I** and **B** must be on, with the flag set to off.

Figure 11.3 illustrates the operation of an ART system through the first two steps: pattern presentation and bottom-up processing, and top-down template matching.

How does ART know if the match is good enough? That is the crux of the entire system, and introduces the need for the orienting subsystem.

11.2.3.3   Basic ART Dynamics — The Orienting Subsystem   The orienting subsystem monitors the *goodness of fit* of the matches by accessing the magnitudes of both the **I** and **X** vectors. It uses a ratio of the two, $|X|/|I|$. This ratio is compared to a parameter which the user can set for the ART system called ρ. The value of ρ controls the degree to which the ART system insists on *goodness of match*; the higher the value of ρ, the better the fit has to be. The values of ρ can be between 0 and 1. If the value is set at 1, that means that the fit between stored template and input pattern must be 100% accurate.

Let's see what happens when we compare $|X|/|I|$ against a fairly usual value for ρ, say 0.85. When the input pattern is first given to the system, **X** is a direct copy of the input vector **I**. That means that their magnitudes are the same, so that the ratio $|X|/|I|$ is 1. So long as this ratio is greater than or equal to ρ, the system accepts the match. If the ratio is less than ρ, the orienting subsystem becomes active. Thus, the action of the ART system depends on the comparison of the ratio $|X|/|I|$ with ρ.

If the ratio is greater than or equal to ρ, then the match is accepted. A little learning takes place, which refines the values of the connection weights, but no new categories are created. The pattern is categorized as the category coded by the ith node in F_2.

If the ratio is less than ρ, the match is not accepted, and the orienting subsystem becomes active. This process is illustrated in Figure 11.4. In this figure, an 11-element pattern is presented to the network which has already been trained to respond to the 9-element diagonal cross. The ratio of X/I is 0.82, which is less than the 0.85 value for ρ. The match is not accepted. The first thing the orienting subsystem does is to suppress

Two Stages of Pattern Recognition in an ART Network

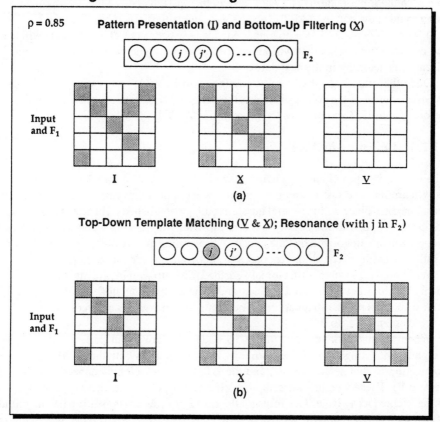

Figure 11.3   The first two steps of ART1 processes. The input is shown as a 2-dimensional matrix, but would actually be encoded as a 1-dimensional array. Vectors **I**, **X**, and **B** have corresponding elements. (a) A noisy cross pattern is input into **I**. **B** has not yet been activated. The units in F_2 compete for the "best match" to the presented pattern. The flag (attentional gain control) is on, so **X** is an exact copy of **I**. (b) The "winning" F_2 unit, j, stores a cross pattern. It sends a "template" of its stored pattern into **B**. The flag is off. Only those **X** elements which have both corresponding **I** and **B** elements "on" are turned "on" themselves. The ratio of $|X|/|I|$ is 9/11 (0.89), which is greater than ρ (0.80), so the match is accepted.

the activity of the jth node in F_2 which has just sent the top-down pattern to F_1. Then, the entire process of having the input vector code into X in F_1, and send a pattern up to F_2 is repeated.

Another node, say j', in F_2 may respond to that pattern. This cycle will repeat until all nodes in F_2 which can potentially respond to the input pattern have tried and failed. Each time a node tries and fails to match the input pattern, its activity is suppressed by the orienting subsystem so that it can not compete with the efforts of the next F_2 node which will attempt a match.

Finally, there will be no *assigned pattern category* nodes left in F_2 which can attempt to match the input pattern. The ART system then assigns one of the unassigned nodes in F_2 to this new pattern. (ART can store as many patterns as it has nodes in F_2. When it runs out of F_2 nodes, it can no longer encode new patterns.) This new node will learn the connection weights that will allow it to accurately encode the input pattern. We discuss this learning in the next subsection.

Once the orienting subsystem has assigned an F_2 node to respond to the input, it releases control of the ART system and the pattern recognition process can continue as before. Table 11.3 summarizes the processes in ART 1.

11.2.4   Learning in ART

There are two types of learning which take place in an ART network: fast learning and slow learning. The fast learning is invoked when a new category is established. This learning takes place within several iterations, all accomplished right after a new F_2 node has been assigned to the new F_1 pattern. Slow learning occurs whenever a pattern is recognized or categorized by an F_2 node. This learning allows the weights to adapt just a little, reflecting recognition responses to subtle changes in the input pattern.

When an ART system is first initialized, the **LTM_up** weights are randomly assigned to small values between 0 and 1. This randomness allows F_2 nodes to differentiate in their responses to the input patterns. The **LTM_down** weights are very small, and can be conveniently set to 0.

When the network first assigns an F_2 node to a new input pattern, all the **LTM** weights connecting to that node learn rapidly. This includes both the **LTM_up** weights connecting to that selected node, and the **LTM_down** weights connecting that node back to F_1. These weight values approach limits. Carpenter and Grossberg refer to this as the Weber Decay Rule. The limits for the **LTM_up** weights depend on the magnitude of on elements in a pattern. We can describe the limits of these weights, where i refers to a node in F_1, and j refers to the active node in F_2:

LTM_up:
For active i, active j: $\lim (t \rightarrow \infty)$ **LTM_ij** = $1/|X|$

For inactive i, active j: $\lim (t \rightarrow \infty)$ **LTM_ij** = 0

For active i, active j: $\lim (t \rightarrow \infty)$ **LTM_ji** = 1

For inactive i, active j: $\lim (t \rightarrow \infty)$ **LTM_ji** = 0

As a result of these limits, connection weights for very small patterns (relative to the size of the input matrix) will be much larger than connection weights learned for large patterns. This allows ART to distinguish small patterns as being unique, even when they use the same input pixels as might be used by larger patterns. This is a very valuable quality of the ART system. For example, after the ART network shown in Figures 11.3 and 11.4 has learned the new pattern, a simple diagonal cross is presented again. (See Figure 11.5.) Both F_2 nodes, j and j', respond. The strength of each of the j **LTM_down**

Creation of a New Category in an ART Network

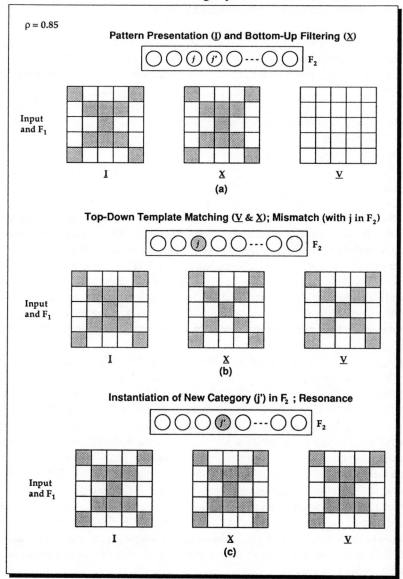

Figure 11.4   ART can create a new category to respond to novel input. (a) An input is presented to F_1, and stored in I. It is copied over into X. (b) The top-down template from the "best matching node" in F_2 is stored in B. The attentional gain control flag is off, and when B matches against I, only a 7 out of the 9 "on" elements in I find a match. X is reduced to those matching elements. (c) The ratio $|X|/|I|$ is less than ρ $(0.77 < 0.8)$, and so the match is not accepted. The orienting subsystem "shuts down" the active node in F_2. A new node in F_2, j', "learns" the new pattern in X.

Table 11.3   Art 1 Processes:

The attentional subsystem has three major processes:

- Bottom-up adaptive filtering and contrast enhancement (STM),
- Top-down template matching and stabilization of learning, and
- STM reset and search. (This process is invoked when template matching doesn't work).

The orienting subsystem has two major processes:
- Vigilence, and
- Attentional gain control and priming.

connections is 1/9, and the sum of those connections is 9/9, or 1. The strength of the j' **LTM_down** connections is 1/11, and their sum for this pattern is 9/11. Thus, j responds to the pattern, even though the pattern is a subset of the pattern encoded by j'.

Note that this involves non-Hebbian learning. **LTM_up** weight vectors that may have been greater than 0 decay down to 0. Carpenter and Grossberg refer to this as the Associative Decay Rule. Note also that learning involves only the connections to and from the selected F_2 node. This means that learning takes place only with regard to a specific input pattern and a specific F_2 node at one time, and does not spill over into the weights for other learned patterns.

Because of this, learned categories in ART can change over time. Suppose the five-element pattern shown in Figure 11.6 is presented to a network trained on the previous examples. Node j in F_2 will respond with a top-down template of a nine-element diagonal cross pattern. Only five of those elemets match those in the iput pattern, but this suffices to give a ratio of $|X|/|I|$ equal to 1.0. The match will be accepted. During the long-term learning and adaptation which follows, the weights between the five "on" elements in F_1 and node j in F_2 will not change. However, the weights between the four unused elements of the original diagonal cross template in F_1 and node j in F_2 will decrease. If this pattern presentation sequence were to be repeated many times, node j would be recoded; it would take on a template of the new five-element pattern and lose the template of the original diagonal cross. Then, if the diagonal cross were to be presented again, it would no longer find a match with node j. (The ratio $|X|/|I|$ would be 5/9, or 0.56, which would be less than $\rho = 0.85$.) Node j' in F_2 would have to respond to the diagonal cross pattern.

11.2.5   Performance and Capabilities of the ART Network

The ART networks (both ART 1 and ART 2) have excited much interest among neural networks researchers because they are the first truly potent self-organizing pattern recognition networks that we have had. Nevertheless, there are some major difficulties when we consider using ART networks for pattern recognition applications. These boil down to the dependence of the network's categorization (and hence its performance)

Competition in F_2 for Response to Input Patterns

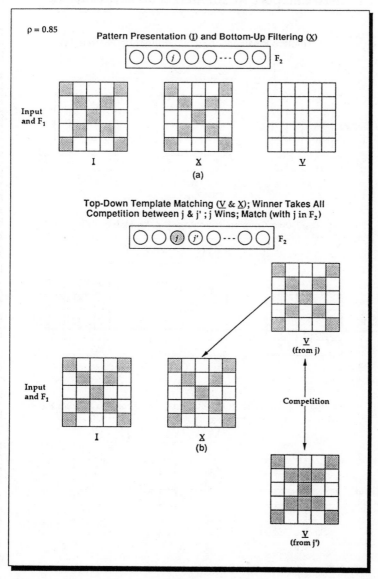

Figure 11.5 (a) A cross pattern is presented to the ART network, previously trained on the patterns shown in Figures 11.3 (cross and noisy cross), and in 11.4. (b) The two trained F_2 nodes, j and j, compete to respond to the input pattern. Connection strengths to node j and stronger, it wins the competition.

Pattern Category Adaptation in an ART Network

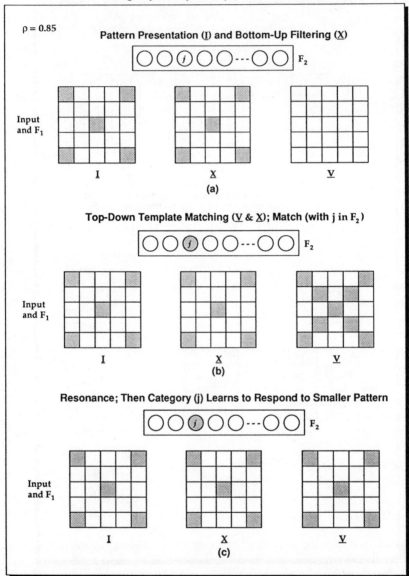

Figure 11.6 ART network pattern category nodes can adapt to slowly learn new patterns over time. This is particularly likely to occur if the new patterns are somewhat smaller than the pattern originally presented.

on both the type and order of patterns learned, and the extent to which even small changes in ρ can have massive influences on the number of pattern classes formed.

Because an ART network responds to the on/off characteristics of each individual element in the input and F_1 nodes, they are not well suited for processing raw data. They cannot handle data which is shifted or distorted. However, when data is preprocessed (as is discussed in the chapter on sonar signal processing), ART is a good high-level classifier [Carpenter and Grossberg, 1987(c)]. ART networks can do a good job of categorical perception, in which an object is classified based on its high-level or descriptive features. Further, systems ART networks can yield hierarchical classification; or identifying an object within multiple classes. Because of these possibilities, several major government and industry research laboratories have been investigating ART networks for different applications.

11.2.6 Applications of the ART Network

The major applications of ART networks thus far have been for speech recognition and generation, recognition of visual patterns, and radar image detection.

ART networks are not the easiest to implement for real-time applications. Several researchers and teams are addressing this issue. Michalson and Heldt [1990] have described a hybrid analog/digital implementation of ART 2. Levine and Penz [1990] have described an "ART 1.5" architecture which they believe is appropriate for radar target classification.

As it stands right now, ART networks are typically too labile to be trusted with certain real-world applications. However, with more development effort, and especially if ART networks were to be embedded into intelligent systems that could supervise behavior (e.g., watch what categories are being formed, detect significant changes in category "templates" from those held earlier by the same category), then ART networks can become the most powerful networks available. The sophistication of the ART network's processes, its ability to discriminate among many different pattern classes, and its ability to respond appropriately to novel input make the ART network the network of the future.

11.2.7 Further Reading About ART

Space limitations prevent a very detailed description of the full complexity of the ART system. For those who want further reading, there are two useful tutorials: one by Gail Carpenter and Steve Grossberg [1988] and another by David Stork [1989]. Two articles written by Steve Grossberg in 1976 form part of the conceptual underpinnings of the ART system. In 1983, Michael Cohen and Steve Grossberg wrote a paper analyzing the stability of patterns formed by neural networks. This is an important work on autoassociators, and is also a significant precursor to the ART system.

The most fundamental and comprehensive paper in which Carpenter and Grossberg described their (now fairly elaborate) ART system was published in *Computer Graphics, Vision, and Image Processing* in 1987. This paper is the major work available on adaptive resonance theory. While not very easy to read, it allows the reader to develop a clear and comprehensive understanding of the ART architecture, learning methods, and performance characteristics. With diligent work, a reader can use this paper as a basis for developing an ART system.

The ART 1 architecture described in the previous reference works on binary patterns. Another 1987 paper, which Carpenter and Grossberg published in *Applied Optics* shows how the ART concept may be extended for systems allowing continuously-valued input [ART 2]. They make it clear that their own work in this area is still experimental, and that they have a variety of architectures under consideration. Although they have achieved some success with their ART 2 implementations, few other laboratories have had such success. This is due to both the complexity of the network and to the fact that performance of that network can vary dramatically with even small changes in the few governing parameters.

Recently, they have extended their work to creating hierarchies of ART systems [Carpenter and Grossberg, 1990, 1899].

11.3   BIDIRECTIONAL ASSOCIATIVE MEMORIES AND RELATED NETWORKS

Bart Kosko has developed another form of resonating, heteroassociative networks called *Bidirectional Associative Memory* (BAM) networks. There are numbers of variants and evolutions on this theme, including continuous and discrete BAMs, *Adaptive* BAMs (ABAMs), *Temporal Associative Memories* (TAMs), *Random* ABAMs (RABAMs), and others. In this section, we focus on simple, discrete BAMs. References to Kosko's works (at the end of this chapter) will put you on the track to BAM variations.

11.3.1   The Underlying Concept for the BAM Network

The underlying motivation for BAM networks is similar to the one underlying development of ART networks. Carpenter and Grossberg refer to it as the stability/plasticity dilemma, and Kosko calls it the stability/convergence dilemma. Kosko focuses on developing a network which stably stores pattern associations, so that any presented input pattern leads to convergence to one of the known associated patterns. Like ART networks, Kosko's BAM networks can learn new patterns.

The basic idea of a BAM is straightforward. It has much in common with the simple linear heteroassociators developed independently by Kohonen and Anderson in the 1970's. The unique characteristic of BAMs is the resonance; passing information back and forth between the layers until both input and output patterns stabilize. This repeated action also has something in common with Anderson's Brain-State-in-a-Box (an autoassociative network which converges after multiple steps) and the Adaptive Resonance Theory network.

Like ART networks, BAMs create their pattern categories as they receive input. The number of categories they will create depends on the number of different types of input patterns they receive. The memory limitations, and possibilities of error when given non-orthogonal patterns, are somewhat like those found in the Hopfield network.

11.3.2   The BAM and ABAM Architectures

Because it is a heteroassociator, there are two layers to a BAM (or any of the BAM variants): an input layer and an output layer. The neurons can be in one of two states, on (+1), or off (-1 or 0, depending on how you want to code it). Typically, they have hard-limit thresholds set at a value of 0.

BAMs use two connection matrices, one feedforward, the other feedback. Kosko calls his input layer F_x and his output layer F_y. The feedforward matrix is M, and the feedback matrix is M^T (the transpose of M). If F_x is of dimension n, and F_y is of dimension p, then M has dimensionality nxp, and M^T has dimensionality pxn. Figure 11.7 illustrates the BAM architecture.

11.3.3   The Dynamics of BAM-Type Systems

In the usual manner, a pattern in the first layer excites a pattern in the second layer using its feedforward matrix of synaptic connections. The novel twist is that this pattern now uses the feedback connection matrix (which is a transpose of the feedforward matrix) to excite a pattern in the input matrix. This new pattern in the input matrix now excites a pattern in the second matrix, and this process repeats until patterns in the input and

**The Discrete and Adaptive
Bidirectional Associative Memory Network**

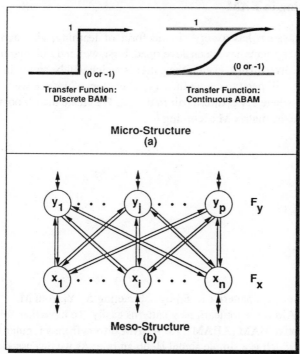

Figure 11.7   The architecture of a BAM network consists of two layers with feedforward and feedback connections.

output matrices stabilize. This is the resonance in the BAM, and is similar to resonance in the ART networks.

Let's walk through a BAM pattern association process. Let's assume that we are running a BAM with 0 and 1 as the two possible states for each neuron. When a BAM operates, a pattern is presented to either layer, F_x or F_y. Suppose that it is presented to layer F_x. This input pattern multiplies matrix M, to yield a p-dimensional pattern. For each element in this pattern, we make one of three possible choices. If the value of this element is greater than the threshold (0), then we store a 1 in the corresponding position in F_y. If the value is less than 0, we store a 0 in the corresponding position. If the value is 0, we leave the currently stored value unchanged. We need to keep track of how many changes we make when we store this new pattern.

Next, the BAM does a feedback stage which is just like the feedforward one. The pattern in F_y multiplies M^T, and the results are bound and stored in F_x in same manner as was just described for storing results in F_y. The feedforward stage is repeated, and so is the feedback stage, until there are no changes in either layer. The input pattern will have converged to the one of the patterns used to train the BAM, and the output pattern will have converged to the pattern associated with the input pattern used for training. Note that since patterns are encoded into the BAM as an associated pair, you can start by inputting a pattern into either F_x or y, and the BAM will produce the pattern closest to the input pattern in the layer you started with, and its associated pattern in the other layer.

11.3.4   Learning in BAM

A simple BAM system has a very simple form of learning, although much more complicated BAM systems have been developed. First, we need a bipolar representation of the patterns involved. Thus, even if the patterns which will be stored in the BAM are (1,0) patterns, we will need to create their (1,-1) analogs. Suppose we have a set of P patterns, (A_i, B_i), where i is a pattern pair in the set. We create their bipolar analogs, X_i and Y_i. We create the matrix M according to:

$$M = \sum_{i=1}^{P} X_i^T Y_i$$

and M^T according to

$$M^T = \sum_{i=1}^{P} Y_i^T X_i$$

Note that we can erase a pattern (A_i, B_i) by subtracting $X_i^T Y_i$ from M.

This simple BAM does not learn new patterns easily. To have that ability, we need to create an Adaptive BAM (ABAM). An ABAM differs from a regular BAM only in its learning rule, which is a simple signal Hebbian rule. As we discussed in Chapter 6, Hebbian learning occurs only when two neurons (a presynaptic and a postsynaptic, or in our case F_{xi} and F_{yj}, are both on at the same time.)

11.3.5   Performance and Capabilities of BAM

An obvious question about such a system is, "Does it work?" Can we be sure that given an input pattern, the right output pattern will show up, and the input pattern will stay close to what it originally was? This is a stability question, and is a very valid question to ask. Because of this, many of Kosko's papers address the stability of the BAM network and its various metamorphoses. We will not present his stability arguments here, but they can be found in his papers cited in the bibliography and at the end of this chapter.

However, the most important consideration about any network is its usefulness — its performance characteristics. BAMs can be confused if like inputs are associated with different outputs, or vice versa.

Some investigators have modified the BAM architecture to attempt to overcome its limitations. Schurmann [1989], Hagiwara [1990], and Lee and Kil [1989] have each proposed variant architectures with greater capability for generalization. Simpson [1989, 1990 (a)] has proposed an architecture consisting of a system of BAMs which allows for greater memory of states. He suggests that a partitioning of patterns into orthogonal sets may facilitate recall. Haines and Hecht-Nielsen [1988] have proposed a non-homogeneous BAM in which neurons may take on values between -1 and 1. They show how this BAM variant may have higher memory capacity and greater fidelity at pattern association.

High-order BAMs [Simpson, 1990(b & c); and Wu et al., 1990] have substantial potential for overcoming the memory problems of simple BAMs. Wu et al. [1990] have shown how a higher-order BAM (one encoding triple and higher connections as well as the more common paired connections) can be applied to frequency classification.

11.3.6   Applications of BAM

There have only been a few applications of BAM networks. This is a direct result of the storage limitations of this network type. The storage of a BAM network can be roughly expressed as P (the maximum number of patterns which can be stored) min(n,p).

11.3.7   Further Reading on BAM-Type Systems

Kosko has written a number of interesting articles enhancing his original BAM system [Kosko, 1987a, 1987b, 1988a, 1988b, 1989]. The most crucial ones are listed at the end of this chapter, and include his 1987 paper on Adaptive BAMs (ABAMs) and his 1989 paper on Random ABAMs (RABAMs). Pat Simpson has shown how use of several sets of BAM-type networks can overcome the memory limitations of a single BAM [Simpson, 1990a]. Wang et al [1989] have also explored enhanced BAMs.

11.4   SUMMARY

To summarize, we have explored three types of heteroassociative networks: an early type with a simple feedforward association matrix, and the resonating ART and BAM systems. The early type of network is not in much use, and suffers from the same types of limitations (memory, need for orthogonal patterns, etc.) as affect the Hopfield network.

Both the ART and BAM classes of networks appear promising. The BAMs are conceptually simpler, but are subject to some of the pattern storage and orthogonal constraints that limit performance of some similar autoassociative networks, such as the Hopfield and BSB networks. ART is not similarly constrained, and is more subtle and complex.

In terms of pattern recognition capabilities, both ART and BAM respond to fixed pattern types, whereas a back-propagation network can be trained to recognize more general or abstract pattern classes, or patterns occurring in different positions in an input layer. There are other weaknesses affecting ART and BAM networks. However, they are the only ones to date which can create a new pattern category (or form a new heteroassociation) when presented with novel input. But as Anderson [1987] has said, "It is hard to avoid the conclusion that a learning network is a dangerous thing." We have a lot of work to do before we can safely and usefully incorporate these networks into usable systems.

REFERENCES

Amari, S. (1967). "A theory of adaptive pattern classifiers," *IEEE Trans. on Electronic Computers, EC-16*, 299-307.

Amari, S. (1972). "Learning patterns and pattern sequences by self organizing nets of threshold elements," *IEEE Trans. Computers, C-21*, 1197-1206.

Anderson, J.A. (1972). "A simple neural network generating an interactive memory," *Math. Biosciences, 14*, 197-220, and reprinted in J.A. Anderson and E. Rosenfeld (Eds.), *Neurocomputing* (1988), MIT Press, Cambridge, MA, 181-192.

Anderson, J.A. and Rosenfeld, E. (1988). "Introduction" (to separate articles by Kohonen and Anderson), in J.A. Anderson and E. Rosenfeld (Eds.) (1988). *Neurocomputing* Cambridge, MA, 171-173.

Carpenter, G. and Grossberg, S. (June 18-22, 1989). "Search mechanisms for Adaptive Resonance Theory (ART) architectures," *Proc. Int'l. Joint Conf. Neural Networks* Washington, D.C., I-201–I-205.

Carpenter, G.A. and Grossberg, S. (March, 1988). "The ART of adaptive pattern recognition," *Computer*, 77-88.

Carpenter, G.A., and Grossberg, S. (1987a). "A massively parallel architecture for a self-organizing neural pattern recognition machine," *Computer Vision, Graphics, and Image Processing, 37*, 54-115.

Carpenter, G.A., and Grossberg, S. (December 1, 1987b). "ART 2: Self-organization of stable category recognition codes for analog input patterns," *Applied Optics*.

Carpenter, G., and Grossberg, S. (June 21-24, 1987c). "Invarient pattern recognition and recall by an attentive self-organizing ART architecture in a nonstationary world," *Proc. First IEEE Int'l. Conf. on Neural Networks* San Diego, CA, II-737–II-745.

Carpenter, G., and Grossberg, S. (1990). "ART 3: Hierarchical search using chemical transmitters in self-organizing pattern recognition architectures," *Neural Networks*, *3*, 129-152

Cohen, M., and Grossberg, S. (1983). "Absolute stability of global pattern formation and parallel memory storage by competitive neural networks," *IEEE Trans. Systems, Man, and Cybernetics, SMC-13* , 815-826.

Grossberg, S. (1976). "Adaptive pattern classification and universal recoding: I. Parallel development and coding of neural features," *Bio. Cybernetics*, *23*, 121-134, and reprinted in *Neurocomputing*, J.J. Anderson and E. Rosenfeld (Eds.) (1987). MIT Press, Cambridge, MA, 245-258.

Grossberg, S. (1976). "Adaptive pattern classification and universal recoding: II. Feedback, Expectation, Olfaction, Illustions," *Bio. Cybernetics*, *23*, 187-202.

Grossberg, S. (1987). "Competitive learning: From interactive activation to adaptive resonance," *Cognitive Science*, *11*, 23-63.

Grossberg, S. (Ed.) (1988). *Neural Networks and Natural Intelligence* MIT Press, Cambridge, MA.

Grossberg, S. (Ed.) (1987). *The Adaptive Brain, Vol. 1: Cognition, Learning, Reinforcement, and Rhythm*, New York, Amsterdam, North Holland.

Hagiwara, M. (January 15-19, 1990). "Multidimensional associative memory," *Proc. Second Int'l. Joint Conf. Neural Networks*, Washington, D.C., I-3-I-6.

Haines, K., and Hecht-Nielsen, R. (July 24-27, 1988). "A BAM with increased information storage capacity," *Proc. Second IEEE Int'l. Conf. on Neural Networks*, San Diego, CA, I-181-I-190.

Kohonen, T. (1972). "Correlation matrix memories," *IEEE Trans. Computers, C-21*, 353-359, and reprinted in J.A. Anderson and E. Rosenfeld (Eds.) (1988) *Neurocomputing* MIT Press, Cambridge, MA, 174-180.

Kohonen, T. (1989). *Self-Organization and Associative Memories*, 3rd Ed., Springer-Verlag, New York, NY.

Kohonen, T. (1987). *Content-Addressable Memories*, 2nd. Ed., Springer-Verlag, New York, NY.

Kosko, B. (September, 1987). "Constructing an associative memory", *Byte*, 137-144.

Kosko, B. (1986). "Adaptive bidirectinal associative memories," *Applied Optics*, *26*, 4947-4960.

Kosko, B. (July 23-27, 1988). "Feedback stability and unsupervised learning," *Proc. Second IEEE Int'l Conf. on Neural Networks*, San Diego, CA, I-141-I-152.

Kosko, B. (1988), "Bidirectional associative memories," *IEEE Trans. Systems, Man, Cybernetics, SMC-L8*, 49-60.

Kosko, B. (June 18-22, 1989). "Unsupervised learning in noise," *Proc. Int'l. Joint Conf. Neural Networks*, Washington, D.C., I-7-I-17.

Lee, S., and Kil, R.M. (June 18-22, 1989). "Bidirectional continuous associator based on Gaussian potential function," *Proc. First Int'l. Joint Conf. Neural Networks*, Washington, D.C., I-45-I-53.

Levine, D.S., and Penz, P.A. (January 15-19, 1990). "ART 1.5 — A simplified adaptive resonance network for classifying low-dimensional analog data," *Proc. Second Int'l. Joint Conf. Neural Networks* , Washington, D.C., II-639-II-642.

Michaelson, W.R., and Heldt, P. (January 15-19, 1990). "A hybrid architecture for the ART 2 neural network," *Proc. Second Int'l. Joint Conf. Neural Networks*, Washington, D.C., II-167–II-170.

Nakano, N. (1972). "Associatron: A model of associative memory," *IEEE Trans. Systems, Man, and Cybernetics, SMC-2*, 381-388.

Shurman, B. (June 18-22, 1989). "Generalized adaptive BAM," *Proc. First Int'l. Joint Conf. Neural Networks*, Washington, D.C., I-91–I-97.

Simpson, P. (1989). "BAM systems," *Heuristics, 1*, 50-59.

Simpson, P. (June 18-22, 1990a). "Associative memory systems," *Proc. Second Int'l. Joint Conf. Neural Networks*, Washington, D.C., I-468–I-471.

Simpson, P. (1990b). *Artificial Neural Systems: Foundations, Paradigms, Applications, and Implementations*, Pergamon, New York, NY.

Simpson, P. (May, 1990c). "Higher-order and intraconnected bidirectional associative memories," *IEEE Trans. Systems, Man, and Cybernetics, SMC-20*.

Stork, D.G. (1989). "Self-organization, pattern recognition, and adaptive resonance networks," *J. Neural Network Computing*, (Summer), 26-42.

Wang, Y.-F., Cruz, J.B., and Mulligan, J.H. (June 18-22, 1989). "An enhanced BAM," *Proc. Int'l. Joint Conf. on Neural Networks*, Washington, D.C., I-105–I-110.

Wu, C.-H., Tai, H.-M., Wang, C.-J., and Jong, T.-L. (January 15-19, 1990). "High-order associative memory and its application to frequency classification," *Proc. Second Int'l. Joint Conf. Neural Networks*, Washington, D.C., I-31–I-34.

12

MULTILAYER COOPERATIVE/COMPETITIVE NETWORKS

Alianna J. Maren

12.0 OVERVIEW

This chapter addresses a different kind of architectural concept from those previously considered — that of using cooperative/competitive networks. In these networks, neurons both cooperate with and compete with each other in order to carry out some task. System architectures often involve layers of networks, as with the basic competitive learning network. Sometimes, the cooperative and competitive processes are split among the different layers. An example of this is Grossberg's Boundary Contour System and Fukushima's Neocognitron.

A typical task for a cooperative/competitive network is to define some sort of mapping relationship between input data and neuron activation in the network. That is, the cooperative/competitive relationship helps resolve which neurons in a set get to respond to some input pattern, whether it is characterized by certain features or by spatial location, or both. The mappings which result from cooperative/competitive networks are useful for pattern recognition sensory mappings, motor control, optimization, and other tasks.

12.1 INTRODUCTION

Cooperative/competitive processes allow us to build up very subtle and sophisticated feature recognition and pattern recognition networks. By using layers of cooperating, competing, and thresholding neurons, we can achieve high-level transformations of input data into very useful representations. This can lead to pattern classification (e.g., character recognition), high-level pattern completion (as with finding the boundaries of regions), and hierarchical clustering.

179

The idea of using a cooperative/competitive network architecture has a very sound biological basis. Neurons in mammalian vision systems use the cooperative/competitive principle with great success. For example, an on-center, off-surround cell in the retina is excited by light stimulus striking the innermost area of the receptive field, and is inhibited by light in the periphery of its small receptive field.

Because cooperative/competitive networks are among the most powerful single-network architectures available, they are worth the investment of study time and experimentation. The networks which we discuss in this chapter include Rumelhart and Zipser's competitive learning network, the Cohen/Grossberg masking fields network, Fukushima's Neocognition, Grossberg's Boundary Contour System, and Minsky and Maren's Hierarchical Scene Structure system. Each is a complex network which exhibits a different and interesting capability of a cooperative/competitive network system.

12.2 COMPETITIVE LEARNING NETWORKS

The competitive learning approach is a variation of a feedforward architecture in which lateral inhibitory connections are allowed. This allows a winner-takes-all mechanism in which neurons on the intermediary and upper layers selectively respond to features in the input and inhibit the response of certain other neurons on their same layer to that input. This leads to a strong specialization in response to pattern features.

The simple competitive learning network (described by Rumelhart and Zipser [1985]) has not found widespread application. However, the basic concepts behind this network have motivated development of much more complex networks. (Note that although these concepts appear in different networks, researchers worked in near-isolation during the early stages of neural network development in the 1970's. Thus, von der Malsburg [1973], Grossberg [1976], Fukushima [1975], and Kohonen [1982] each independently developed neural networks which used lateral inhibition to achieve different goals.) The important thing to be gained from this section is an understanding of a basic architectural concept which has found a home in many more sophisticated networks.

12.2.1 The Concept Underlying the Competitive-Learning Network

The concept underlying the competitive-learning network is that neurons in the higher levels of a multilevel network can learn to recognize features in input patterns by using an architecture which has lateral inhibitions combined with a learning method that reinforces the connections to winning neurons. This leads to a network which can recognize different types of input patterns based on the features of these patterns.

The feature discovery and generalization mechanisms of a competitive learning network are different from the feature generalization abilities of a back-propagation network. The back-propagation network undergoes supervised learning, which forces (given an appropriate architecture) the hidden neurons to generalize about features in the input population. The competitive-learning network undergoes unsupervised learning, and so creates its own feature classes, which leads to its own categories of output.

Architecture of the Competitive Learning Mechanism

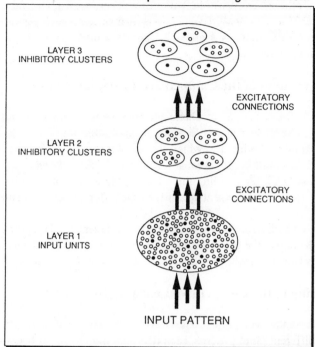

Figure 12.1 The competitive-learning network described by Rumelhart and Zipser [1985] has three layers. The input was presented to the first layer. The second and third layers consisted of "clusters" of neurons. Active neurons in the figure were represented by filled-in circles, and inactive ones by open circles. Neurons had inhibitory connections to other neurons in their own cluster, and excitatory connections to neurons at the next highest layer. The neurons in Layer 1 did not have any lateral connections, and no neurons had connections to any neurons on its layer who were not within its cluster. Figure adapted from D. Rumelhart and D. Zipser, "Feature discovery by competitive learning," *Cognitive Science*, *9*, p. 85. Reprinted with the permission of Ablex Publishing Corporation.

12.2.2.   The Architecture of the Competitive-Learning Network

In the architecture described by Rumelhart and Zipser [1985], each neuron could have an activation value between and inclusive of 0 and 1. The activation of each node was proportional to the sum of all its inputs. The strength of input into a node is governed by the constraints on the total amount of weight on the input to each node. Each node is allowed a fixed amount of incoming weight (all weights are positive) which is distributed among its input lines.

The network architecture had multiple layers. The first was for input and was not partitioned or treated in any special manner. The second and any subsequent layers consisted of clusters of nodes or neurons. Each neuron within a cluster had inhibitory connections to all other neurons within that cluster. There were no lateral connections outside of a cluster. The feedforward connections (from a neuron in layer i to all neurons in layer i+1) were all excitatory. Figure 12.1 shows a three-layer version of such a competitive learning network.

12.2.3 The Dynamics of the Competitive-Learning Network

The system used a winner-take-all mechanism to select a maximally responsive neuron from within each cluster to respond to a given input stimulus. The winning neuron represents the feature in the input to which that cluster has the greatest response. The winning neuron inhibits all other neurons in its cluster, and sends an excitatory signal to the next highest level. By iterating this process, the excited neuron in the highest layer responds to a set of features sent upwards by the different clusters in the preceding layer.

The features to which neurons respond at different layers can be at different levels of abstraction. The higher the layer, the more abstract or generalized the feature can be. (This process has been used with substantial success in the Neocognitron.)

12.2.4 Learning in the Competitive Learning Network

The connections to the winner neurons undergo Hebbian learning. After repeated presentation of different input patterns, each neuron within a cluster becomes specialized to respond to different features in the input. Neurons in higher-level clusters learn to respond to higher-level features; features which could be abstracted from the multiple features active in the immediately preceding layer.

12.2.5 Performance and Capabilities of the Competitive Learning Network

The competitive-learning network described by Rumelhart and Zipser functioned effectively as a regularity detector, and could be interpreted as a pattern classifier. Bairaktaris [1990] used a competitive-learning network with modifiable thresholds for visual pattern (character) recognition, but found substantial limitations with this approach.

Because the competitive-learning network undergoes unsupervised learning, it does not have a separate learning phase. The connection weights are subject to constant change. Grossberg [1987] showed that the network described by Rumelhart and Zipser [1985] could have convergence problems; that is, the classes which it created might not be stable. He suggested that the ART network had greater robustness in terms of the stability of its classes. (Pattern classes in the ART network are still somewhat labile, so the problem isn't solved yet. Nevertheless, the use of the vigilance parameter and the orienting subsystem within the ART system does help its stability. In this regard, the ART concept is more refined than the competitive-learning network.)

Reggia [1989] has recently proposed an alternative to the competitive learning method in which inhibitory interactions are produced by a controlled spread of activation rather than by direct inhibitory connections. He applied this approach to two moderately large-scale problems: print-to-sound mapping and fault diagnosis. Results in both cases were encouraging, and suggest that this modification of the competitive learning method may prove useful.

12.3   MASKING FIELDS

Masking fields are a way to adapt pattern recognition systems (especially ART systems) to recognize complex sequences or patterns of data. They have been applied to speech recognition tasks. They were first introduced by Grossberg [1978], and further developed by Michael Cohen and Steve Grossberg [1987].

12.3.1   The Concept Underlying the Masking Field Network

The essence of the masking field concept is that recurrent self-connections are used in the pattern recognition layer F_2 of an ART system. This enables a node which responds to an input pattern to stay activated even though the stimulus pattern is no longer present.

12.3.2   The Architecture of the Masking Field Network

The masking field network is an adaptation of the Adaptive Resonance Theory (ART) network, discussed in the previous chapter. There are three major differences between a masking field network and an ART network.

The first major difference is that each node in the upper layer of a masking field network has an excitatory self-connection as well as lateral inhibitory connections to other nodes in the same layer. That means that once a neuron wins any kind of competitive process, it can continue its (slowly decreasing) activation through the next several pattern presentations. This is shown in Figure 12.2.

The second difference between a masking field and an ART network is that there may be additional nodes in the F_2 layer of the masking field. These nodes will be triggered by sequential activations of two or more different input patterns. Thus, these nodes respond to temporal patterns, rather than just a single static input pattern.

The third major difference between the two network types is that another pattern recognition layer of nodes (or even another network) needs to be used in order to recognize and categorize the pattern which appears in the F_2 layer. Cohen and Grossberg [1987] have also described further architectural developments leading to a smoother response to pattern sequences in the F_2 field.

12.3.3   The Dynamics of the Masking Field Network

The basic dynamics of a masking field are the same as those of the ART system. The only difference is in the temporal persistence of the activation of the recognition node in F_2. Because of the self-excitatory feedback connections in F_2, the winning F_2 node remains active even when the stimulus originally presented at the F_1 layer is removed or changed. Thus, after several nodes in the F_2 layer recognize patterns, there is an activation pattern across the F_2 nodes. This can be used to encode representations of temporal pattern sequences.

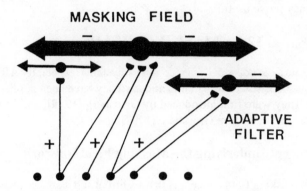

Figure 12.2   A masking field is similar to an ART network, except that the neurons in the second layer have excitatory self-connections, which permit them to remain temporally active after they have responded to an input pattern. Figure taken from M. Cohen and S. Grossberg, "Masking fields: a massively parallel neural architecture for learning, recognizing, and predicting multiple groupings of patterned data," *Applied Optics*, 26, p. 168, May 15, 1987. Copyright by the Optical Society of America. Reproduced by permission.

12.3.4   The Learning Method of the Masking Field Network

The connections between the F_1 and F_2 layers in a masking field learn in a similar manner as the corresponding connections in an ART network. The connection strengths between the F_2 nodes are dependent upon the number of connections to each F_2 node made by different sets of active F_1 nodes.

12.3.5   Performance and Capabilities of the Masking Field Network

Cohen and Grossberg [1987] have presented the results of several different computer simulations of masking fields using synthetic data. Cohen et al. [1987] described how a masking field plays a role in a more complex system which allows for feedback-dependent speech articulation. The masking field facilitates context-sensitive *chunking* of emerging discrete linguistic units. Nigrin [1990] has developed a similar approach which is useful for discovering significant chunks of information in a temporal pattern of inputs.

12.4   THE BOUNDARY CONTOUR SYSTEM

The Boundary Contour System (BCS), developed by Steve Grossberg and Ennio Mingolla [1985 (a) & (b)], is a complex multilevel, cooperative/competitive method for biologically based image segmentation. It is capable of connecting edge segments over a distance of several pixels. It mimics a number of phenomena observed in studies of visual psychophysics, and is probably the most complete and sophisticated approach to low-level vision processing available.

The BCS is actually a subsystem within a more complex architecture for pre-attentive vision. This overall architecture involves the BCS, a Feature Contour System (FCS), and an Object Recognition System (ORS). Because we want to explore cooperative/competitive processes as a design tool, we will concentrate on one subsystem, the BCS, which illustrates both cooperative and competitive interactions.

12.4.1   The Concept Underlying the BCS

The BCS/FCS/OR system is strongly based on the neurophysiology of the lower levels of the mammalian visual system. It illustrates how multiple levels of interacting modules can yield effective processing of complex input information. The network nodes are topologically isomorphic, but their functions correspond to simple, complex, hypercomplex, and unoriented and oriented cooperative cells.

A key design aspect in creating the BCS is the use of what Grossberg calls the *CC loop*; a cooperative/competitive feedback network. In a CC loop, an incoming stimulus excites certain neurons, which then send out cooperative (excitatory) signals to similar neurons (or to neurons whose activation can strengthen their own activation), and send out competitive or inhibitory signals to neurons whose activation would interfere with their own activation, or would compete with their activation in a later stage of processing. The CC loop involves feedforward, feedback, and lateral connections. It is useful for sharpening distinctions when several different neurons are excited in response to the same initial stimulus. Figure 12.3 illustrates the operation of a simple CC loop.

12.4.2   The Architecture of the BCS

The combined BCS/FCS/ORS system is composed of six layers of neurons. Each layer contains one or more networks each having the same number of nodes. However, each layer performs a very different task. The first layer is the input, and the next four layers (Layers 2-5) are used for the BCS. A CC loop is used between layers 3, 4, and 5 [Grossberg, 1987 (a)]. Figure 12.4 illustrates a schematic representation of the BCS/FCS system.

12.4.3   The Dynamics of the BCS

Each neuron in Layer 2 receives an on-center, off-surround stimulus from a neighborhood of neurons (pixels) in Layer 1. The on-center, off-surround processes generate a response which is fed to Layer 3.

Layer 3 of the BCS/FCS/ORS architecture mimics the simple cells in the visual cortex. It is composed of several different neural networks, each of which is tuned to a particular orientation and direction-of-contrast from its input cells in Layer 2. Thus, each neuron (or group of neurons) in Layer 2 connects with each of the different subnetworks of Layer 3.

A Cooperative-Competitive Feedback Exchange

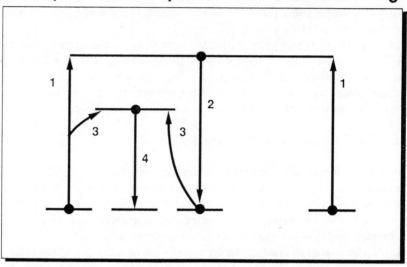

Figure 12.3   The cooperative/competitive loop (CC Loop) defined by Stephen Grossberg allows certain neurons in a system to preferentially strengthen their response to a stimulus and decrease the activity of neighboring neurons which have a similar (but lesser) response to the input stimulus. The CC loop thus exerts a sharpening and defining effect on a neural network's response to stimulus. Figure from S. Grossberg, E. Mingolla, and D. Todorovic, "A neural network architecture for preattentive vision," *IEEE Trans. on Biomedical Engineering, 36* (January 1989), p. 78. © 1989 IEEE. Reproduced by permission.

Active Layer 3 cells cooperate with those in different Layer 3 nets corresponding to the same retinal positions and orientation, but different direction-of-contrast. (Each Layer 3 net is tuned to a certain orientation and direction-of-contrast specificity. The simulations by Grossberg and his colleagues used 12 nets at Layer 3.) These cooperative interactions produce activations of Layer 4 cells which are sensitive to the same orientation and position, but which insensitive to direction-of-contrast. Thus, if twelve networks are used in Layer 3 to store twelve different orientation and direction-of-contrast specific activations, only six subnetworks are needed in Layer 4. This represents *short-range cooperation*, since interactions between cells are corresponding to the same local area in the retinal plane.

The active Layer 4 cells engage in two types of short-range competitive interactions. In one type, an active neuron inhibits nearby neurons which respond to the same orientation. This thins and sharpens edges. In the other type of competition, cells with different types of orientation sensitivities inhibit each other. This prevents extension of boundary edges beyond the region which generated them.

Layer 5 combines thresholded outputs from all of the orientation-specific networks in Layer 4, and has a long-range form of cooperation which is a major component of the CC loop. This long-range cooperation between neurons with similar orientation

The Boundary Contour System

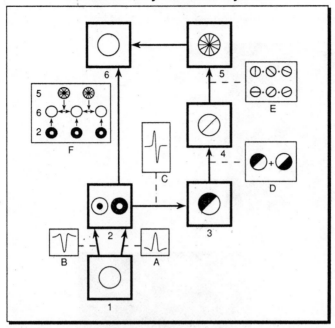

Figure 12.4   The Boundary Contour System (Boundary Contour System) consists of five layers embedded into a more complete Boundary Contour System / Feature Contour System / Object Recognition System. The activities of the five layers are explained in the text. Figure from S. Grossberg, E. Mingolla, and D. Todorovic, "A neural network architecture for preattentive vision," *IEEE Trans. on Biomedical Engineering*, **36** (January 1989), p. 70. © 1989 IEEE. Reproduced by permission.

sensitivities enables boundaries to override small gaps which might otherwise produce disjointed, segmented regions. As part of the CC loop, active Layer 5 cells also send positive feedback to Layer 3 cells of similar orientation. (These Layer 3 cells require input from two or more Layer 5 cells to become active.) Competition at this layer keeps the edge from spreading, but the new feedback-induced activation of more Layer 3 cells excites corresponding cells in Layer 4, and these cells in turn excite (via another positive feedback loop) appropriate cells in Layer 3. This results in a complete yet thin boundary contour. (The complete dynamics are described more fully in the papers by Grossberg and his colleagues listed at the end of this chapter.)

Together, these outputs represent the boundary contour of a region in an image. Overall, this contour has been achieved by using five layers of representation and processing, and multiple networks at Layers 3 and 4 to represent different orientations and directions of contrast. The dynamics involve feedforward, feedback, and lateral connections (both cooperative and competitive), and thresholding at most layers (especially from Layer 4 to Layer 5).

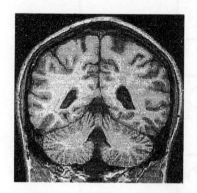

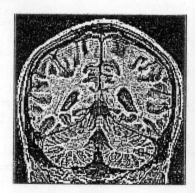

Figure 12.5 The Boundary Contour System (BCS) and associated Feature Contour System (FCS) have been applied to segmenting: (a) Original Image, (b) Image after BCS segementation, (c) Image after FCS fills in the BCS segements. Reproduced by permission of Dr. S. Leher, at Boston University.

12.4.4   The Learning Method of the BCS

The connection strengths in the BCS do not change, so the system does not learn.

12.4.5   Performance and Capabilities of the BCS

The result of all of this work is a region-contour detection scheme which is substantially more subtle and sophisticated than any of the image processing algorithms currently in use. It has the capability of overcoming numerous difficulties which beset current image processing technology (e.g., region boundaries which break where we would like them to jump a few pixels and connect, etc.). Figure 12.5 illustrates the effect of BCS segmentation on a CAT scan image. Note that it links up two offset edges, even though it has to make a diagonal connection over several pixels to do so. The type of robust segmentation which potentially generated by this architecture is one of the better arguments for neurally based engineering.

12.4.6   Applications of the BCS

One of the major deterrents to widespread adoption of the BCS/FCS/OR system has been its computational complexity and its memory requirements. Diamond and Holden [1989] have implemented the BCS on a multi-vector processor iPSC/2 Hypercube, with satisfying results. This type of highly-parallel implementation will undoubtedly lead to greater exploration of the BCS paradigm.

Since the introduction of the BCS, Grossberg and his colleagues have continued to refine the system [Grossberg 1987 (b) & (c); Grossberg & Todorovic, 1988; Grossberg, Mingolla, & Todorovic, 1989]. More recently, Carpenter, Grossberg, and Mehanian [1989] have introduced the CORT-X boundary segmentation network which has only feedforward operations and thus might be easier to implement in hardware. Most recently, Grossberg and Rudd [1989] have extended the BCS system for detecting group and element motion. Daugman [1989] reports using the BCS approach for both motion and texture analysis.

12.5   HIERARCHICAL SCENE STRUCTURES

The important aspect in Part II is not so much the specific details of any neural network system, but rather how the architectural and dynamic concepts used in building a given system can be abstracted and used for other applications. The Hierarchical Scene Structure (HSS) system developed by Veronica Minsky and Alianna Maren [Minsky & Maren, 1989; Minsky, 1990] is an example of this. Its design draws greatly from the principles used in the BCS.

12.5.1   The Concept Underlying the HSS System

The HSS system was developed to provide a hierarchical data structure which represents the perceptual organization of segmented regions in an image. This is necessary because a single object may be segmented into many discrete segments, each of which has its own characteristics or features. Further, due to shadows or other impediments, the segments of a single object may appear in several physically different locations in a image. By grouping together the most closely related and perceptually salient regions, the HSS facilitates scene understanding and object identification.

The HSS system accomplishes its goal of creating a hierarchical data structure by progressively identifying the most *closely related* and *perceptually distinct* regions in an image. There are many possible types of relationships between each pair of regions. Examples include similarity of orientation, similarity of intensity, and boundary line continuation. The proximity of two regions also plays a role in region grouping.

To identify which regions should be grouped together, the HSS system calculates values for six different types of pairwise relationships for each possible combination of pairs in a segmented image. Cooperative processes are used to increase the likelihood that *highly-related* sets of regions will be grouped. Competitive processes discourage the involvement of strongly-related pairs of regions in other possible groupings, and also enhance the dominance of *perceptually salient* (the most perceptually distinguishing) relationships. Thresholding is used to ensure that only the most strongly related regions survive to form new groups or clusters of regions.

12.5.2   The Architecture of the HSS System

The architecture of the HSS system has noticeable conceptual similarities to that of the BCS. Figure 12.6 illustrates the architecture of the HSS system. In order to show how cooperative/competitive principles can be abstracted and used in different neural architectures, we present the architecture of the HSS system in terms of its similarities and differences from the BCS.

First, the similarities. Both networks use multiple levels of neurons, and have multiple networks at certain levels to represent different specific tuning aspects of the same type of phenomena. (For example, the BCS has different networks at Level 3 for different orientations, and the HSS has different networks at Layers 2-3 for different types of relationships between regions.) The "layers" in the HSS pond roughly in organizing layers of activity with the levels in the BCS. Both networks use combinations of cooperation, competition, and thresholding between layers. In each case, cooperative processes preferentially encourage certain responses between strongly activated neurons and their corresponding neurons in other nets on the same layers. Competitive processes discourage activations which would diminish the strength of the most strongly activated processes. Thresholding is used, particularly in going from one layer to another, to extract the most powerful signals for further processing.

Major differences between the BCS and the HSS also exist. The BCS is totally isomorphic to the original input image in its representation. That is, every neuron in each layer of the BCS corresponds to one of the image neurons. In contrast, the HSS begins operation at a more abstract representational level. Input to the HSS is a list of segmented regions along with their features. The individual neurons in the HSS represent pairwise relationships between these regions, and the cooperative/competitive processes strengthen or diminish the relationships that are relative to each other. The output is a set of relationships which are strongest or most perceptually salient.

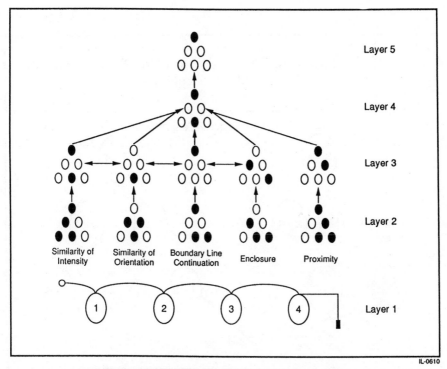

**A cooperative/competitive architecture for
creating the perceptual organization of an image.**

Figure 12.6   The architecture of a Hierarchical Scene Structure system consists of five layers. The first layer is a list of segmented regions and their features. The nodes on the remaining layers correspond to pairwise relationships between regions. There are several nets at each of layers 2 and 3 to correspond to different types of pairwise relationships. Lateral connections are not shown in this figure.

12.5.3   The Dynamics of the HSS System

There is no attempt to have the neurons in the HSS correspond to biological neurons. However, the processing method in the HSS is somewhat like the BCS. The relationship values from the input layer are stored in Layer 2 of the HSS, and passed up to Layer 3. In Layer 3, active neurons both inhibit neurons in other, lateral nets that would compete with the relationship indicated by the active neurons, and excite the neurons in the lateral nets that encourage the same pairwise grouping. This is shown in Figure 12.7. The results of this processing are thresholded and passed up to Layer 4, which consists of a single net. Each neuron in this net collects inputs from the corresponding neurons in each of the nets in Layer 3. Again, thresholding is used before the strongest values are passed up to Layer 5.

As a result of these processes, the active nodes in Layer 5 represent the regions with the strongest pairwise relationships. More than two nodes may be identified for

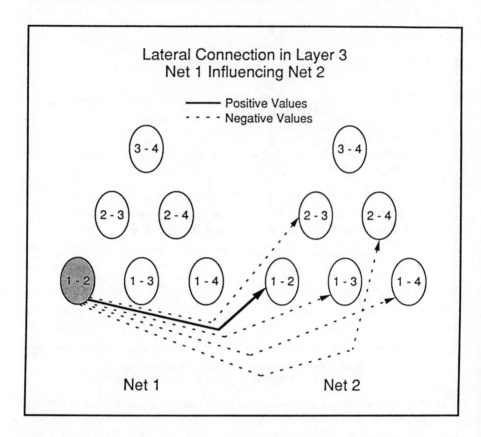

Figure 12.7 In the example shown, the strong relationship between nodes 1 and 2 (shown as node 1-2) in the net on the left side of the figure excites a similar 1-2 node in the neighboring net on the right. This node also inhibits nodes 1-3, 1-4, 2-3, and 2-4 within the net on the right, as each of these nodes represents a possible competition for the 1-2 pairing into a perceptual cluster.

grouping at the same time; they may belong to the same group, as when the relationships between regions 1 and 2, and 2 and 3, and 1 and 3 all are active in the fifth layer. Also, different groups may be indicated. The next task is to use this information to group together these regions which are closely related. This will reconfigure the original input list into the first stages of a hierarchical structure. In order to group these regions together, the HSS system uses an algorithmic subsystem which operates on the input linked list, grouping together the identified regions and threading them as children under a new parent node inserted in the original list.

12.5.4 The Learning Method of the HSS System

The parameters of the HSS system are generally fixed at the outset, and not changed during the course of processing. Thus, the system does not learn. It would be possible to introduce learning to *tune* the system to a certain type of inter-region relationship, e.g., boundary line continuation. However, promising results have been achieved without such tuning.

12.5.5   Performance and Capabilities of the HSS System

The current system, which uses cooperative and competitive processes, is very robust. An example of a HSS for a complex image is shown in Figure 12.8), which addresses neural networks for machine vision applications.

The HSS approach has also been adapted to create Hierarchical Data Structures (HDSs) from one-dimensional data which varies through time, as with single-sensor readouts [Maren et al., 1989 (a)]. This approach is useful for representing complex events, such as seismic occurrences. An extremely useful aspect of HSS and HDS representations is that they provide a higher-level representation for multisensor information fusion and knowledge-based data interpretation [Maren et al., 1989 (b)].

12.6   THE NEOCOGNITRON

The Neocognition, developed by Kunihiko Fukushima, is an outgrowth of about 20 years of work on the problem of biologically based methods for pattern recognition. It is specialized for character recognition, but can be applied to recognizing patterns expressed as binary-coded edges in a pixel-like array. The Neocognition is widely acknowledged as one of the most complex neural networks in existence, but it is worth the effort of understanding and building to obtain the shift, size, and rotational invariant pattern recognition abilities of this network.

12.6.1   The Concept Underlying the Neocognitron

The Neocognitron (an extension of the earlier Cognitron) was developed to aid computer recognition of Japanese handwritten characters. It is one of the few neural network systems to directly address issues such as position, rotation, and scale invariance in character recognition. It does this by using an architecture which interleaves feature-selection cells with cells which can respond to the same feature (or activated cell) in different positions. This leads to position-invariant feature recognition. However, the Neocognitron is not a general-purpose image processing system. It is specialized for recognizing characters or objects which can be depicted in terms of their edges.

The Neocognitron, like the BCS, attempts to emulate certain characteristics of the mammalian visual system. However, its design relies on applying the same structure and dynamics at each level of its architecture. The goal is to have the replicated structure perform the same function but at higher levels of abstraction at each level. This is in sharp contrast with the design of the BCS, in which each level is functionally and architecturally unique.

12.6.2   The Architecture of the Neocognitron

The Neocognition is unique when compared with most neural networks in that it contains four distinct types of artificial neurons. These four types are S-cells, C-cells,

Gun
Barrel

Extreme
Top of Tank

Cylinder Shape
at Rear of Tank

Upper
Tank Area

Tread   Tread Cover   Dark Middle Area

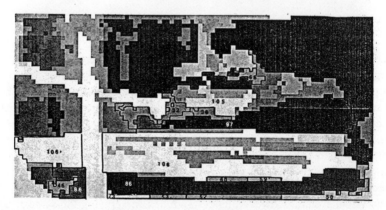

Figure 12.8 The Hierarchical Scene Structure (HSS) applied to grouping disjoint segments with an image of a tank. (a) Original image of a Russian tank in snowy woods, from *Soviet Military Power*, 1988. Certain components of the tank have been manually labeled to facilitate tracking and understanding of the grouping process. (b) Region-based segementation of the image produced 78 regions. (c) Manual resegmentation and initial stages of the HSS process. New *super-regions* have been formed, logically joining two or more segmented regions.

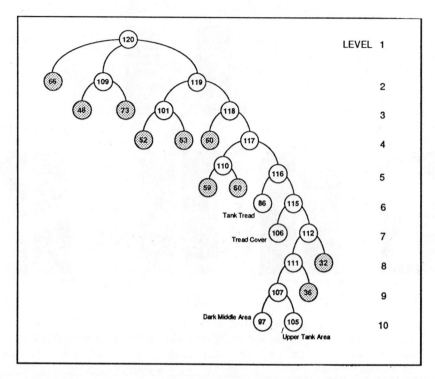

Figure 12.8 (continued)   (d) Final HSS of the tank. Shaded circles represent regions from initial segmentation, clear circles represent super-regions. Grouping is done using perceptual principles, e.g., adjacent and similarly oriented regions 97 and 105. Reproduced courtesy of V. Minsky.

V_S-cells, and V_C-cells. The S-cells are specialized for feature extraction. The C-cells assist with position-invariant object recognition. The V_S- and V_C-cells provide linkages between the S- and C-cells.

The Neocognitron is organized into a hierarchical series of levels. Each level has a similar structure to all other levels, and each level contains each of the four different types of cells, organized into layers. The signals traverse these layers in the same order from level to level. Connections are feedforward.

Figure 12.9 shows the organization of levels and layers in the Neocognitron. This figure illustrates the architecture in terms of S- and C-cells. A level is divided into two layers, one of S- cells and one of C-cells. Each layer of S-cells or C-cells is divided into subgroups (called *cell-planes*) according to the features to which they respond. Each subgroup of the S-cells responds to the same feature. Each C-cell receives signals from a group of cells that extract the same feature, but from slightly different positions. This allows the C-cell to form (within the range of its receptive field) a position-invariant response to a certain feature stimulus.

A layer of V_S- and V_C-cells exists at each level, but is omitted from the figures for the sake of simplicity. These cells are used to provide inhibitory connections which

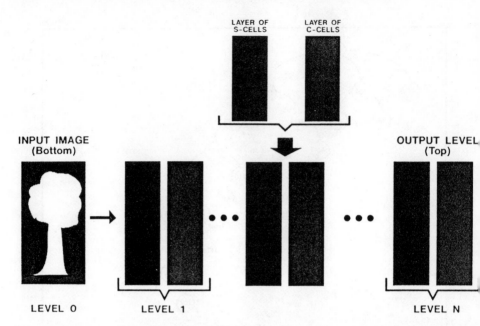

Figure 12.9 The Neocognitron meso-structure can be expressed as N levels, where each level consists of a distinct layer of S-cells, C-cells, V_s-cells, and V_c-cells. For clarity, only the layers of S- and C-cells are shown. Reprinted with permission from *Neural Networks*, *1*, M. M. Menon and K. G. Heineman, "Classification of patterns using a self-organizing neural network." © 1988, Pergamon Press, Inc.

focus the response of S-cells, so that each S-cell only responds if the particular feature to which it has been trained is presented.

12.6.3 The Dynamics of the Neocognitron

The input layer connects directly to the first S-cell layer. S-cells are feature-extracting cells. The connections to each S- cell are refined by learning, and each cell-plane of S-cells learns to respond to a different feature. Thus, when a pattern is presented, each cell-plane attempts to recognize the feature for which it is specialized.

Each C-cell receives signals from a group of S-cells that extract the same feature. A C-cell is activated if at least one of its connecting S-cells is active. The output from a single C-cell will be fed to many different S-cell planes in the next layer. Each of these S-cell planes will attempt to recognize a new, *higher-order* feature. In this manner, local features are gradually integrated into more global and position-independent features. This process is illustrated in Figure 12.10.

The topmost layer in the network is a C-cell layer. Each cell in this layer is specialized to recognize a certain whole pattern which is built up through the progressively more complex and abstract features extracted by the different layers. The interleaving of feature selection S-cells with C-cells which provide position invariance is a key factor in enabling the network to recognize complex patterns, even when they are distorted.

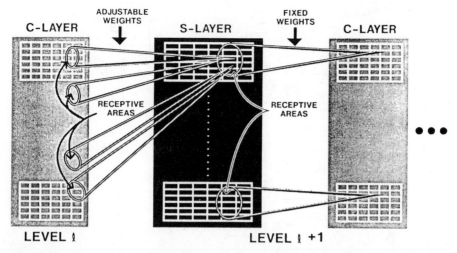

Figure 12.10   The interconnection architecture of the neocognitron specifies different receptive fields for the S-cells and the C-cells. Reprinted with permission from *Neural Networks, 1*, M. M. Menon and K. G. Heineman, "Classification of patterns using a self-organizing neural network." © 1988, Pergamon Press, Inc.

This interleaving allows positional error to be tolerated a little at a time, and allows simple features to be aggregated into complex features which define a unique pattern. Figure 12.11 illustrates the pattern-recognition action of the Neocognitron.

12.6.4   The Learning Method of the Neocognitron

The Neocognitron can learn under two different training approaches: learning with and without a teacher. In each case, only the connections from the C-cells or the V-cells to the S- cells are variable, all others are fixed.

In unsupervised learning, the S-cells undergo a cell-plane variant of winner-takes-all signal Hebbian learning. The active connections to the winning cell undergo the greatest learning, but all the other cells which share the same cell-plane also learn to respond to the same feature. As a result, all the S-cells in a cell-plane grow to have input connections of the same spatial distribution as the original winning cell. However, each S-cell in the cell-plane learns to respond to the feature in a slightly different position. The goal is that with sufficient cells in the cell-plane, at least one cell in that plane will respond to the presence of a given feature anywhere in the preceding layer.

In supervised learning, the teacher assigns certain features to different cell-planes. This is useful when we want the Neocognitron to classify patterns not only on similarity of shape, but also on certain user-defined conventions which the system might not learn independently.

12.6.5   Performance and Capabilities of the Neocognitron

Fukushima [1988 (a)] has described how a four-layer Neocognitron with feedforward and feedback connections learned to distinguish among eleven Arabic digits. (The

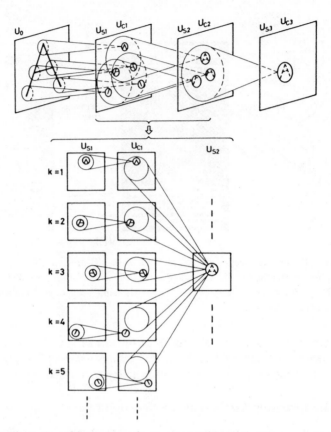

Figure 12.11 Illustration of the process of pattern recognition in the Neocognitron. Figure from K. Fukushima, *Neural Networks*, ... Reproduced by permission.

number 4 was written two different ways, and two recognition categories were assigned to it.) The system was able to recognize digits even when their positions in the input plane shifted, or their size changed, and even when they were distorted with different writing styles and slants. In another work, Fukushima [1988 (b)] shows how the attentional gain in the Neocognitron allows it to recognize first one character and then another, even when the two characters are superimposed or blended. He has also analyzed the performance of the Neocognition [Fukushima, 1989]. Ito and Fukushima [1990] have extended the basic Neocognitron system to recognize spatio-temporal patterns with some success.

Fukushima demonstrated the effectiveness of the Neocognitron on fairly small images, typically on the order of 16x16 pixels. Recently, Menon and Heinemann [1989] have explored how well the Neocognitron performs on much larger (128x128 pixel) images. They studied how the Neocognitron responded to shifted, rotated, and noisy images. Their implementation of the Neocognitron was able to discern between three pattern categories (using edge drawings of vehicles as patterns) when the drawings were rotated up to 15 degrees. They found that system performance with noisy data became unreliable when the noise level reached 30%. They found that their four-level system was not able to recognize patterns when the pattern was shifted to a very different location from the area in which the system was trained.

12.6.6   Applications of the Neocognitron

Jakubowics [1989] has used the Neocognitron concept for situation analysis. This required a very large network, coded for both spatial localization and different feature types corresponding to symbolic categories (e.g., tank, troops, etc.). This application is particularly interesting because it can deal with configurations expressing different battlefield situations, even when their spatial representation is somewhat distorted, or is shifted or rotated from specific training configurations.

A basic problem with using the Neocognitron for large-scale applications is its memory requirements. Medawar and Noetzel [1990] have developed a feature-detection based model which may reduce hardware complexity and facilitate implementation.

12.6.7   Configuring the Neocognitron

There are two issues to consider when configuring a Neocognitron. The first is the number of cell planes per layer, and the second is the number of levels (consisting of a two layers each) needed. Each cell plane becomes specialized for recognizing a certain type of feature. Thus, there needs to be sufficient cell planes available that will recognize enough features to carry out character recognition. Since the network is self-organizing, and selects the features which it will recognize as a result of training, it is not easy to specify the number of cell planes which will be needed. However, the number of cell planes needs to be increased when the number of pattern categories cited for recognition is increased. This increase is somewhat less than linear, because the local features to be extracted at the lower stages are usually shared by several patterns of different categories. Minimally, the number of cell planes in the top-most layer must be equal to the number of categories to be recognized.

Menon and Heinemann [1988] performed experiments to test the shift invariance pattern recognition abilities of a four-level Neocognitron, and found that four levels were not adequate. They conclude that in order to achieve good shift-invariant recognition, the Neocognitron must be designed with a high degree of overlap between adjacent regions of input. They estimate that a Neocognitron of about 15 levels would be adequate for shift-invariant object outline recognition in 128x128 pixel imagery.

12.7   SUMMARY

Cooperative/competitive processes are a powerful tool for achieving high levels of network performance on complex pattern recognition tasks. These processes may be implemented via feedforward, lateral, and sometimes feedback connections in multi-layered neural architectures. Example networks which have successfully used this structural and dynamic principle include the competitive learning network (for creating self-organized classes), the masking field network (for creating representations of and recognizing spatio-temporal patterns), the BCS (for low-level image processing, fusing edge detection with regional feature-filling-in), the HSS network (for creating perceptually based data structures to represent the organization of regions in a scene), and the Neocognitron (for size, location, and rotation-invariant character recognition).

REFERENCES

Bairaktaris, D. (1990). "Competitive learning with modifiable thresholds for visual pattern recognition," *Proc. Second Int'l. Joint Conf. Neural Networks*, Washington, D.C.; Jan. 15-19, I-357–I-364.

Carpenter, G.A., Grossberg, S., and Mehanian, C. (1989). "Invarient recognition of cluttered scenes by a self-organizing ART architecture: CORT-X boundary segmentation," *Neural Networks*, 2, 169-181.

Cohen, M.A., and Grossberg, S. (1987). "Masking fields: A massively parallel neural architecture for learning, recognizing, and predicting multiple groupings of patterned data," *Applied Optics*, 26, 1866-1891.

Cohen, M.A., Grossberg, S., and Stork, D. (1987). "Recent developments in a neural model of real-time speech analysis and synthesis," *Proc. IEEE First Int'l. Conf. Neural Networks*, San Diego, CA; June 21-24, IV-443–IV-453.

Daugman, J. (1989). "Networks for image analysis: Motion and texture," *Proc. First Int'l. Joint Conf. Neural Networks*, Washington, D.C.; June 18-22, I-189–I-193.

Diamond, A., and Holden, D. (1989). "Implementing the Boundary Contour System on a multivector processor iPSC/2 Hypercube," *Proc. First Int'l. Joint Conf. Neural Networks*, Washington, D.C.; June 18-22, I-189–I-193.

Fukushima, K. (1975). "Cognitron: A self-organizing multilayered neural network," *Biological Cybernetics*, 20, 121-136.

Fukushima, K. (1988a). "Neocognition: A hierarchical neural network capable of visual pattern recognition," *Neural Networks*, 1, 119-130.

Fukushima, K. (1988b). "A neural network for visual pattern recognition," *IEEE Computer*, (March), 65-75.

Fukushima, K. (1989). "Analysis of the process of visual pattern recognition by the Neocognitron," *Neural Networks*, 2, 413-420.

Grossberg, S. (1976). "Adaptive pattern classification and universal recoding, I: Parallel development and coding of neural feature detectors," *Biological Cybernetics*, 23, 121-134.

Grossberg, S. (1978). "A theory of human memory: Self-organization and performance of sensory-motor codes, maps, and plans," in R. Rosen and F. Snell, Eds. *Progress in Theoretical Biology, Vol. 5*, Academic, New York, NY.

Grossberg, S. (1987). "Competitive learning: From interactive activation to adaptive resonance," *Cognitive Science*, 11, 23-63.

Grossberg, S. (1987). "Cortical dynamics of three-dimensional form, color, and brightness perception: Monocular theory," *Perception and Psychophysics*, 41, 87-116.

Grossberg, S. (1987b). "Cortical dynamics of three-dimensional form, color, and brightness perception: Binocular theory," *Perception and Psychophysics*, 41, 117-158.

Grossberg, S. and Mingolla, E. (1985a). "Neural dynamics of form perception: Boundary completion, illusory figures, and neon color spreading," *Psychological Review*, 92, 173-211.

Grossberg, S., and Mingolla, E. (1985b). "Neural dynamnics of perceptual grouping: Textures, boundaries, and emergent segmentations," *Perception and Psychophysics*, 38, 141-171.

Grossberg, S., Mingolla, E., and Todorovic, D. (1989). "A neural network architecture for preattentive vision," *IEEE Trans. Biomedical Engineering*, 36, 65-84.

Grossberg, S., and Rudd, M.E. (1989). "A neural narchitecture for visual motion perception: Group and element apparent motion," *Neural Networks*, 2, 421-450.

Grossberg, S., and Todorivic, D. (1988). "Neural dynamics of 1-D and 2-D brightness perception: A unified model of classical and recent phenomena," *Perception and Psychophysics*, 43, 241-277.

Ito, T., and Fukushima, K. (1990). "Recognition of spatio-temporal patterns with a hierarchical neural network," *Proc. Second Int'l. Joint Conf. Neural Networks*, Washington, D.C.; January 15-19, I-273-I-276.

Jakubowicz, O.G. (1990). "Multilayer multi-feature map architecture for situation analysis," *Proc. First Int'l. Joint Conf. Neural Networks*, Washington, D.C.; January 15-19, II-23-II-30.

Kohonen, T. (1982). "Clustering, taxonomy, and topological maps of patterns," *Proc. 6th Int'l. Conf. Pattern Recognition* IEEE Computer Society Press, Silver Spring, MD, (October, 1982), 114-128.

Maren, A.J., Harston, C.T., and Pap, R.M. (1989a). "A hierarchical data structure representation for fusing multisensor information," *Proc. SPIE Technical Symposia on Aerospace Sensing, Sensor Fusion Section*, Orlando, FL; March 27-31.

Maren, A.J., Harston, C.T., and Pap, R.M. (1989b). "A hierarchical structure approach to multisensor information fusion," *Proc. Second Nat'l. Symposium on Sensors and Sensor Fusion*, Orlando, FL; March 27-31.

Medawar, B. and Noetzel, A. (1990). "Towards reducing the hardware complexity of a feature detection-based models," *Proc. Second Int'l. Joint Conf. Neural Networks*, Washington, D.C.; January 15-19, I-440-I-442.

Menon, M.M. and Heinemann, K.G. (1988). "Classification of patterns using a self-organizing neural network," *Neural Networks*, 1, 201-215.

Minsky, V.A. (1990). *A Multilayer Cooperative/Competitive Method for Grouping and Organizing Related Image Segments*, Master's Thesis, The University of Tennessee Space Institute, Tullahoma, TN.

Minsky, V.A., and Maren, A. (1989). "Representing the perceptual organization of segmented images using hierarchical scene structures," *J. Neural Network Computing*, 1, (Winter), 14-33.

Nigrin, A.L. (1990). "The real-time classification of temporal sequences with an adaptive resonance circuit," *Proc. Second Int'l. Joint Conf. Neural Networks*, Washington, D.C.; Jan. 15-19, I-525-I-528.

von der Malsburg, C. (1973). "Self-organizing of orientation sensitive cells in the striate cortex," *Kybernetik*, 14, 85-100, and reprinted in J. A. Anderson and E. Rosenfeld (Eds.) (1988). *Neurocomputing*, MIT Press, Cambridge, MA, 212-228.

Reggia, J.A. (1989). "Methods for deriving competitive activation mechanisms," *Proc. Int'l. Joint. Conf. Neural Networks*, Washington, D.C.; June 18-22, I-357–I-363.

Rumelhart, D.E., and Zipser, D. (1985). "Feature discovery by competitive learning," *Cognitive Science, 9* , 75-115.

13

HYBRID AND COMPLEX NETWORKS

Alianna J. Maren

13.0 OVERVIEW

A single neural network is useful for a single task. To carry out complex tasks, more complex networks are needed. Solutions range from hybrids of two or more existing networks (generally offering some small degree of improvement in learning or performance) to creating systems of interacting networks. We can define several different system macro-architectures, and relate these architectures and their corresponding dynamics to different applications. These systems contain two or more networks, which might interact. Such systems may operate in logical or real parallelism.

Network systems may be tightly or loosely coupled. Tightly coupled systems are often hybrids of pre-existing network types, such as the Hamming network (a hybrid of the Perceptron and MAXNET), or the counter-propagation network (a hybrid of a Kohonen adaptive vector quantization network with a Grossberg outstar). Loosely coupled systems of networks may arrange similar networks in parallel or in hierarchies for rapid evaluation of different types of information from the same data, or to yield increasingly higher levels of data abstraction. Some loosely coupled systems may use one network as a preprocessor or filter for another network. Another approach system is to use one network to influence or set the weights in another. This type of system has been recommended for neurocontrol applications.

13.1 INTRODUCTION

One of the most important issues which is beginning to emerge for neural networks systems designers is simply that — how to design systems of neural networks. This is not just a matter of selecting and configuring a single network (e.g., the back-propagating Perceptron or a Hopfield network). It is a matter of creating an entire macro-architecture consisting of multiple interacting neural networks, each performing some unique and vital task in solving a complex problem.

The topics we have covered in the preceding chapters position us to take on this more global issue. In the previous chapters, we have investigated five different types of architectures. These were feedforward networks (for pattern classification), laterally-connected networks (for autoassociation), vector-matching networks (for topographic maps), feedforward/feedback networks (for pattern heteroassociation), and cooperative/competitive networks (for mapping, clustering, and other complex tasks). Now we are ready to put all of this background together, and take on a bigger issue: that of creating network systems.

We will make a distinction between two major ways in which we can integrate neural networks to create a system. One is to create a hybrid network by tightly fusing two existing neural network architectures. The Hamming net and the counter-propagation network are examples of this approach. In these networks, it is not readily possible to separate out individual networks. The limitations of this approach is that we are still relying on a single network to accomplish a task which may be greater than the capabilities any single network may offer.

The alternative approach is to create systems of individual but interacting networks, expanding laws described by Edelman [1987]. Within the last year, this area has emerged as a major research area. The reasons for this are that people are discovering that complex problems cannot usually be solved by a single network. One network will usually only do one task, or make one type of classification. Often, complex problems require multiple stages of processing. This requires multiple networks.

There are other reasons for exploring systems of modular networks. The computational complexity of fully connected networks scales as the square of the number of neurons involved. A large network can take up a great deal of memory. Even with the prospect of using neural network chips to store weight connections, it is still useful to explore ways to cut back on the memory required. Modular network design is one way of doing this.

Yet, another reason to adopt a modular approach to network system design is that functions can be decomposed and associated with specific networks. This will make it easier to test and debug (and if necessary, repair) different functional networks. Although nothing like structured programming principles have yet invaded the province of neural network system design, it is only logical to assume that the rules we learned in our first year of structured computing will carry over to designing systems of interacting networks. A structured, modular system design will make it easier to verify and validate network systems. This will be very important as we seek to apply systems of neural networks to projects sponsored by DoD, NASA, and other government agencies which have critical fielding requirements.

There are several major types of loosely coupled systems of networks which have been developed in the past year. These are macro-structures of neural networks. We illustrate the basic types of systems of networks in Figure 13.1. One common approach (Figure 13.1a) is to create a hierarchy of networks of a similar type. In such systems, multiple networks might be used to partition a multiscale pattern recognition problem into separate problems. For example, ART 3 is a hierarchy of ART 1 (or ART 2) networks. Similar systems have been built using exclusively back-propagation networks or self-organizing topology-preserving maps.

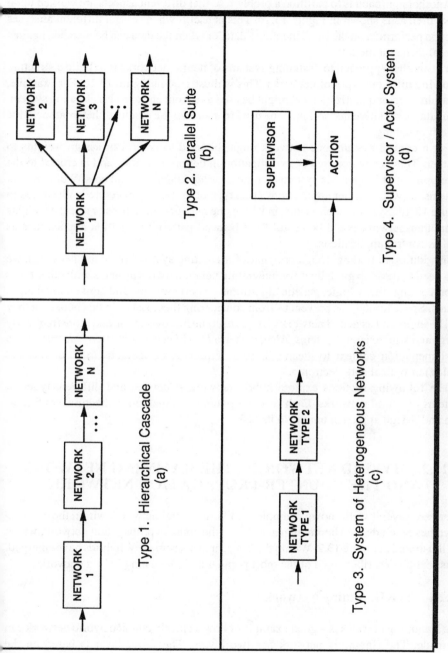

Figure 13.1 Four basic types of system architectures. (a) A hierarchical system of similar networks. (b) A system of networks operating in parallel. (c). Systems of dissimilar networks, each of which performs a different type of task. (d). A control system architecture, in which one network sets or influences the weights of another network.

Another approach is to partition a problem so that different aspects are worked upon in parallel, as is shown in Figure 13.1b. This is useful when several different analyses must be performed on the same incoming data, or when the data can be matched against several different models.

Yet another approach to designing systems of neural networks is to build a structure involving different types of networks. This is illustrated in Figure 13.1c. For example, a learning vector quantization net might be used as a front end to a classification system, or an autoassociative network may be used to *clean up* the results of heteroassociative pattern matching.

For a control system, one network might be used to control or assign weights to another. This is a major area of investigation in neural networks, and is crucial to the development of large-scale systems for monitoring, diagnosis, and control of complex systems, as well as other applications. This type of system architecture is illustrated in Figure 13.1d. Systems such as this, in which one network influences or sets the weights of another network could be useful for advanced pattern recognition applications as well as control applications.

In addition to finding design principles for creating systems of neural networks, we also need to develop guidelines for integrating neural network systems with other types of processing: fuzzy logic, genetic algorithms, expert systems, and conventional algorithms. Space limitations prevent us from considering these issues in this book, but they will be important as neural network systems become more common and more frequently integrated with existing systems. Also, the term *hybrid network* is often construed in an implementation context to mean a hybrid of processing methods (e.g., digital and analog) or optical and electronic.

The following sections explore hybrid network structures and different types of systems of neural networks. Each type of system has a macro-architectural configuration and design approach unique to its task.

13.2 HYBRID NETWORKS: THE HAMMING NETWORK AND THE COUNTER-PROPAGATION NETWORK

There are several well-known examples of hybrid neural networks within the community; these include the Hamming network and the counter-propagation network, which are illustrated in Figure 13.2. Within the last year or so, some new hybrids have emerged. This section describes some of the most popular and interesting hybrid networks.

13.2.1 The Hamming Network

The Hamming network is a good example of how a closely coupled hybrid network can be made. The Hamming network has two layers. The lower layer is based on the Perceptron, and the upper layer (called MAXNET), is like a Hopfield network. The Hopfield-like layer in the Hamming network is very similar to the previously described Hopfield network, with a few key differences. It uses threshold-logic rather than hard-limit nodes, the interactive-convergence layer is fully connected with each node feeding back to itself as well as to all of its neighbors on that level, and the weight

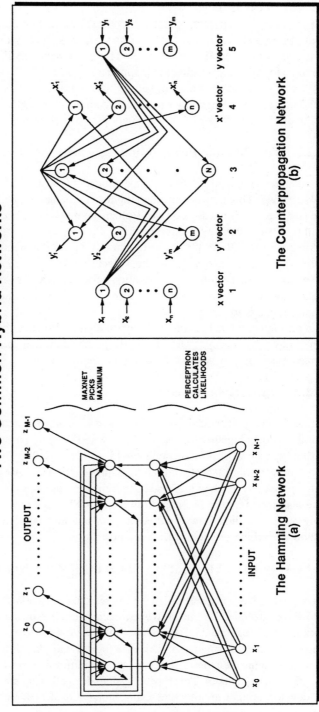

Figure 13.2 Two well-known hybrid networks. (a) The Hamming net, and (b) the counter-propagation network. Reproduced with permission of Dr. Richard Lippmann and Dr. Robert Hecht-Nielson, respectively.

specifications are different. Like the Hopfield network, the weights for the node interconnections are fixed.

The Hamming network operates by calculating a Hamming distance between the input pattern and each of the exemplar patterns. In a 1987 report, Lippmann developed the concept of the neural network method for implementing a Hamming distance metric as a basis for pattern classification [Lippmann et al., 1987]. This distance is equal to the number of elements in the input pattern which are different in value from the corresponding node in the exemplar pattern. Lippmann showed that the Hamming network can be used as an efficient optimal binary pattern classifier.

13.2.2   The Counter-Propagation Network

The counter-propagation network, developed by Robert Hecht-Nielsen, acts as a look-up table. This network can be taught to recognize inputs to different categories. The performance is not as good as a Boltzmann machine or back-propagation network, but this network converges more readily.

Maureen Caudill [1988] has written a tutorial on this network which is useful for those who want to explore it further. Hecht-Nielsen's 1987 article should enable a person to construct a counter-propagation network. In a following 1988 article, Hecht-Nielsen discusses applications. He also reviews basic concepts and structure for counter-propagation nets.

In an interesting extension to the basic counter-propagation network, Xu and Oja [1990] have explored a modified counter-propagation map as a means of establishing correspondences between two topology-preserving maps.

13.2.3   Other Hybrid Networks

Within the last year, researchers have begun to explore more ways of fusing different neural networks together to create hybrids with interesting properties. A good example of this type of work is that done by Li and Wee [1990], who have modified the Neocognitron using aspects of both back-propagation and ART networks. Their resultant network has resulted increased generalization in learning. Rajapakse et al. [1990] have also modified the Neocognitron using ART and a multilayer multifeature map. They state that their new architecture can automatically incorporate new patterns without disturbing previously trained information.

13.3   NEURAL NETWORKS OPERATING IN PARALLEL

There are two basic ways in which we can organize networks to operate in parallel. One is to have several different networks, each operating in logical or real parallelism, working on the same data. Each network will be trained to extract different features or to make different distinctions. This is a logical and straightforward way to design a network system when there are several types of information which need to be extracted from the same data pool. This type of system design is very similar to the way in which we develop system architecture using conventional programming; neural networks simply replace some of the algorithmic units.

This approach has been adopted in some systems which have been developed for and fielded into real applications tasks. Gevins and Morgan [1988, and references contained therein] have developed systems of interacting neural networks to analyze multiple EEG signals recorded from the brain. They use neural networks for detecting transient contaminants, where different networks are specialized in detecting different types of contaminants. They have exhaustively searched for subsets of plausible features for use in signal classification. As a result, their network has replaced one of two human scorers who previously hand marked the data they collected. Routine use of their system over the past two years has led to a 15 man-hour reduction of the time needed to edit contaminated data from recordings with up to 64 channels.

Casselman and Acres [1990] have developed a similar system (DASA/LARS) which uses multiple back-propagation networks in parallel to solve different diagnostic tasks based on satellite spectral data. Their system has been field tested, and is slated for permanent installation.

More interesting system design issues arise when we seek to develop systems which can make subtle and complex distinctions from a large body of incoming data. Sometimes, a complex distinction may require that several different types of features need to be extracted from the data, and we may not know in advance what features will be needed. This is particularly true in the many cases in which we use back-propagation networks. As the network size grows, the representation of feature information can become more distributed, or needed generalizations may fail to be made.

Sometimes, when a set of subtle or complex distinctions are desired, the best solution is to break the problem up into a number of subtasks, each solved by a separate network. A system of networks can be created. The results of the different networks can be fused or correlated to obtain desired results. Hering et al. [1990] illustrate how this can be done in a couple of different configurations. They found that by separately training and combining a system of networks, they were able to develop a network system that could make distinctions that were difficult to encode into a single equivalently sized network. Rossen and Anderson [1989] have similarly used two modules of hidden layer nets within a larger network system for speech recognition.

13.4   HIERARCHIES OF SIMILAR NETWORKS CAN BUILD LEVELS OF INFORMATION PROCESSING

One straightforward and effective way of creating a neural network system to handle complex pattern recognition tasks is to hierarchically link together a number of networks of the same type. Fukushima's Neocognitron (which we discussed in the previous chapter) is an example of this approach. This network is viewed as a single one, and not a composition of separate networks. Because of that, we discussed it in the context of cooperative/competitive networks. Nevertheless, it also provides a good exemplar of the type of network we'll consider in this section.

Gail Carpenter and Steve Grossberg have recently published [1989, 1990] work showing how systems of ART 1 networks can be combined for complex, high-level pattern recognition. This is both an exciting and natural development of their work with ART. Although ART is probably not the network of choice for low-level pattern recognition, it is a strong contender for being the best network for high-level pattern recognition and for forming categories based on features extracted from lower-level primitives.

When ART networks are combined into a system, the top-down priming capability built into ART comes into play. This could be used, even by an expert system or knowledge-based system operating in concert with the ART network system, to reevaluate the features presented as input.

This approach, or a similar one which could be done using systems of back-propagation networks, is not the only approach we can take to building a hierarchical system of neural networks. Helge Ritter [1989] has presented an interesting approach to decomposing the function which a mapping network would learn into smaller functions, each of which is recognized or mapped by an individual network. The functions do not need to be known or specified in advance. This allows us to keep one of the major advantages of using neural networks for mapping (that is, we do not have to know their mapping functions, just have a good set of teaching examples). At the same time, we can build a system which can break down a very complex function into manageable units.

A major issue which we will have to address in developing systems of neural networks will be to find some formal means of describing system properties. Morris Hirsch [1989] has recently done some work along these lines, by exploring the convergence properties of cascades (hierarchical series) of networks. An example of this type of approach is the work done by Anikst et al., who have used two binary-tree structured neural networks for encoding speech information at two different levels: the feature level (extracted directly from frames of raw data) and the segment level (composed of time-contiguous blocks of frames).

Lee et al. [1990] have explored decomposing a multilayered network into pairwise linkages of single layer nets. Each pairwise linkage of two nets is individually trained using the coulomb energy law for creating clusters. Then, the nets are assembled into the full multilayer system and the back-propagation law is used to train the output classes to correspond to the desired class recognition. Their approach seems a little stilted, as the network designer must select the different classes for each level of the network. This cuts across the ability of the back-propagation method to discover features which extract and generalize properties from the presented input. Nevertheless, their approach may have some merit, and it is useful to think of ways of decomposing and separately training modules within a large multilayer system.

The idea of using hierarchies of neural networks may be useful in building systems of networks out of neural network chips, as well as designing software systems. Jackel et al. [1987] illustrate how a hierarchy of image vector quantizing network chips can rapidly search a database for a codeword. This approach uses a system of neural network chips for rapid and effective image data compression, with compression ratio of about 8:1.

13.5   SYSTEMS OF DIFFERENT TYPES OF NEURAL NETWORKS CAN HANDLE COMPLEX TASKS

We can't always expect that one type of neural network will be sufficient to solve all of the tasks associated with a single complex problem. But we need to find some rules

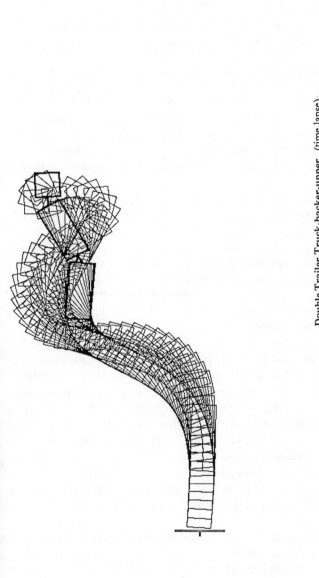

Double Trailer Truck-backer-upper (time lapse).
The controller is a layered neural network with 5 inputs (the state variables), 25 hidden units, and 1 output (the steering angle). The controller is adapted using the backpropagation through time algorithm where the error gradient is backpropagated directly through the dynamic plant (i.e. without the use of a neural network emulator).

Figure 13.3 Francois Beausays and Dr. Bernard Widrow have used a dual-network simulator/controller system to simulate backing up a truck with two trailers. This is an extension of earlier work done by Nguyler and Widrow [1989]. Reproduced courtesy of Drs. Beausays and Widrow, Department of Electrical Engineering at Stanford University.

or guidelines that will help us wisely select the right types of networks and put them together in a useful architecture. Little formal work has been done in this area (see, e.g., [Coleman et al., 1990]). Most guidelines will have to be obtained from empirical studies.

One important task will always be that of creating the right type of input for the main problem-solving unit of a system. Kadaba et al. [1990] have done some interesting work which could be an example of a procedure we can follow. They have used two self-organizing back-propagation networks to encode and decode data sent to a main back-propagation generalization network. Their self-organizing back-propagation networks (based on work by Myers, et al.) is something of a conceptual hybrid between a regular back-propagation network and a BAM; the same pattern vector is used as both input and desired output for a five-layer network. The innermost hidden layer has substantially fewer nodes than the input and output layers, and this forces a generalization of the input into a smaller representation space. This is a useful technique for data compression, and could be adopted for many different applications. Kadaba et al. use the compressed data (as represented by the output of the hidden nodes) as input to their main network.

Rossen and Anderson [1989] have demonstrated another useful approach. They use a back-propagation network (with two modules of hidden layers) for training a speech recognition system. Their output or categorization layer has recurrent connections as well as receiving input from the hidden layer. This means that the output classification is dependent on previous classifications as well as on the temporal unit being currently processed. They use a Brain-State-in-a-Box (BSB) algorithm for setting the recurrent connection weights. This is an example of extending the basic BSB approach for autoassociation into the temporal realm. It acts as a *clean-up filter* for assisting categorization of a temporally-varying input.

Sometimes, it is not easy to identify a set of appropriate features for a pattern recognition and classification tasks. Neumann et al. [1990] illustrate how a system of two neural networks can be used for classifying insect courtship songs. Their approach can be adapted for similarly complex frequency-temporal or spatio-temporal pattern recognition tasks. They described each pulse in a song with nine zero-crossing features, which have been shown to encode frequency information. This information was used to train an Kohonen Learning Vector Quantization (LVQ) network. The nodes in the LVQ network took on a distribution corresponding to the distribution of pulse types in the training set, and let the nine zero-crossings be represented by a single scalar value. This network was used as a front end to input the data from multiple pulses into a back-propagation network for insect genotype classification. Their preliminary results are encouraging.

Jakubowicz et al. [Jakubowicz & Ramanujam, 1990; Samarabandu & Jakubowicz, 1990] have developed a very similar approach to classifying time-varying signals such as temporally-varying fault states and speech. In their approach, a three-layer network has a topographically-organizing layer as the second layer. This layer is created from input presented to the first layer. The third layer functions as a back-propagation-type classifier. Information is stored in the Kohonen net over time (as with the network developed by Neumann et al.). In the work with Samarabandu [1990], this net also feeds back into the input, allowing further activations in the Kohonen net to be dependent on temporal context.

Nishikawa et al. [1990] have also developed a very similar approach, in that they use a Kohonen process to partition the hidden layer in a back-propagation network into separate subnets. The resultant architecture is similar to the networks described in Section 13.2 which have partitioned hidden layers.

13.6 SYSTEMS OF NETWORKS ARE USEFUL FOR ADAPTIVE CONTROL

Neurocontrol applications seem to intrinsically require systems of neural networks. This area is one which will grow in importance, and a chapter in Part IV of this book is devoted to neurocontrol. This section addresses some of the architectural issues involved in creating a system for neurocontrol. These systems are likely to exhibit the greatest complexity, subtlety, and criticality in design of the neural network systems which will be built and implemented over the coming decade.

Information flows in neurocontrol systems are more complex than in most systems of neural networks. This is because neurocontrol systems both monitor and affect some ongoing process, and often involve using both a model (to provide guidance as to the desired state) as well as a network to actually control or influence the process. Two, sometimes even three or more, networks may be involved in a neurocontrol system. [See, e.g., Bar-Kama and Guez, 1990.]

K.S. Narendra and K. Parthasarathy [1990] have recently overviewed neural networks for control of dynamical systems. They describe several system architectures (similar to those described here) which can be used for control systems. Their point of view is from engineering control systems, and may not be an easy introduction for those not familiar with that field.

Barto, Sutton, & Anderson [1983] developed an early neurocontrol scheme. Their system involved two networks, both of which were simple table look-ups. Their work is significant because they were among the first in the neurocontrol community to make the conceptual distinction between action and critic networks consciously by using two subnetworks. Their adaptive critic approach has formed a basis for a substantial line of research in the neurocontrol community. [See, e.g., Schmidhuber, 1990.]

Later, Lapedes and Farber [1986, 1987] expanded on the work of Barto, Sutton, and Anderson by developing a two-network system which consisted of a *master* network which set the weights for a *slave* network. Both of these networks were Hopfield networks. The key problem they addressed was how to use one network to set or control the weights of another. In their 1986 paper, they explored finding optimal weights. In their 1987 paper, they extended their initial approach to programming nets that contain interneurons (hidden neurons). This allowed them to use their system to address more complex problems. They have explored using their network system to model time-dependent behavior.

Paul Werbos [1989 and Part IV, this book] has described the general principles of neurocontrol, with specific descriptions of different adaptive critic architectures. His description of the major methodologies for neurocontrol serve to bring some conceptual order to a field that is undergoing tumultuous growth common to any emergent area. Lukes, Thompson, & Werbos [1990] have recently expanded on the adaptive critic idea by developing an *Action Dependent Heuristic Dynamic Programming* system. Their approach may be very useful for dynamic systems where the costs are incurred by taking actions, instead of being intrinsic to a state.

Werbos has also developed the concept of back-propagation-through-time. This approach requires an adaptive part (schedule of actions) and model part. For example, in applying the back-propagation through time method to predicting profits from the natural gas industry [Werbos, 1989b], the model is a description of natural gas industry and how profits are determined. The back-propagation-through-time is done in this case using functional equations, not neural networks. Werbos explains back-propagation (of utility) through time, adaptive critic networks, and other neurocontrol architectures in Part IV.

Recently, Nguyen and Widrow [1989] have developed a neurocontrol system (the *truck backer-upper*) which consists of an emulator network and controller (action) network. First, they trained the emulator network to mimic truck and trailer kinematics. Next, they trained the neural controller to control the neural emulator. The supposition is that once the neural controller can effectively control the emulator, it will then be able to control the actual trailer truck. Beaugays and Widrow have extended this work. Their results are illustrated in Figure 13.3.

Michael Jordan [1989, and references cited therein] has developed a system for neurocontrol that incorporates three networks into one. These are the model, action, and utility functions. The main content of his work deals with how to design utility networks. The utility network is not an adaptive network, but it is a crucial part of the system. Jordan concentrates on formulating a useful utility mapping. His network architecture has several layers, two of which have recurrent and/or feedback connections.

Rabelo et al. [1990] have replaced a hardwired controller with several neural control networks in an architecture based on Jordan's. Their moderately complex system illustrates a possible evolutionary direction for neural systems.

Mitsuo Kuwato [Kuwato et al., 1988, and with Miyamoto et al., 1988] has been developing systems of networks for neurocontrol of robot motion. His approach, that of using a hierarchical neural network system, uses both feedforward and feedback torques (motor commands) in a learning scheme. As a result, his network acquires an inverse dynamics model to guide arm motion. This model generalizes its capabilities once it has learned basic motions. His work is solidly based on studies of the neurophysiology of sensory data association and neural feedback loops for voluntary motion. Kawato and his colleagues have applied his approach to learning trajectory control of an industrial robot manipulator.

Recently, we [Parten et al., 1990] have explored using systems of networks for neurocontrol when only information about joint positions and velocities is available (no torque feedback). We have developed a design of networks feeding each other. Our concept for control of the telerobotic arm features a three-level design. The optimization network plans the path trajectory. The planning neural network performs nonlinear mapping for the end effecter position to the desired joint angles. These serve as input to the various joint angle neural networks which drive the motors to produce the desired angles.

13.7 SUMMARY

Hybrid networks, such as the Hamming network or the counter-propagation network, may offer slight improvement in classification or learning speed over simpler networks. For truly difficult tasks, we need to investigate ways of creating systems of interacting networks. There are at least four basic categories of system architecture; similar networks in hierarchy, similar networks in parallel, systems of different network types, and network control systems. Recently, there has been a shift in attention from the performance of single networks to an emphasis on how systems of networks can effectively handle complex pattern recognition and control problems.

REFERENCES

Anikst, M.T., Meisel, W.S., Newstadt, R.E., Pirzadeh, S.S., Schumacher, J.E., Shinn, P., Soares, M.C., and Trawick, D.J. (1990). "A continuous speech recognizer using two-stage encoder neural networks," *Proc. Second Int'l. Joint Conf. on Neural Networks* (Washington, D.C., Jan. 15-19, 1990), II-306–II-309.

Bar-Kana, I., and Guez, A. (1990). "Neuromorphic computing architecture for adaptive control," *Proc. Second Int'l. Joint Conf. on Neural Networks* (Washington, D.C., Jan. 15-19, 1990), II-323–II-326.

Barto, A.G., Sutton, R.S., and Anderson, C.W. (1983). "Neuron-like adaptive elements that can solve difficult learning control problems," *IEEE Trans. on Systems, Man, and Cybernetics, SMC-13*, 834-846.

Carpenter, G., and Grossberg, S. (1989). "Search mechanisms for Adaptive Resonance Theory (ART) networks," *Proc. First Int'l. Joint Conf. on Neural Networks* (Washington, D.C., June 18-22, 1989), I-201–I-206.

Carpenter, G., and Grossberg, S. (1990). "ART 3: Hierarchical search using chemical transmitters in self-organizing pattern recognition architectures," *Neural Networks, 3*, 129-152.

Casselman, F. and Acres, J.D. (1990). "DASA/LARS, a large diagnostic system using neural networks," *Proc. Second Int'l. Joint Conf. on Neural Networks* (Washington, D.C., Jan. 15-19, 1990), II-539–II-542.

Caudill, M. (1989)."Neural Network Primer, Parts IV-VI," *AI Expert*, (August, 1989), 61-67, (November, 1988), 57-85, and (February, 1989), 61-67.

Coleman, W.P., Sanford, D.P., De Gaetano, A., and Geisler, F.H. (1990). "Modularity of neural network architecture," *Proc. Second Int'l. Joint Conf. on Neural Networks* (Washington, D.C., Jan. 15-19, 1990), I-51–I-54.

Edelman, G. (1987), *Neural Darwinism: The Theory of Neuronal Group Selection*, Basic, New York.

Gevins, A.S., and Morgan, N.H. (1988). "Application of neural-network (NN) signal processing in brain research," *IEEE Trans. Acoustics, Speech, and Signal Processing, 36*, 1152-1166.

Hecht-Nielsen, R. (1987). "Counter-propagation networks," *Applied Optics, 26*, 4979-4984.

Hecht-Nielsen, R. (1988). "Applications of counter-propagation networks,"*Neural Networks*, *1*, 131-140.

Hering, D., Khosla, P., and Kumar, B.V.K.V. (1990). "The use of modular neural networks in tactile sensing," *Proc. Second Int'l. Joint Conf. on Neural Networks* (Washington, D.C., Jan. 15-19, 1990), II-355-II-358.

Hirsch, M.W. (1989). "Convergence in cascades of neural networks," *Proc. First Int'l. Joint Conf. on Neural Networks* (Washington, D.C., June 18-22, 1989), I-207-I-208.

Jackel, L.D., Howard, R.E., Denker, J.S., Hubbard, W., and Solla, S.A. (1987). "Building a hierarchy with neural networks: an example — image vector quantization," *Applied Optics*, *26*, 5081-5084.

Jakubowicz, O., and Ramanujam, S. (1990). "A neural network model for fault-diagnosis of digital circuits," *Proc. Second Int'l. Joint Conf. on Neural Networks* (Washington, D.C., Jan. 15-19, 1990), II-611-II-614.

Jordan, M.I. (1989). "Generic constraints on underspecified target trajectories," *Proc. First Int'l. Joint Conf. on Neural Networks* (Washington, D.C., June 18-22, 1989), I-217-I-225.

Kadaba, N., Nygard, K.E., Juell, P.L., and Kanga, L. (1990). "Modular back-propagation neural networks for large domain pattern classification," *Proc. Second Int'l. Joint Conf. on Neural Networks* (Washington, D.C., Jan. 15-19, 1990), II-551-II-554.

Kawato, M., Uno, Y., Isobe, M., and Suzuki, R. (1988). "Hierarchical neural network model for voluntary movement with application to robotics," *IEEE Control Systems*, (April), 8-16.

Lapedes, A. and Farber, R. (1986). "A self-optimizing, nonsymmetrical neural net for content addressable memory and pattern recognition," *Physica 22D*, 247-259.

Lapedes, A. and Farber, R. (1987). "Programming a massively parallel, computation universal system: Static behavior," *Proc. IEEE Second Conf. on Neural Information Processing*, ed. by D. Anderson, Snowbird, CA, 283-298.

Lee, W.D., Lee, K., and Jang, J. (1990). "Modular neural networks: Combining the coulomb energy network algorithm and the error back-propagation algorithm," *Proc. Second Int'l. Joint Conf. on Neural Networks* (Washington, D.C., Jan. 15-19, 1990), I-651-I-654.

Li, D., and Wee, W.G. (1990). "A new Neocognitron structure modified by ART and back-propagation," *Proc. Second Int'l. Joint Conf. on Neural Networks* (Washington, D.C., Jan. 15-19, 1990), I-420-I-423.

Lippmann, R.P., Gold, B., and Malpass, M.L. (1987). *A Comparison of Hamming and Hopfield Neural Nets for Pattern Classification*, MIT Lincoln Lab. Tech. Report 769 (Lexington, MA, May 21, 1987).

Lukes, G., Thompson, B., and Werbos, P. (1990). "Expectation driven learning with an associative memory," *Proc. Second Int'l. Joint Conf. on Neural Networks* (Washington, D.C., Jan. 15-19, 1990), I-521-I-524.

Miyamoto, H., Kawato, M., Setoyama, T., and Suzuki, R. (1988). "Feedback-error-learning neural network for trajectory control of a robotic manipulator," *Neural Networks*, *1*, 251-265.

Myers, M., Kuczewski, R., and Crawford, W. (1987). *Application of New Artificial Information Processing Principles to Pattern Classification*, Final Report, U.S. Army Research Office, Contract DAAG-29-85-C-0025.

Narendra, K.S., and Parthasarathy, K. (1990). "Identification and control of dynamical systems using neural networks," *IEEE Trans. Neural Networks, 1*, 4-27.

Neumann, E.K., Wheeler, D.A., Burnside, J.W., Bernstein, A.S., and Hall, J.C. (1990). "A technique for the classification and analysis of insect courtship song," *Proc. Second Int'l. Joint Conf. on Neural Networks* (Washington, D.C., Jan. 15-19, 1990), II-257-II-262.

Nishikawa, Y., Kita, H., and Kawamura, A. (1990). "NN/I: A neural network which divides and learns environments," *Proc. Second Int'l. Joint Conf. on Neural Networks* (Washington, D.C., Jan. 15-19, 1990), I-684-I-687.

Nguyen, D., and Widrow, B. (1989). "Truck backer-upper: An example of self-learning in neural networks," *Proc. First Int'l. Joint Conf. on Neural Networks* (Washington, D.C., June 18-22, 1989), II-357-II-363.

Parten, C.R., Pap, R.M., Harston, C.T., Maren, A.J., Rich, M.L., and Thomas, C. (1990). "Applications of neural networks for telerobotics as applied to the space shuttle," to appear in *Simulation*.

Rabelo, L, Kim., D., and Erdogan, T. (1990). "Integrating digital and artificial neural networks using neurocontrollers: An intermediate step toward the universal computer," *Proc. Second Int'l. Joint Conf. on Neural Networks* (Washington, D.C., Jan. 15-19, 1990), II-191-II-194.

Rajapakse, J.C., Jakubowics, O.G., and Acharya, R.S. (1990). "A real time ART-1 based vision system for distortion invarient recognition and autoassociation," *Proc. Second Int'l. Joint Conf. on Neural Networks* (Washington, D.C., Jan. 15-19, 1990), II-298-II-301.

Ritter, H. (1989). "Combining self-organizing maps," *Proc. First Int'l. Joint Conf. on Neural Networks* (Washington, D.C., June 18-22, 1989), II-499-II-502.

Rossen, M.L., and Anderson, J.A. (1989). "Representational issues in a neural network model of syllable recognition," *Proc. First Int'l. Joint Conf. on Neural Networks* (Washington, D.C., June 18-22, 1989), I-19-I-26.

Samarabandu, J.K., and Jakubowicz, O.G. (1990). "Principles of sequential feature maps in multi-level problems," *Proc. Second Int'l. Joint Conf. on Neural Networks* (Washington, D.C., Jan. 15-19, 1990), II-683-II-686.

Schmidhuber, J., (1990). "Recurrent networks adjusted by adaptive critics," *Proc. Second Int'l. Joint Conf. on Neural Networks* (Washington, D.C., Jan. 15-19, 1990), I-719-I-722.

Werbos, P.J. (1989a). "Backpropagation and neurocontrol: A prospectus," *Proc. First Int'l. Joint Conf. on Neural Networks* (Washington, D.C., July 18-22, 1989(a)), I-209-I-216.

Werbos, P.J. (1989b). "Maximizing long-term gas industry profits in two minutes in Lotus using neural network methods," *IEEE Trans. on Systems, Man, and Cybernetics, SMC-19* (March, 1989(b)), 315-333.

Xu, L., and Oja, E. (1990). "Vector pair correspondence by a simplified counter-propagation model: A twin topographic map," *Proc. Second Int'l. Joint Conf. on Neural Networks* (Washington, D.C., Jan. 15-19, 1990), II-531-II-534.

14

CHOOSING A NETWORK: MATCHING THE ARCHITECTURE TO THE APPLICATION

Dan Jones and Stanley P. Franklin

14.0   CHAPTER OVERVIEW

This chapter provides general guidelines for the design and development of a useful neural network system. Such a project is not for the unadventurous. Although neural network technology has left the neonatal incubator, it remains very much in its infancy. The theory of neural networks is not yet sufficiently developed to allow a network application to be precisely engineered in the manner that, say, a bridge or a radio is engineered. Development of a successful system will require a combination of careful research and planning, educated guesswork, and outright trial- and-error.

The following material is aimed at assisting in the de novo development of a useful neural network. If one prefers not to get extensively involved in neural network development, it may be possible to find an existing network for a similar application. It is often easier to modify an existing application than to create one from scratch. Whether the plan is to develop a network from scratch, or to adapt an existing product, a review of the current literature will prove invaluable.

In general, neural network design remains a rather ad hoc, empiric process. Sufficient theoretical and experimental data is available to provide general guidelines. But it is wise to obtain or develop a system that is flexible, allowing all network parameters to be empirically optimized for any particular application. For those who are new to neural network development, consider the use of consultants. Several companies and individuals now offer neural network consulting services. Look for advertisements in recent neural network journals. Also, seek out any local companies or university departments that are involved in neural network research and development. Interest in neural networks is becoming widespread, and one may not have to look far to find the needed expertise.

14.1　WHEN TO USE A NEURAL NETWORK

Algorithmic, statistical, and more traditional AI ("Artificial Intelligence") approaches should be considered before deciding on neural network technology. DARPA (the Defense Advanced Research Projects Agency) sponsored an in-depth study of neural networks [DARPA, 1988]. Their report concluded that neural networks (at that time) were generally comparable, but not superior, to the best algorithmic/statistical methods for a variety of applications. These included associative memories, pattern classifiers, robotic control, signal processing, and speech and vision applications. Since then, neural networks have demonstrated superiority over classical methods for certain applications. Waibel [1989], for example, has demonstrated neural network phoneme recognition to be superior to the best statistical methods. The developing superiority of neural networks for many problems is demonstrated by the rapidly growing number of fielded applications. See Miller, et al [1990] for a discussion of approximately 300 existing applications. For any particular application, the appropriateness of a neural network solution can only be determined by assessing the state-of-the-art of both neural network and more traditional solutions.

Statistical approaches (Gaussian classifiers, hidden Markov models, etc.) have been the mainstay for problems involving large amounts of data that cannot be well described symbolically or mathematically. Such problems include visual, radar, and sonar pattern recognition, and speech. Due to their great flexibility and adaptability, neural networks appear promising for rapid development of products in these areas. Both statistical and neural network approaches are computationally intensive. (Indeed, some claim that neural networks are merely an especially elegant, automated way to measure, compute, and apply statistics.)

A less computation-intensive, algorithmic solution may be preferable, if an appropriate one is available. The problem here is that algorithmic approaches have generally not demonstrated the robustness required to deal with problems involving perception and control under real-world conditions (i.e. high noise & ambiguity, variable perspective, natural distortion and variation, large numbers of interdependent variables, etc.).

In some cases, a traditional AI or rule-based solution may be more efficient. In general, an AI approach may be more appropriate when dealing with a relatively simple or closed system, where the characteristics and behavior of all components are well understood, and can be accurately expressed in the form of logical or mathematical rules. Unfortunately, many practical problems are not that simple, and have proved intractable to AI or rule-based approaches.

In general, neural networks are superior for dealing with more complex or open systems, which may be poorly understood and which cannot be adequately described by a set of rules or equations. Tasks requiring fault tolerance or coping with noise, or involving pattern detection or recognition, diagnosis, abstraction, and generalization, are all good candidates for a neural network approach. Any situation in which the input data is incomplete or ambiguous may also be suitable.

14.2   WHAT TYPE OF NETWORK?

Once there has been a decision to use a neural network approach, one has to choose from almost two dozen different types of neural networks. The choice of a neural network depends on the application task. We briefly identify some networks which have been found useful in specific applications. More detail is given later in Part IV.

14.2.1   Pattern Recognition and Signal Filtering

The network of choice for most pattern recognition, signal processing, and similar applications is a multilayered feedforward network—the back-propagation network [Werbos 1974, Rumelhart 1986]. The chapter on feedforward delta rule networks in Part II briefly reviews several successful back-propagation applications. Back-propagation is probably the best approach to use if the input array is reasonably small (probably a few hundred items or less), and if the patterns to be learned do not vary greatly in their size or position in the input array.

Limitations of the back-propagation network include a long training time for large networks [Waibel, 1989], a propensity not to train at all due to local minima in the error surface, and/or to flat spots at the extremes of the sigmoidal activation function ("network paralysis," see Part II), and limited ability (for networks of practical size) to deal with input patterns that are not translationally, rotationally, and size invariant. However, with proper conditioning of the inputs, and by using recent improvements to the back-propagation algorithm, these limitations can often be surmounted. We discuss such methods in the next chapter.

Huang and Lippmann [1988] have compared traditional and neural netwokr classifiers. They find that neural networks form a flexible framework which can be used to construct many different classifiers, such as Gaussian and k-means as well as the traditional Perceptron. Performance is comparable to traditional classifiers. Neural networks can be implemented to operate more rapidly than most traditional classifiers using neural networks hardware. Huang and Lippmann expressed concern about the relatively long training time of the back-propagation method versus the training time needed for traditional classifiers, however this concern is unwarranted in light of the many options available to speed up the training and convergence of a back-propagation network, as discussed in the next chapter.

The basic back-propagation algorithm has been modified in myriad different ways, each attempting to circumvent one or more of these limitations. Some of these techniques have demonstrated dramatic superiority to standard back-propagation. Notable examples include Scott Fahlman's [1988] quickprop algorithm; Montana and Davis's [1989] combination of genetic algorithms with back-propagation; and Wasserman's [1990] back-propagation-Cauchy machine hybrid.

If the network is to be used for temporal or spatio-temporal pattern recognition, it may be important to use one of the more recent architectural modifications which has recurrent or time- delay connections. We discuss these in Part IV.

An alternative approach for some spatial pattern recognition tasks is the Neocognitron [Fukushima, 1988, 1989], which is a pattern recognition architecture modeled after the cerebral cortex. Its success tends to support its underlying assumptions about the functioning of the cerebral cortex. And to the extent that it incorporates essential characteristics of the cerebral cortex, this network has great potential for additional intelligent applications.

The major advantage of the Neocognitron is that it, like the brain, is able to recognize patterns that vary greatly in size, orientation, location in the input array, and degree of deformation. This is accomplished by using a large number of layers, and by alternating feature-detection layers with position-tolerance layers. Others (e.g., Hecht-Nielsen, Kurtzwiel AI, Bell Labs) have demonstrated similarly robust recognition performance, but have been less generous in publishing the details of their methods.

Unlike the back-propagation network, the Neocognitron has not been tested in a wide variety of applications. Also, at least as presented by Fukushima for handwritten character recognition, the Neocognitron requires that each layer be trained with a separate set of input patterns. Finding the right training patterns and the optimum number of nodes for each layer appears to be a trial-and-error process, requiring extensive analysis of the input patterns. Unsupervised training is also possible, but is less well explored.

Although it shows great promise, the Neocognitron is simply not as mature as the back-propagation algorithm. Nonetheless, if an application involves 2-dimensional pattern classification, and requires a high tolerance for deformity or size and position variance in the input patterns, the Neocognitron may be a good choice. Be prepared for a greater investment in training and fine-tuning if this route is taken.

If it is important that the network be able to recognize and create pattern classes for novel patterns, the only network type worth considering is the Carpenter-Grossberg Adaptive Resonance Theory (ART) network [Carpenter and Grossberg, 1988]. Its primary advantage is that learning and performance both occur simultaneously and continuously; it is an explicit attempt to deal with the "plasticity-stability dilemma" of natural intelligent systems. Feedback and a "vigilance" parameter are used to determine when a new pattern is sufficiently different from previously learned patterns to warrant the creation of a new category. See Carpenter et al, 1989, for a discussion of visual scene recognition using ART.

Orlando et al. [1990] have compared traditional and neural classifiers for the problem of using radar to classify sea-ice. They compared a Bayesian (Gaussian) classifier, a back- propagating Perceptron, and a Kohonen Learning Vector Quantization (LVQ) network, using the results of the Bayesian classifier to benchmark the two neural network classifiers. Each classifier produced similar decision regions and similar overall performance, differing on the average by -0.1% for the best configuration of the LVQ network (20 by 1 map) and +0.6% for the best configuration (20 hidden units) of the back-propagating Perceptron, where the Bayesian classifier produced an average of 82% correct classifications. They conclude that the only advantage of neural network classification in this type of task might be its potential for real-time operation.

Hush and Salas [1990] have studied the performance of four neural network classifiers on a 1-class classifier task, in which the classifier had to create a decision boundary that completely surrounded one class of data. In comparisions, a localized receptive field network (e.g., a radial basis function network, such as discussed in the next chapter) gave the best results, followed by a Higher-Order Neural Network (HONN) [Giles and Maxwell, 1987], followed by a back-propagating Perceptron and a Learning Vector Quantization network.

There are also a number of other complex networks which have potential for spatio-temporal pattern recognition. Some of the cooperative/competitive networks are useful; we discuss others in the appropriate chapter in Part IV.

14.2.2   Pattern Autoassociation

There are several networks which are useful for pattern autoassocation, allowing a complete pattern to be reconstructed when only a partial or degraded pattern is used as input. They have been explored mainly for the reconstruction of relatively small, binary patterns. Two of the most common autoassociators are the Hopfield/Tank network and Anderson's Brain-State-in-a-Box (BSB). They work well on small pattern sets, but cannot store large numbers of patterns without interference. The Hopfield network is particularly susceptible to generating spurious states in its output. Kohonen's Learning Vector Quantization network is a useful autoassociator, producing vector outputs whose distribution represents the probabilistic distribution of the training data.

These networks are most useful for cleaning up or completing noisy or degraded input patterns, and can also be combined with other architectures for more sophisticated applications. For example, Rossen and Anderson [1989] obtained improved syllable-recognition by using a BSB for the output layer of a 3-layer feedforward network that was trained by the back-propagation method.

Tsoi [1990] studied three networks; the Hopfield, the Hamming, and the back-propagating Perceptron; on an autoassociative retrieval task. He found that the Hamming network operated best when given noisy data, and the Perceptron was the worst. He also attempted a generalization study, but the method he chose (data concatenation) was not able to probe the generalization capabilities of the back-propagating Perceptron, and yielded a spurious minimum in the case of the Hopfield network.

14.2.3   Pattern Heteroassociation

Although there are several networks which are prototypically heteroassociators (e.g., Kosko's Bidirectional Associative Memory, or BAM networks), the best choice might again be to use a back-propagation network. This is due to the memory limitations of BAM-like networks; only a limited number of pattern-pairs can be stored without interference. Also, without the middle layer of "feature-recognizing" neurons, BAM-like networks lack the ability to generalize about pattern features. (This is also true of ART, and any network which lacks some method for encoding a feature-based pattern representation. The only two-layer networks which are able to form generalized pattern representations have some special means of encoding richer representations at the neuronal level, such as using radial basis functions or functional links. We discuss both of these alternatives in the next chapter.)

14.2.4   Data Compression and Mappings

A varient use of the back-propagation network allows for data compression and/or dimensionality reduction. The Learning Vector Quantization network can be used as a codebook. The self- organizing Topology-Preserving Map can perform complex mappings for dimensionality reduction. We discuss networks for these applications in more detail in Part IV.

Both the back-propagation network and the Topology-Preserving Map have been used for mappings. In a parameter estimation mapping study, Shadmehr and D'Argenio [1990] compared the back-propagation network to two statistical techniques; a Maximum Likelihood (ML) estimator and a Bayesian Maximum A Posteriori probability (MAP) estimator. The task involved estimating model parameters in a sparse and noisy environment. The neural network consistently performed better than the ML estimator, and approached the performance of the MAP estimator.

14.2.5   Optimization

There are only a few choices for optimization using neural networks. Theoretically, the Hopfield network can perform optimization. Because of the difficulties in using a "vanilla" Hopfield network (spurious states, etc.), this network is not recommended. The Boltzmann machine can do optimization, even though it has a feedforward structure. This is because it learns using an (statistical) optimizing method. However, the learning rule for the Boltzmann machine is complex, and the network takes a long time to learn. Kohonen's Topology-Preserving Map has demonstrated some very interesting capabilities for optimization (see discussion in Part II). Certain business applications (e.g., scheduling) require optimization. We discuss these in Part IV.

14.2.6   Speech and Vision

Speech, vision, and similar applications are very complex tasks. Approximately one-third of our cerebral cortex is devoted to speech and language processing, and another one-third to processing visual information. Therefore, we should not expect any simple neural network to take on such demanding tasks. For these applications, context is important. Previous research in both areas indicates that we need to have transform data from one representation level to another, moving progressively from the data level to the feature level, the structure level, and the symbol level. Most investigations of neural networks for these applications has focused on use of networks at the data and feature levels.

Even at these levels, the networks are typically quite complex (see, e.g., Lippmann, [1989] for a comprehensive review of neural networks for speech.) Because of the contextual dependency and the need for multiple representation levels, the networks used often involve cooperative/competitive connections, and may also employ feedback. Their dynamics are therefore more complex, and may more closely approximate the neural dynamics of the brain. Examples include Grossberg's Boundary Contour System for finding scene boundaries, and Kohonen's speech analyzer. In Kohonen's speech analyzer, a "Mexican hat" distribution of lateral excitation and inhibition is used to convert temporal phonetic sequences into 2-dimensional spatial maps [Kohonen, 1988].

These cooperative-competitive models show great promise, in the long run, for emulating brain function. To date they have been most useful for handling sub-tasks of large, multi-stage problems, such as scene analysis and speech recognition. Therefore, these approaches should probably be reserved for longer range, multi-stage research and development projects.

14.2.7 Neurocontrol

Neural networks for control applications require either a system of interacting networks or complex networks involving recurrent connections. This is because control processes involve building up a model for the desired state or output of the system, and influencing some ongoing process to bring its state or output more into accord with what is desired. The last chapter in Part II described some systems of networks which could be used for control, and a significant chapter by Paul Werbos in Part IV of this book addresses this topic.

14.3 DEBUGGING, TESTING, AND VERIFYING NEURAL NETWORK CODES

14.3.1 Debugging Neural Networks

Debugging the user interface of a neural network system or the network itself is no different from debugging other kinds of software. Debugging the training portion of a neural network system is quite another matter. Here it is most advantageous to rely on some form of numerical or, preferably, graphical output showing how the weights and activations change during the training process.

Useful statistics include the maximum, minimum, and average weights for each node and layer; the individual activations, as well as their maximum, minimum, and average; and the individual output errors as well as their maximum, minimum, and average sum of squares. This information will make it easier to recover from problems such as local minima, logical errors, and "network paralysis" resulting from the flat spots of the activation function. Commercial neural network software often contains graphical tools for monitoring training, although they may only be capable of displaying small networks.

14.3.2 Efficiency Considerations for Optimal Network Performance

Whatever network paradigm is chosen, significant efficiencies can be gained through careful implementation and minor modifications. It may be helpful to model the network code on one or more of the networks available in source code (e.g., McClelland and Rumelhart's Explorations in Distributed Parallel Processing [1988]). Existing implementations generally contain optimizations which have been tried and tested. And it is sometimes easier to begin with a functioning network simulation and modify it, than to code one from scratch. Although the network algorithms themselves are quite simple, the tasks of data structure management, and of controlling, training, and monitoring the network are not trivial.

In general, the safest course is to begin with the smallest and simplest possible configuration and then, once the system is working, experiment with various options. Once a small version has been perfected, scale-up can be attempted.

14.3.3 Testing and Verifying Neural Networks

There are, as yet, no theoretically sound methods for verification of neural network systems, although this is a topic for which considerable research funding is available. Trial and error testing, such as that often employed for more traditional programs, for symbolic expert systems, and indeed, for humans, must be employed.

Some fraction (typically about 20-25%) of the training cases should be reserved for testing the fully trained network. This is essential to verify the ability to generalize to novel inputs. Successful training, with failure to generalize, suggests the need to reconfigure the network using fewer hidden nodes. This, paradoxically, often improves the ability of the network to generalize accurately.

Verification should always include testing with boundary cases. A careful analysis should be done to construct the "hardest," "trickiest," or most unusual inputs likely to be encountered by the system in the field. Inadequate responses to these cases may require retraining with a more complete set of training cases.

Conventional algorithmic software is notorious for its propensity to exhibit catastrophic failure when presented with even mildly deviant unanticipated inputs. Due to their inherent ability to generalize from trained inputs, neural networks may be more resistant to this sort of failure.

Their complexity and ability to generalize also imply that neural networks cannot be adequately tested with a small set of specific inputs. Like the humans they attempt to emulate, networks must be tested on a broad sampling of the inputs they are required to cope with. The rigor and expense of this testing must be proportional to the risk of operational failure, just as physicians and airline pilots are more rigorously tested than cosmetologists and letter-carriers. Nonetheless, they sometimes fail.

14.4 IMPLEMENTING NEURAL NETWORKS

One can estimate the computational resources needed to run a specific neural network in a specific environment. Different environments can be used for developing and implementing neural networks.

14.4.1 Computing Requirements

In general, computing requirements will be determined by the size of the neural network. Once the network configuration has been determined, the number of interconnects should be calculated. For example, in a fully connected multilayer feedforward network with 2 hidden layers, this number will be: $IC = ([IP+1] * HL1) + ([HL1+1] * HL2) + ([HL2+1] * OP)$ where IC=number of interconnects, IP=number of inputs, HL1=number of 1st hidden- layer nodes, HL2=number of 2nd hidden-layer nodes, and OP=number of outputs. A 1 is added to one factor of each term for the biases. The formula for a single hidden-layer net is: $IC = ([IP+1] * HL) + ([HL+1] * OP)$. For example, for a network with 200 inputs and 10 outputs with 60 nodes in HL1 and 20 in HL2, the number of interconnects will be: $IC = (201 * 60) + (61 * 20) + (21 * 10) = 13490$.

The next question is "how fast does the network need to run?" This will determine the number of interconnects per second (IPS) needed. If the above network needs to run in 0.25 seconds, then 13490/0.25 = 53960 IPS will be needed.

For technical reasons, back-propagation is generally performed with floating point arithmetic. (Some commercial packages are switching to integer arithmetic for more speed.) Therefore, processing power requirements are calculated in FLOPS or MFLOPS (million floating point operations per second). In most networks, each interconnect amounts to a multiply-and-sum operation, such that each interconnect requires 2 floating point operations (although an increasing number of floating point processors are "pipelined" to accomplish the multiply-and- accumulate in a single operation). In this example: FLOPS = (2 * IPS) = (2 * 53960) = 107920 FLOPS. The actual number of FLOPS required will be somewhat higher because this estimate does not include computation of the activation functions.

So far, it has been assumed that the hardware will be used with 100% efficiency. In reality, a great deal of overhead is required to organize and execute the floating point operations. With custom-optimized assembly language routines, at least 50% efficiency is attainable; with a high-level language, 10-25% efficiency is probably more realistic. Therefore, the final estimate of the computing power requirement is: FLOPS(actual) = FLOPS(ideal)/0.50 = 107920/0.5 = 215840 for the assembly language case, or FLOPS(actual) = FLOPS(ideal)/(0.2) = 107920/0.2 = 539600 for the high-level language case.

This discussion has pertained to a determination of the computing requirements to run a network. Computing requirements for development and training will depend on the amount of time available for training. Training the network is likely to require on the order of 10,000 times as long as running the network once. During development, a considerable amount of time is likely to be spent waiting for the system to train or test. Therefore, the preceding method for making estimates should be regarded as a bare minimum and the maximum available computing power should be used for rapid development.

14.4.2 Neural Network Computations in a PC Environment

Neural network research and development is currently being performed on a wide variety of hardware platforms, ranging from supercompters to desktop personal computers (PCs).

When working on a PC, a numeric coprocessor is essential (80x87 or Weitek). Recently, several manufacturers have come out with pin and software compatible 80x87 clone chips that purport to be several times faster than the Intel product. Currently, fast coprocessor equipped PCs can perform in the 1 MFLOPS range. With careful software design, these machines can support small- to-moderate sized realtime networks, and larger networks at slower speeds. For larger networks, and to speed development by reducing training times, a coprocessor board with a DSP (digital signal processor) is a good option. 10-100 fold speed improvements can be obtained with currently available boards. These are discussed in the next section.For the most demanding neural network applications, a true parallel processing machine should be considered. Examples include the Intel Hypercube, the Quadputer, and the Connection Machine, from Thinking

Machines, Inc. A fully configured Connection Machine is capable of 2-4 GFLOPS (giga, or billion, FLOPS). Although opinions vary, this may be approaching the computational power of the human brain.

14.4.3　Coprocessing Boards

A variety of DSP coprocessor boards are available for PCs, as well as for workstations and for the Macintosh. We can mention only a few of these products. This is a rapidly evolving market, so review current journals for up-to-date information.

HNC (Hecht-Nielsen Neurocomputing; San Diego, CA; 619- 546-8877) offers the ANZA neural network coprocessor board for PCs and Sun workstations. This board is available with 10 megabytes RAM, and can perform 10 million IPS. HNC also offers a variety of development tools, including many network paradigms and AXON, a neural development language.

SAIC (Science Applications International Corporation; San Diego, CA; 619-546-6290) offers the Delta II neural network floating point coprocessor board. This card is rated at 22 MFLOPS and can perform extensive I/O at 20 MBytes per second. Again, a variety of network paradigms and development tools are offered.

For those who prefer to program their own networks on a fast DSP coprocessor board, several companies offer boards based on the TMS32030 DSP chip from Texas Instruments. This chip is rated at 33 MFLOPS. As of this writing, prices on these boards range from about $2500 to $10,000, depending on the manufacturer and memory configuration. Companies supplying TMS32030-based coprocessor boards include: Atlanta Signal Processors, Inc., Atlanta, GA (404-892-7265); Spectrum Signal Processing, Inc., Blaine, WA (800-663-8986); and Sonitech International, Inc., Wellesley, MA (617-235-6824).

The boards mentioned above simulate neural networks in software with a fast DSP chip. They are not true hardware implementations of neural network architectures. The real promise of neural networks will not be fulfilled until affordable massively parallel hardware is available for them to run on. It is likely that within a few years parallel coprocessor boards and optical computers with performance in the GFLOPS range will become available [Morton, 1989; see the discussion of Intel's ETANN chip].

True neural network chips are now being tested, and offer the prospect of quantum improvements in network size and speed. For example, Intel's ETANN (electrically trainable analog neural network) chip permanently stores its weights on-chip, and is capable of 3 billion multiply-and-accumulate operations per second. PC plug-in boards containing one or more of these chips should be available in the near future. We discuss neural networks hardware later in Part III.

It should also be mentioned that several manufacturers of neural network development environments now supply proprietary "neural development languages." These languages have high-level constructs and procedures to facilitate the rapid development of network applications. For example GM and TA Korn (Tuscon, AZ) offer "Desire/Neunet," a flexible programming environment that allows, for example, a very fast back-prop network to be created with less than a dozen lines of code. The resulting software may not be portable to other hardware or environments, however. Keep in mind, however, that software created with a proprietary language may not be portable across a wide variety of hardware/software platforms.

14.5   NEURAL NETWORK DEVELOPMENT TOOLS

Once the network configuration and hardware requirements have been determined, the next issue is what software to use. Many options are available, and the best choice may depend somewhat on the experience of the programmers involved.

14.5.1   Neural Network Shells and Environments

One of the quickest ways to get a custom network up and running is to use one of several software products available for this purpose. (For a listing of vendors, see Byte, August 1989, pp. 244-245.) Some very inexpensive products are available that are useful for education and experimentation, such as the software accompanying the book Explorations in Parallel Distributed Processing, by McClelland and Rumelhart [1988]. These are generally unsuitable for application development, however, without extensive modification of the code, and separate development of a user interface.

Nestor, Inc. (Providence, RI; 401-331-9640) advertises a "complete developer's environment" for creating neural network applications on PCs and workstations. Their products include training, software, and complete neural network workstations. Source code is not available. NeuralWare, Inc. (Sewickley, PA; 412-741-5959) supplies neural network development systems for PCs, Macs, workstations, and the N-Cube. HNC and SAIC, mentioned above, provide software development environments with their neural network coprocessor boards.

Neural network development environments can provide a fast track for application development, but they also have potential drawbacks. First of all, most will not have the tools needed to create a top-notch user interface for the final product. Once the network is working, it may be necessary to develop the user interface in another environment and then integrate the two. Another drawback is decreased flexibility. Some of these systems restrict the user to whatever paradigms are supplied, making it difficult or impossible to experiment with novel networking ideas. For this reason, it is desirable to use a product for which the source code is available.

Portability is another concern with neural network development environments. If the development environment includes a coprocessor board, the final product may be locked to that board. Also, some of these products come with proprietary "neural programming languages" that may reduce product portability.

A final concern regarding neural network environments is product quality. Most of these products are relatively new. It is easy end up spending more time testing and debugging bad development software, than using it to develop a network application. Others who are developing similar products should be consulted to determine which development products are capable and reliable.

14.5.2   Neural Network Development Languages

Due to the computational demands of neural networks, the choice of a programming language is more critical than in other situations. Whatever language is used, it is advisable to seek a version that has been optimized for numeric data processing (NDP).

The C language has become somewhat of a de facto standard for network programming. Most of the published source code examples are in C (e.g., Explorations in Parallel

Distributed Processing). Working in C may facilitate the use these examples as templates or models.

Only with painstakingly coded assembly language is it possible to approach the theoretical efficiency of the hardware. Drawbacks of assembler include increased programming and debugging time, as well as decreased portability. In the short run, it may be cheaper to program in a high level language and use faster hardware.

Object oriented languages are conceptually ideal for neural network implementations, but for the moment are too slow for production networks. They can be profitably used to rapidly produce early prototypes. Another option is to develop the user interface in a suitable object oriented language, and then link in the network written in C or assembler.

14.6   CHAPTER SUMMARY

In summary, the following steps are recommended in the design and development of a neural network application:

1.   Verify the competitiveness of a neural network solution relative to more traditional approaches.
2.   Determine the most appropriate neural network architecture for the application.
3.   Ascertain the most efficient means of representing the network inputs. Optimum preprocessing of the inputs can greatly improve network performance.
4.   Estimate the number of layers and hidden nodes required, allowing a generous safety factor.
5.   Select a delivery hardware platform adequate to run the network at design speed. For development and training, use the fastest hardware available.
6.   Choose a language or development environment conducive to the implementation of fast network computation.
7.   Design a flexible network that allows network size and all other parameters to be easily adjusted.
8.   Begin with the smallest possible network, for rapid verification of training and performance. Then scale up.
9.   For large networks it may be advisable to scale up in stages, or to use a modular approach, combining multiple smaller networks in an integrated system.
10. Experimentation with various optimizations may be required in order to obtain optimum performance.

REFERENCES

Carpenter, G.A., & Grossberg, S. (1988). "The ART of adaptive pattern recognition," *IEEE Computer,* March, 77-88.

Carpenter, G.A., Grossberg, S., & Mehanian, C. (1989). "Invariant recognition of cluttered scenes by a self-organizing ART architecture: CORT-X boundary segmentation," *Neural Networks, 2,* 169- 182.

DARPA Neural Network Study, October 1987, February, 1988 Final Report. Chap. 19, Comparison with Other Technologies. AFCEA Int'l. Press (AIP).

Fahlman, S.E. (1988). "Faster-learning variations on back-propagation: An empirical study," *Proceedings of the 1988 Connectionist Models Summer School.*

Fukushima, K. (1988). "Neocognitron: A hierarchical neural network capable of visual pattern recognition," *Neural Networks, 1,* 119-130.

Fukushima, K. (1989). "Analysis of the process of visual pattern recognition by the Neocognitron," *Neural Networks, 2,* 413-420.

Giles, C.L, & Maxwell, T. (1987). "Learning, invariance, and generalization in higher-order neural networks," *Applied Optics, 26,* 4972-4978.

Huang, W.Y., & Lippmann, R.P. (1988). "Neural net and traditional classifiers," in D.Z. Anderson (Ed.), *Neural Information Processing Systems,* AIP, New York, 387-396.

Hush, D.R., & Salas, J.M. (1990). "A performance of neural network classifiers for the 1-class classifier problem," *Proc. Second Int'l. Joint Conf. Neural Networks* (Washington, D.C., Jan. 15-19, 1990), I-396-I-399.

Kohonen, T. (1988). "The 'neural' phonetic typewriter," *IEEE Computer* (March), 11-22.

Lippmann, R.P. (1989). "Review of neural networks for speech recognition," *Neural Computation, 1,* 1-38.

McClelland, J.L., & Rumelhart, D.E. (1988). *Explorations in Parallel Distributed Processing,* MIT, Cambridge, MA.

Miller, R.K., Walker, T.C., and Ryan, A.M. (1990), *Neural Net Applications and Products,* SEAI Publications and Graeme Publications.

Montana, D.J., & Davis, L. (1989). "Training feedforward neural networks using genetic algorithms," *Proceeding of the Int'l. Joint Conf. Artificial Intelligence,* 762-767.

Morton, S.G. (1989). "Intelligent memory chips: Practical neural network building blocks," *J. Neural Network Computing, 1 (2),* 39-53.

Orlando, J., Mann, R., & Haykin, S. (1990). "Radar classification of sea-ice using traditional and neural classifiers," *Proc. Second Int'l. Joint Conf. Neural Networks* (Washington, D.C., Jan. 15-19, 1990), II-263-II-266.

Parker, D.B. 1982, Learning logic, Invention Report S81-64, File 1, Office of Technology Licensing, Stanford University, Stanford CA.

Rumelhart, D.E., Hinton, G.E., & Williams, R.J. (1986). "Learning internal representations by error propagation," in Rumelhart, D.E., McClelland, T.L., & the PDP Research Group (Eds.), *Parallel Distributed Processing: Explorations in the Microstructure of Cognition: Volume 1: Foundations,* MIT, Cambridge, MA.

Rossen, M.L., and Anderson, J.A. (1989). "Representational issues in a neural network model of syllable recognition," *Proc. First Int'l. Joint Conf. Neural Networks* (Washington D.C., June 18-22, 1989) I-19-I-25.

Shadmehr, R., & D'Argenio, D.Z. (1990). "A comparison of a neural network based estimator and two statistical estimators in a sparse and noisy data environment," *Proc. Second Int'l. Joint Conf. Neural Networks* (Washington, D.C., Jan. 15-19, 1990), I-289-I-292.

Tsoi, A.C. (1990). "Comparison of the performance of three popular neural networks," *Proc. Second Int'l. Joint Conf. Neural Networks* (Washington, D.C., Jan. 15-19, 1990), II-707–II-711.

Waibel, A., Sawai, H., Shikano, K. (1989). "Modularity and scaling in large phonemic networks." *IEEE Trans. Acoustics, Speech, and Signal Processing, 37,* 1888-1898.

Wasserman, P.D. (1990). "A combined back-propagation/Cauchy achine." *J. Neural Network Computing, 1* (Winter), 34-40.

Werbos, P. 1974. Beyond regression: new tools for prediction and analysis in behavioral sciences, Ph.D. thesis in applied mathematics, Harvard Univ.

15

CONFIGURING AND OPTIMIZING THE BACK-PROPAGATION NETWORK

Alianna J. Maren, Dan Jones, and Stanley Franklin

15.0 OVERVIEW

Multilayer feedforward networks are good for pattern classification, signal filtering, complex mapping, and other tasks, but they often require long training times. Network performance is also strongly affected by the configuration of the network structure and the type of preprocessing and representations of input data used. Use of novel microstructural elements, such as radial basis functions or functional links, can enhance network performance for some applications. Both theoretical and experimental results now guide network configuration, and there are many ways in which training time can be reduced. These allow for optimal learning and performance in a back-propagation or similar network.

15.1 ISSUES IN OPTIMIZING AND GENERALIZING FEEDFORWARD NETWORKS

During the past several years, researchers have explored many ways to enhance the performance, maximize likelihood of convergence while learning, and improve the overall learning rate of the back-propagation (and related) neural networks. There are several factors which have provided an impetus for this work. These include the fact that the basic (*vanilla*) back-propagation network often takes a discouragingly long time to learn, and that sometimes it does not converge at all for a given random set of starting weights. Also, as we use the back-propagation for more demanding applications (such as classifying sonar returns), we need to have ways of optimizing performance to obtain the best possible results. We need to know how to train a network to achieve these best results. Thus, the major issues in designing, optimizing, and training a back-propagation (or similar) network are:

- How can we decrease the learning time?
- How can we select and configure different architectures for different types of applications?
- How can we increase the synergy between associative memory networks and back-propagation networks in order to achieve more complete and robust generalization across a broader variety of domains?
- How can we apply neural networks to the problem of tracking dynamic systems?
- How can we assess the capabilities of the back-propagation (or similar) network vis-a-vis other alternative methods?

Recently, there has been a lot of research in developing methods for faster learning. No single method has emerged as *the* learning paradigm, and the *Proceedings* of the different neural networks conferences have papers on improved learning algorithms.

The most interesting architectures involve more than a simple back-propagation network. Typically, some sort of preprocessing is still necessary to achieve optimal results. For example, Glover [1989] uses Fourier preprocessing of images before using a back-propagation network for automated inspection.

Werbos [1988] has addressed some of the issues involved in generalizing the performance of the back-propagation network to recurrent systems. This would enable a link between associative networks (such as Hopfield or Grossberg networks) and feedforward networks. He also shows how the generalized back-propagation method can be used for forecasting over time.

It is typically more difficult to apply feedforward neural networks to time-varying problems than to static pattern recognition. Derrick Nguyen and Bernie Widrow [1989] have recently applied the ADALINE to the control problem of backing up a truck, and have demonstrated their results in simulation. In Part IV, we discuss neural networks for recognition of different types of time-varying signals.

Werbos [1989] has also shown how back-propagation can be used as one element of a more comprehensive system to control dynamic processes, such as robotic arms (or muscles). He presents a more expanded version of his ideas in Part IV of this book.

There are three major ways to improve performance of the back-propagation (or similar multilayer feedforward) network: through modifying the structure, the dynamics, or the network training and learning rules. In this chapter, we consider some of the most promising options in each of these areas.

15.2   MICRO-STRUCTURAL CONSIDERATIONS FOR OPTIMIZING PERFORMANCE OF FEEDFORWARD NETWORKS

We can modify the structure of the basic feedforward network in several different ways. At the micro-structural level, we can use a new transfer function or create new types of connection weights. There are several ways in which changes to the transfer function or the connection weight capability can have dramatic affects on the network performance.

15.2.1   Rescaling the Transfer Function

The activation function most commonly used with back propagation is the logistic function, $1 / (1 + \exp[-\alpha * \text{net\_input}])$. The output of this function ranges over the open interval 0 to 1. Certain limitations in training and performance can occur with this function, due to the fact that it is not symmetric about 0. Simply by subtracting 1/2 from the logistic function, thereby shifting the output range to the open interval -1/2 to + 1/2, Stornetta and Huberman [1987] reduced training times by 30-50 percent. (Don't neglect to also change the output function derivatives from x(1-x) to (x+1/2)(1/2-x) = 1/4-x2!) Fahlman [1988a] found this activation function significantly faster (on the average) and more reliable for training encoder-decoder networks.

15.2.2   Radial Basis Functions

The micro-structure of a Perceptron-like network can be modified by using a radial basis function as the transfer function in the hidden layer [Broomhead and Lowe, 1988]. Radial basis functions are particularly useful for complex mapping tasks where the mapping is continuous. (For example, parity problems are not continuous, and are not a good radial basis function application.) Network architectures using these functions only require a single hidden layer, as they allow disjoint regions in space to be satisfactorily represented by different hidden layer nodes.

The radial basis function, shown in Figure 15.1 (a), has two parameters in the transfer function itself: c (the center of each function) and sigma {small Greek sigma}, the variance. Each hidden layer node receives a weighted input from all the neurons in the first layer of the network. The activation of each hidden layer neuron is

$$\text{hidden\_activ}_j = \Phi\,(\,|\,x_j - c_j\,|\,) = \exp\left[\,-\sum_{i=1}^{\infty}(x_j - c_j)^2 / 2\sigma_{ji}^2\right] \qquad \textbf{Eq. 15.1}$$

Where n is the total number of input neurons,

indices i and j refer to the ith element of the input layer and the jth element of the hidden layer, respectively,
c_j is the center of the jth radial basis function,
σ_j is the *spread* around the center c_j of the jth function,
x_j is the input into the jth function.

The hidden nodes are linearly and fully connected to the output nodes. The output nodes may or may not have a nonlinear transfer function.

Tsoi [1989] has presented a full derivation of the radial basis function technique as applied to multilayer feedforward networks. He states that when radial basis functions are used, the number of hidden nodes should be at least equal to the number output classes. They conclude that the field width scalars can be safely set to 1.

The Radial Basis Function Network

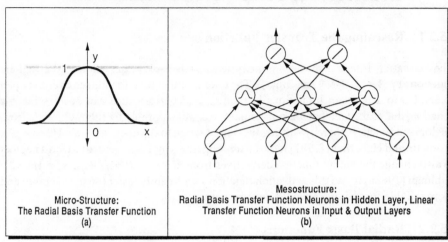

Micro-Structure:
The Radial Basis Transfer Function
(a)

Mesostructure:
Radial Basis Transfer Function Neurons in Hidden Layer, Linear
Transfer Function Neurons in Input & Output Layers
(b)

Figure 15.1 The radial basis function is useful as a transfer function for feedforward networks doing continuous mappings.

Moody and Darken [1989] and Saha and Keeler [1990] have explored how to adaptively set the centers of the radial basis functions. They have also developed improved methods for learning in radial basis function systems. Saha and Keeler have also adaptively set the receptive field width of the different centers. Moody and Darken applied their network to phoneme recognition, and compared results of their network with the results of k-nearest neighbor, Gaussian classifiers, a two-layer back-propagation network, and a Kohonen feature map classifier. Percent error on the four comparison networks ranged from 18.0 percent (for the k-nearest neighbor system) to 22.8 percent (for the Kohonen network). With 100 Gaussian units, their radial basis function network achieved the same accuracy as the k-nearest neighbor, 18 percent. They attribute the improved accuracy of their network (as compared to the Kohonen network) to the fact that the Gaussian response functions yielded smooth interpolations of classification regions, rather than sharp discontinuities from one cluster region to the next.

Poggio and Girosi [1990] have also developed a rigorous formalization of neural networks using radial basis functions with adaptive center locations. They show that their network is the feedforward version of regularization and are equivalent to generalized splines. Mel and Koch [1990] show how this type of network maps naturally on to cortical hardware, and gives coherence to a number of aspects of learning observed in cortical anatomy.

15.2.3 Higher-Order Connections

The high-order connection networks, sometimes called *sigma-pi networks*, achieve greater processing power by using more complex connections. Various combinations of inputs are multiplied together. These products are then multiplied by their respective

The High-Order (Sigma-Pi) Network

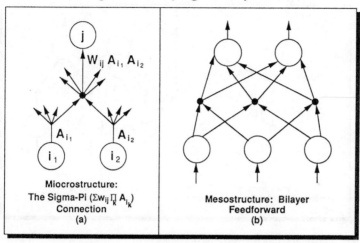

Microstructure:
The Sigma-Pi $(\Sigma w_{ij} \prod_k A_{i_k})$
Connection
(a)

Mesostructure: Bilayer
Feedforward
(b)

Figure 15.2 High-order networks have neurons with the usual (e.g., sigmoid) transfer functions, but have connections which pair two or more neurons together for a single connection to another neuron.

weights and summed to give an output node's net activation. A single-layer HONN is therefore capable of computing non-linear functions. For a sizable input vector, however, the possible number of input products that can be formed is enormous, precluding the use of fully connected HONNs. Spirkovska and Reid [1990] used a restricted HONN (no products of more than 3 inputs) to achieve translation, rotation, and scale-invariant pattern recognition. In their application, the HONN was superior to a back-propagation network in terms of training time, accuracy, and training set size. An illustration of a HONN is shown in Figure 15.2.

15.2.4   Functional-Link Networks

The functional-link network [Pao, 1989] achieves nonlinear responses using a single-layer net without hidden nodes. This is accomplished by applying nonlinear functions to some or all of the inputs before they are fed into the network. This can be conceived as a network implementation of a superposition approximation of a nonlinear function, such as Fourier synthesis or polynomial curve-fitting. For example, Kraiss and Küttelwesch [1990] successfully trained a functional-link network to guide a vehicle through an obstacle course by inputting the distance datum (D) in $2n+1$ different ways: D, $\sin(\pi * D)$, $\cos(\pi * D)$, $\sin(i * \pi * D)$, $\cos(i * \pi * D)$,....., $\sin(n * \pi * D)$, $\cos(n * \pi * D)$. Good results were obtained with an n of 10 or less. Although the total number of inputs is increased by the application of the nonlinear functions, the functional-link net converges during training more rapidly than back-propagation, due to the absence of hidden nodes. Drawbacks appear to be the considerable amount of preprocessing required, and the necessity of ad hoc or empiric selection of the appropriate functions to be applied to the inputs. Clearly, a thorough understanding of the nature of the input data and the

The Functional Link Network

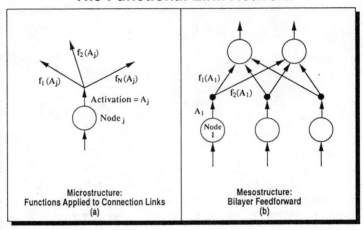

Microstructure:
Functions Applied to Connection Links
(a)

Mesostructure:
Bilayer Feedforward
(b)

Figure 15.3   Functional link networks have connections which are functions instead of scalars.

desired network transformation is essential to selection of the optimal input functions. Figure 15.3 illustrates a functional link network.

15.3 MESO-STRUCTURAL CONSIDERATIONS FOR OPTIMIZING NETWORK PERFORMANCE

At the meso-structural level, we can explore basic design issues, such as: how many layers, how many nodes, what type of connectivity. Sometimes, when we use a new micro-structural feature (e.g., more complex connections), this influences how we make choices at the meso-structural level (e.g., how many layers).

15.3.1 Choosing the Right Number of Layers

When using a back-propagation network, one of the most important configuration issues is to select the optimal number of hidden layers. Hecht-Nielsen [1987] has given a neural network interpretation of Kolmogorov's Theorem, proving that a network with one hidden layer can compute any arbitrary function of its inputs. This result is of dubious utility, however, because it relies on neurons with different output functions, but gives no constructive method of obtaining them.

Cybenko [1988a] shows that one hidden layer (with homogeneous sigmoidal output functions) is sufficient to compute arbitrary decision boundaries for the outputs, and that two hidden layers are sufficient to compute an arbitrary output function of the inputs. Since simple sigmoidal output functions are used, these results are much more useful. Lippmann [1987] has provided a very lucid and intuitive geometric argument that two hidden layers are sufficient to compute any decision boundary. Although more comprehensible to the non-mathematician, Lippmann's result is weaker than

Cybenko's, requiring two hidden layers, rather than one, to compute arbitrary decision boundaries.

Empirical tests using the back-propagation network for diagnosis (primarily a decision boundary problem) have not demonstrated a significant advantage for 2 hidden layers over 1 in a relatively small and simple diagnostic network [Bounds and Lloyd, 1988]. This is in agreement with Cybenko's results [Cybenko, 1988, 1989]. The best advice based on current knowledge is probably this: for a classification (decision boundary) problem, where the output node with the greatest activation will determine the category of the input pattern, one hidden layer will most likely be sufficient. On the other hand, if the outputs need to be continuous functions of the inputs (e.g., if the outputs need to represent relative probabilities of possible categories, or graded motor responses to sensory inputs), then plan on using two hidden layers or different transfer function, such as the radial basis function. Even for continuous outputs, a single hidden layer may be sufficient, depending on the nature of the problem. In linearly separable cases, NO hidden layers may be required.

15.3.2   Designing the Meso-Structure of a Feedforward Network: Choosing the Right Number of Nodes

Selecting the number of nodes or neurons for each layer of a back-propagation network is a crucial task, and has a great deal of impact on later network performance. There are three major considerations: the number of output nodes, the number and arrangement of the input nodes, and the number of hidden node layers and the number of nodes per hidden layer.

15.3.2.1   Selecting the Right Number of Output Nodes   The number of output nodes will equal the number of scalars required in the output vector (e.g., the number of possible diagnoses, characters, or other patterns to be recognized; assuming local representation in the output layer).

Sometimes, if we have many pattern classes, a pattern of activations across a number of output nodes will be used instead of having a single output node for each pattern. As an example, if we wanted to represent eight different pattern classes, we could either have eight output nodes, or have three nodes and use a binary encoding pattern. However, use of encoding patterns forces additional work onto the hidden nodes, which may require an additional hidden layer.

15.3.2.2   Selecting the Right Number of Input Nodes   Usually, the number of input nodes will equal the number of data items in the input array (assuming local representation of the inputs, by convention, these are noncomputing fanout nodes).

If a 1-D, binary pattern will be used as input, it can be represented as a mapping into an equivalent number of nodes. If a 1-D, discretely sampled analog signal is to be used as input, it can be passed into a set of nodes which are configured to receive real-valued input. Alternatively, a 1-D analog signal or a 2-D binary pattern can be represented by a set of input nodes configured as a 2-D array. The partitioning of the signal into appropriate horizontal and vertical intervals may require experimentation with different alternatives.

15.3.2.3 The Number of Hidden Nodes Choosing the right number of hidden nodes is the most interesting and challenging aspect of configuring a back-propagation network architecture. We can approach this issue by studying an example. Figure 15.4 shows four possible configurations of a back-propagation network to solve the X-OR problem. In each case, there are two input nodes (which each can receive a "0" or a "1"), and one output node (which should yield either a "0" or a "1"). Figure 15.4 (a) shows a network with one hidden node, Figure 15.4 (b) has four hidden nodes, and Figures 15.4 (c) and (d) have two and three hidden nodes, respectively. The questions we need to ask ourselves now are which of these configurations will work, and which of these configurations is optimal.

In Figure 15.4 (a), the two most probable values for output node (0 or 1) can be mapped to the two most probable values for the hidden node. Thus, the output node is redundant and the network will not work.

Figure 15.4 (b) illustrates the opposite extreme. Each hidden node can learn to respond to a different input pattern. It would be relatively easy to set the connection weights so each of the hidden nodes learns to respond to one of the input patterns. This design works, but is not optimal for two reasons. First, back-propagation networks should respond to the features in the input rather than to the exact input pattern.[1] Second, keeping the number of hidden nodes to a minimum reduces the computational time needed for training. This means keeping the number of hidden nodes to be substantially less than the number pattern types which could be presented.

If the number of hidden nodes is too small, it may be difficult to obtain convergence during training. Because of these difficulties, it may be wise to keep the number of hidden nodes low, but to allow for a little more than is absolutely necessary.

For these reasons, Figure 15.4 (c) probably has the optimal, compressed feature extraction architecture, but the chances of finding a set of working connection weights when beginning with random value weights are not 100 percent. (The probability of convergence from a given set of starting weights is about 50-90 percent, depending on the modifications of the learning rule used.) The last architecture, 15.4 (d), may have a less focused feature extraction representation, but the chances of convergence to a working set of connection weights is better. Thus, either 15.4 (c) or 15.4 (d) is a good solution.

Another way to approach setting the number of hidden nodes would be to find a theoretical upper bound on the number of nodes needed, and work back from there. In Hecht-Nielsen's [1987] neural network interpretation of Kolmogorov's Theorem, 2N+1 hidden nodes (where N is the number of inputs) are required for a network with one hidden layer to compute an arbitrary function.

Lippmann [1987] has provided one of the most comprehensible geometrical arguments regarding node requirements. His discussion concerns networks with hard-lim-

---

1 Use of one hidden node per pattern type leads to what is called grandmothering. This term refers to a now-discarded concept used earlier in neurophysiology, which suggested that individual cells could learn to respond to specific input patterns. Thus, a person could have a "grandmother cell," which would recognize his or her grandmother.

How Do We Determine the
Optimal Number of Hidden Layer Nodes?

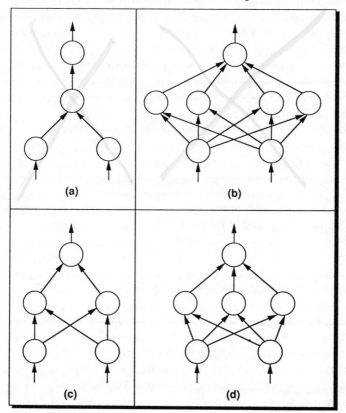

Figure 5.4   Four possible alternative meso-structures for the X-OR problem; only two are good.

iting activation functions and binary outputs, but it is probably pertinent to back-propagation networks with continuous outputs and sigmoidal activation functions. In
Lippmann's scheme, each node of the first hidden layer determines a decision plane (or
hyperplane), bisecting the n-dimensional input space (where n is the number of inputs).
Each node of the second hidden layer adds one or more of the subspaces created by
these decision planes to create a convex subspace or decision region; and each output
layer node adds one or more of these convex subspaces to create a subspace or decision
boundary of arbitrary shape.

As a practical example, consider a top-down estimation of the number of nodes
required to accomplish a pattern recognition task with a two-hidden-layer network,
using Lippmann's approach. This is a decision region determination, not a function
computation. Suppose there are eight medical diagnoses locally represented by eight
output nodes, and 20 input findings locally represented by 20 input nodes. If all the

decision boundaries for the diagnoses are convex, then only one hidden layer would be required. However, assume that the boundaries are of arbitrary shape and that, on the average, each diagnostic boundary is created in the output layer by combining (with an or operation) 2 of the convex subspaces created in the upper hidden layer. In this case, as an upper bound, at most 8 * 2 = 16 nodes will be required in the upper hidden layer.

Empirically, Kudrycki [1988] has found that the optimum ratio of first to second hidden layer nodes remains 3:1, even for higher dimensional inputs. Based on these considerations, our hypothetical diagnostic network should have 20 input nodes, 48 first-hidden-layer nodes, 16 second-hidden-layer nodes, and 8 output nodes. In practice, fewer hidden nodes will often be found to be sufficient.

Many researchers have found experimentally that, for diagnostic problems, good results can be obtained with only one hidden layer [Bounds and Lloyd,1988; Yoon et al., 1989; Saito and Nakano, 1988]. This seems to imply that in problems such as medical diagnosis, the decision regions for the outputs are often convex. Lippmann's reasoning would predict that the maximum number of nodes required for the single hidden layer should equal OP * (N+1) where OP is the number of output nodes and N is the number of input nodes. If Kudrycki's finding holds for the single hidden-layer case, then this maximum should equal OP * 3. In most published accounts of networks the optimum number of hidden layer nodes has been found to be less than the number of inputs. (For some applications, e.g., encoder-decoders, the number of hidden-layer nodes is by definition less than the number of inputs or outputs.) For a single hidden layer classifier or function mapper, an upper bound is about OP*3 or 2N+1 (Kolmogorov) hidden nodes, whichever is larger. In reality, fewer nodes are often sufficient. The relationship between network accuracy and the number of hidden nodes appears to be approximately quadratic [Stubbs 1990].

Gorman and Sejnowski [1988] have found that after a certain point, adding more neurons to a single hidden layer does not yield any improvement in performance. They studied the way in which their network made its classifications, and found that the discrimination process was quite complex. In their experiment, the back-propagation networks which had 12 or 24 hidden nodes had some hidden nodes which performed feature detection, and others which handled "special case" signal returns (in other words, grandmother nodes). For their application, both types of hidden nodes were necessary.

Table 15.1 shows the number of hidden nodes resulting in good performance in several single-hidden-layer networks for medical diagnosis. The geometric mean of the inputs and outputs is a good rough predictor of the optimum number of hidden-layer nodes in a small network with more inputs than outputs. (In the back pain experiment in Table 15.1, diagnostic accuracy with 10 hidden nodes (geometric mean) was 78 percent versus 80 percent with 30 hidden nodes.)

In summary, with too few hidden nodes, the network is unable to create adequately complex decision boundaries. With too many, the decision boundaries may perfectly encapsulate the training points, but have no ability to generalize. A single hidden layer is often adequate, either because the decision regions for the outputs are convex, or because Cybenko's reasoning is more relevant than Lippmann's. A smaller than expected number of first hidden layer nodes is usually sufficient, perhaps because each decision plane created in the first hidden layer can contribute to more than one of the convex decision regions created in the next layer.

Table 15.1   Diagnostic Inputs Hidden Nodes Outputs Reference

Number of hidden nodes required in single-hidden-layer medical diagnostic networks (as determined empirically, and as predicted by Lippman, Kudricki, and the geometric mean of the number of inputs and outputs [the geometric mean of 2 numbers is the square root of their product]).

Diagnostic Task:	Inputs	Hidden nodes				Outputs	Reference
		Emp	Lip	Kud	Mean		
back pain	50	30	102	6	10	2	Bounds 1988
dermatology	96	20	970	30	31	10	Yoon 1989
headache	216	72	4992	69	69	23	Saito 1988

15.3.3   Selecting the Number of Connections

Most feedforward networks are fully connected between layers. It is possible to use a lower or higher degree of connectivity. Higher-order connections (sigma-pi connections) were discussed in the previous section. Lower connectivity, or sparse-connectivity, means that a neuron on layer j is not connected to every neuron on layer j+1. Rumelhart, Hinton, and Williams [1986] used a locally-structured connectivity from the input to the hidden layer in order to discern between spatially rotated and translated versions of the letters T and C as did Sejnowski, Kienker, and Hinton [1986]. Their use of structured connections is a simple form of the basic concept underlying the Neocognitron, in which structured local connectivity forms the basis for shift, scale, and rotation-invariant line-pattern recognition.

15.3.4   Automatic Sizing and Generalization

There have recently been several attempts to circumvent the issue of number of hidden layers and nodes by having them selected by some procedure, rather than the usual ad hoc methods. The network architecture is generated automatically by the procedure. The following examples are all experimental, but may be worth considering.

Harp, Samad, and Guha [1989] have applied genetic algorithms to populations of descriptions of back-propagation networks in order to learn the most appropriate architecture. Each description (genome) contains values of the various network parameters (genes), including number of layers, number of nodes in each layer, etc. The population converges via a selection process to a near optimal network architecture. The selection criteria may include network size, training time, running time, and so forth, so that the developer may vary the meaning of optimal at will.

Fahlman and Lebiere [1990] have experimented with an algorithm which adds hidden nodes one by one as needed during training. After training a no-hidden-layer network to its optimum performance, hidden nodes are added and trained one at a time (via Fahlman's quickprop algorithm), until the target error is achieved. This method appears to generate networks of near-optimum size, while requiring significantly less training time than back-propagation. Rossen and Anderson [1989] have experimented with adding additional groups of hidden-layer units after existing hidden units have been trained.

Le Cun and his colleagues [1989] have produced an algorithm that removes ineffective hidden nodes from an existing network. This results in improved learning and running times, as well as superior generalization.

Mozer and Smolensky [1989] have developed a technique for determining the relevance of each hidden unit after training, and for *skeletonizing* the network by removing redundant hidden units. They assert that the most rapid training can be achieved by using an ample number of hidden units; size and speed can then be optimized using their technique to prune the network. Bartlett [1990, and Bartlett and Uhrig, in preparation] has developed a network which performs "simulated condensation" in determining the number of hidden nodes in a network, adding and subtracting nodes as needed to obtain optimal network configuration.

15.4 OPTIMIZING NETWORK DYNAMICS FOR IMPROVED PERFORMANCE

15.4.1 Optimal Input Representation and Preprocessing

In theory, a multilayer network should be able to produce the desired outputs from any input representation that encodes the relevant information. In practice, careful preprocessing of the inputs is usually required to obtain an efficient network. In general, all inputs should be scaled to a common interval (usually 0 to 1 or -1 to +1). If the inputs represent symbolic concepts (e.g., presence or absence of a feature) or continuous measurements (e.g., temperature or indebtedness), that may be all that is necessary.

Anytime the dynamic range of an input variable is more than a few octaves, the logarithm of the variable should probably be used. This transformation makes small changes in small inputs just as significant as large changes in large inputs. In the study of perception, this sensitivity to proportional changes is known as Weber's Law. It enables biological neurons with very limited dynamic range to encode light and sound intensities that vary over twelve orders of magnitude.

For patterned input arrays (e.g., visual, sonar, or speech patterns), it is important to use an image that is detailed enough for the task at hand, but not excessively detailed. Logarithmic transformation of the individual inputs is often used, as well as a variety of more global image transformations. In general, transformations should be used that inhibit noise and irrelevant variances (e.g., translation, rotation, deformity), and that emphasize relevant information. The Fourier transform or spectragram is often used [Waibel et al., 1989; Rossen and Anderson, 1989; Glover, 1989], and sometimes the cepstrum (Fourier transform of the spectrogram). The Gabor transform has been found to be superior for preprocessing fingerprint patterns for automated feature detection [Leung, Engeler, and Frank, 1990].

Another preprocessing technique that has recently come to prominence is principal components analysis (PCA; see [Vrckovnik et al., 1990]. In this technique, all the training vectors are arranged in a data matrix. The eigenvectors of the autocorrelation matrix of this data matrix are then computed. Usually (depending on the nature of the data), it will be found that the great majority of the energy or information is contained in only a small minority of the eigenvectors. Only the high-energy eigenvectors are kept, in a new (much smaller) matrix, which is used to premultiply the input vectors. This can result in a dramatic reduction in the dimensionality of the input vectors, with minimal loss of information.

Techniques for the detection of edges, contours, and gradients (either within an image or between consecutive images), are often useful. These functions can be accomplished by specialized input layers of the network [Carpenter et al., 1989; Fukushima, 1988], or by algorithmic preprocessing before the image is presented to the network [Rossen and Anderson, 1989; Glover, 1989].

Clues regarding the most effective types of preprocessing may be available from studies of analogous biological systems [Fukushima, 1989; Rossen and Anderson, 1989]. In practice, it is often necessary to experiment with a variety of transformations. For optimum performance, several different transformations may need to be used in parallel [Rossen and Anderson, 1989].

15.4.2   Input-Oriented Computation of the First Hidden Layer

For classification or analysis of time-varying data, pertinent information is often accumulated in an incremental fashion. The difference between two successive values can be the important input to the network. If one is careful to choose an input-oriented algorithm for the computation of the first hidden layer, then a significant speed improvement can be attained. Rather than recompute the entire first hidden layer each time an input item is changed, simply subtract the old value of the changed input (multiplied by the appropriate outgoing weight) from the existing net input of each node in the first hidden layer. Then add the new value for the changed datum (again multiplied by the proper weight) to each hidden layer node to obtain the new net inputs for the first hidden layer.

In this way, the number of multiply-and-accumulate operations required to compute the new net inputs for the first hidden layer (assuming a fully connected network), when a single input item is changed, is reduced from n*m to 2m (n = number of inputs; m = number of first hidden-layer nodes). Since the input layer is often the largest layer, and the first hidden-layer is often the largest computing layer in the network, this method is likely to speed recomputation by a factor of two or better.

15.5   LEARNING RULE MODIFICATIONS FOR IMPROVED PERFORMANCE AND DECREASED TRAINING TIME

There are two basic ways in which we can speed up network learning. The first is to modify the learning rules. An example of this would be to modify the α term, which governs the degree to which a connection weight is adapted proportional to the delta term calculated by the back-propagation method. The other way is to work with the training data itself, presenting it in such a way that the network learns the necessary distinctions more readily. We consider these two related issues in this and the following subsection.

15.5.1   Causes of Non-Convergence During Training

When using an architecture with the minimal number of neurons, there is a chance that the back-propagation learning law will not lead to a convergent set of weights. Non-convergence can occur because the system may move in a direction which produces a lower total error than it previously had, but which does not yield the global error minimization. A useful back-propagation system should have a built-in detector to determine whether the total error is decreasing over a large number of iterations during training. If the error is not decreasing, then corrective action may be necessary (e.g., restarting with new random weights).

In addition to non-convergence, a network may converge more slowly if the outputs of the nodes are very close to 0 or one, that is, produced by the tail ends of a non-linear sigmoid transfer function. This corresponds to very small values of the derivative of the transfer function, as described in Chapter 7.

When a network fails to train satisfactorily, one of the first remedies that should be attempted is varying the number of hidden nodes. As mentioned above, the network error is a quadratic (bowl-shaped) function of the number of hidden nodes. Contrary, perhaps, to intuition, a decrease in the number of nodes is at least as likely to result in improvement as an increase [Kudrycki, 1988].

Another technique worth trying is to recondition the inputs. Transformations such as taking the logarithm, square root, or even the Fourier or Gabor transform, or doing a preliminary principal component analysis, may result in an improvement.

15.5.2   Factors Affecting Training Efficiency

As networks grow, learning speed can quickly become a critical factor. Many faster variants of the basic back-propagation algorithm have been presented [Fahlman, 1988a]. These variants should be explored if training time becomes a limiting factor in a neural network development project.

The learning factor alpha used in the basic back-propagation equations has a significant effect on learning speed and efficiency. A large learning factor results in more rapid learning, but makes the network subject to oscillation. A small learning factor stabilizes the process, but results in slower learning and increased susceptibility to entrapment in local minima. Beginning with a large learning factor, and gradually decreasing it as learning progresses, appears to give the optimum combination of rapid early learning with stability and resistance to local minima. Fahlman's [1988a,b] *quickprop* variant of back-propagation appears to automatically and continuously optimize the training parameter for each weight.

One of the most common methods to speed up learning is to use an additional "momentum" term in the weight change. The change in weight for a given connection then becomes equal to a sum of both the calculated weight change (using the General-ized Delta Rule) and a scalar times the previous weight change for that connection [Rumelhart, Hinton, and Williams, 1986]. Use of a momentum term helps to filter out high-frequency variations in the error surface in the weight space, and allows the effective weight steps to be bigger.

Particularly if the variation in fan-in between the layers is very large, it may be helpful to use a *split-training parameter* technique [Fahlman, 1988b]. In this technique, the training parameter of each node is made inversely proportional to the fan-in of that node. This helps avoid the problem of nodes with high fan-in being driven to extreme values.

The outputs of some nodes may tend to get stuck in the *flat spots* at the extremes of the sigmoid function. This is due to the fact that the derivative of the output function multiplies the error used in computing the weight changes. At the extremes of the sigmoid, the derivative is near 0, so only very small weight changes can occur. Fahlman [1988a] achieved a dramatic improvement in training time simply by adding 0.1 to the sigmoid derivative to prevent it from becoming 0.

Fahlman has also achieved additional improvements in training speed by replacing the error on the output nodes with the hyperbolic arctangent of the error. For small errors, this function is nearly linear, but approaches infinity as the error approaches 1.0. The error was clipped at +/-17 to avoid the approach to infinity. This technique permits large errors to have a disproportionate effect on the training, thereby moving more rapidly toward an approximate solution. This technique is likely to be effective only when the input patterns are free of noise. As White has pointed out (plenary lecture, IJCNN 90 San Diego), when the training data are noisy, large errors are likely due to noise. In such cases training may be improved by applying a function (e.g., the Cauchy function to the error that decreases, rather than magnifies, large errors.

A great deal of work has been done within recent years on devising ways to improve the performance of the back-propagation network.

15.6   MODIFICATIONS TO NETWORK TRAINING SCHEDULES AND DATA SETS CAN INFLUENCE NETWORK PERFORMANCE AND LEARNING SPEED

In order for a supervised feedforward network to make the fine category distinctions of which it is capable, it must be presented with hundreds, sometimes thousands, of training examples. These training examples may be a small set of patterns repeated many times, or a large set of patterns where each (slightly different) pattern may need to be presented less often.

15.6.1   Ways to Improve Network Training

It is important to present the training data in random order. Otherwise, response of the trained network may vary with the order of pattern presentation. Although this has interesting implications for dynamic learning, in most contexts it is undesirable.

Shaping is reported to increase the efficiency of learning by increasing the training data set in batches. Errors are reduced after the addition of each batch of training patterns, before the addition of the next batch. By this technique, the network is able to rapidly learn the rough shape of the decision space on a small data set. The training set is then enlarged to allow the network to refine its knowledge. This process seems beautifully analogous to the common pedagogical technique of teaching students a simple set of axioms or principles first, then gradually elaborating on their significance.

15.6.2   Training Data Requirements

Estimates of the exact number of training examples needed vary, but multiple examples of each pattern should be used [Kudrycki, 1988]. Otherwise the network may not learn to generalize. The number of examples required probably varies, depending on the complexity of the network and of the input patterns. Intuitively, it seems that the larger and more complex the input space of a particular pattern is, the more examples will be required to teach the network the boundaries within that space.

Several authors have given varying estimates of the number of training patterns required for a network to learn a function of the input vector. [See, e.g., Tishby et al., 1989]. This problem is distinct from the pattern recognition (classification) case of the previous paragraph. Stubbs [personal communication] found that the number of training patterns should be at least five times the number of nodes in the network.

In cases where large numbers of training cases are not readily available, some development teams have found it advantageous to generate artificial training examples automatically. One example in the ALVINN system [Pomerleau, 1989] for guiding a truck along a road. The local roadways did not provide sufficient variety of conditions for their video cameras and range finders. Their solution was to generate simulated data via computer. They obtained reasonable performance after training with this computer-simulated data.

The authors have recently obtained good results training a diagnostic network with only a very small number of actual training example patterns. A potentially infinite number of training patterns are randomly generated, and their expected outputs determined by comparison with the small number of example patterns. This technique achieves good generalization from a small number of examples, but training times tend to be long.

REFERENCES

Bartlett, E. (1990). Nuclear Power Plant Diagnostics Using Simulated Condensation — An Adaptive Neural Network Connectionist Technique, Ph.D. Dissertation, Dept. of Nuclear Engineering, The University of Tennessee, Knoxville, TN, August, 1990.

Bartlett, E., and Uhrig, R.E. (in preparation). "Simulated condensation: An auto-adaptive network with a dynamic architecture." For reprints contact Dr. R.E. Uhrig, Dept. Nuclear Engineering, University of Tennessee at Knoxville, Knoxville, TN 37996-2300.

Bounds, D.G., and Lloyd, P.J (1988). "A multilayer perceptron network for the diagnosis of low back pain," *Proc. Second IEEE Int'l. Conf. Neural Networks* (San Diego, CA, July 24-27, 1988), II-481–II-489.

Broomhead, D.S., and Lowe, D. (1988). Radial Basis Functions, Multi-Variable Interpolation and Adaptive Networks. Royal Signals and Radar Establishment Memo. 4148, Worcestire, United Kingdom.

Carpenter, G.A., Grossberg, S., and Mehanian, C. (1989). "Invariant recognition of cluttered scenes by a self-organizing ART architecture: CORT-X boundary segmentation," *Neural Networks, 2,* 169-182.

Cybenko, G. (1988a). "Continuous valued neural networks with two hidden layers are sufficient," preprint, Computer Science Dept., Tufts University.

Cybenko, G. (1989b). "Approximations by superpositions of a sigmoidal function," *Math. of Control, Signals, and Systems, 2 (4)*, 303-314.

Fahlman, S.E. (1988a). "Faster-learning variations on back-propagation: An empirical study," *Proc. of the 1988 Connectionist Models Summer School.*

Fahlman, S. E. (1988b). "An Empirical Study of Learning Speed in Backpropogation Networks. Carnegie Mellon Technical Report, CMU-CS-88-162, June.

Fahlman, S.E., and Lebiere, C. (1990). "The cascade-correlation learning architecture," in D. Touretzky (Ed.), *Advances in Neural Information Processing Systems 2*, Morgan Kaufmann, San Mateo, 524-532.

Fukushima, K. (1988). "Neocognitron: A hierarchical neural network capable of visual pattern recognition," *Neural Networks, 1*, 119-130.

Fukushima, K. (1989). "Analysis of the process of visual pattern recognition by the Neocognitron," *Neural Networks, 2*, 413-420.

Glover, D.E. (1989). "Optical processing and neurocomputing in an automated inspection system," *Journal of Neural Network Computing, 1* (2), 17-38.

Gorman, R.P., and Sejnowski, T.J. (1988). "Analysis of hidden units in a layered network trained to classify sonar targets," *Neural Networks, 1*, 75-89.

Harp, A.H., Samad, T. and Guha, A. (1989). The Genetic Synthesis of Neural Networks, Honeywell Technical Report CSDD-89-I4852-2, June.

Hecht-Nielsen, R. (1987). "Kolmogorov's mapping neural network existence theorem," *Proc. First IEEE Int'l. Joint Conf. Neural Networks* (San Diego, CA; June 21-24, 1987), III-11–III-14.

Kraiss, K.F., and Kuttelwesch, H. (1990). "Teaching neural networks to guide a vehicle through an obstacle course by emulating a human teacher," *Proc. Third Int'l. Joint Conf. Neural Networks,* (San Diego, CA, June 17-21, 1990), I-333–I-337.

Kudrycki, T.P. (1988). Neural Network Implementation of a Medical Diagnosis Expert System, Master's Thesis, College of Engineering, University of Cincinnati.

Le Cun, Y., Denker, J., and Solla, S.A. (1990). "Optimal brain damage," in D. Touretzky (Ed.), *Advances in Neural Information Processing Systems 2*, Morgan Kaufmann, San Mateo, 598-605.

Leung, M., Engeler, W.E., and Frank, P. (1990). "Fingerprint processing using back propagation neural networks," *Proc. Third Int'l. Joint Conf. Neural Networks*, (San Diego, CA, June 17-21, 1990), I-15–I-20.

Lippmann, R.P. (1987). "An introduction to computing with neural nets," *IEEE ASSP Magazine,* (April), 4-22.

Mel, B.W., and Koch, C. (1990). "Sigma-pi learning: On radial basis functions and cortical associative learning," in D. Touretzky (Ed.), *Advances in Neural Information Processing Systems 2*, Morgan Kaufmann, San Mateo, 474-481.

Moody, J., and Darken, C.J. (1989). "Fast learning in networks of locally-tuned processing units," *Neural Computation, 1*, 281-294.

Mozer, M.C., and Smolensky, P. (1989). "Using relevance to reduce network size automatically," *Connection Science, 1*, 3-16.

Nguyen, D., and Widrow, B. (1989). "The truck backer-upper: An example of self-learning in neural networks," *Proc. Int'l. Joint Conf. on Neural Networks* (Washington D.C., June 18-22, 1989), II-357–II-363.

Pao, Y.-H. (1989). *Adaptive Pattern Recognition and Neural Networks,* Chapter 8, Addison-Wesley, Reading, MA.

Poggio, T., and Girosi, F. (1990). "Regularization algorithms for learning that are equivalent to multilayer networks," *Science, 247,* 978-982.

Pomerleau, D.A. (1989). "ALVINN: An autonomous land vehicle in a neural network," in D. Touretzky (Ed.) *Advances in Neural Information Processing Systems 1,* Morgan Kaufmann, San Mateo, CA, 305-313.

Rossen, M.L., and Anderson, J.A. (1989). "Representational issues in a neural network model of syllable recognition," *Proc. First Int'l. Joint Conf. Neural Networks,* Washington D.C., (June 18-22, 1989), I-19-I-25.

Rumelhart, D.E., Hinton, G.E., and Williams, R.J. (1986). "Learning internal representations by error propagation," in D.E. Rumelhart and J.L. McClelland (Eds.), *Parallel Distributed Processing: Explorations in the Microstructure of Cognition, Volume 1: Foundations,* MIT Press, Cambridge, MA, 318-362.

Saha, A., and Keeler, J.D. (1990). "Algorithms for better representation and faster learning in radial basis function networks," in D. Touretzky (Ed.), *Advances in Neural Information Processing Systems 2,* Morgan Kaufmann, San Mateo, 482-489.

Saito, K., and Nakano, R. (1988). "Medical diagnostic expert system based on PDP model," *Proc. Second IEEE Int'l. Conf. Neural Networks,* (San Diego, CA, July 24-27, 1988), I-255-I-262.

Sejnowski, T., Kienker, P.K., & Hinton, G.E. (1986). "Learning symmetry groups with hidden units: Beyond the Perceptron," *Physica 22D,* 260-275.

Spirkovska, L., and Reid, M. (1990). "Connectivity strategies for higher-order neural networks applied to pattern recognition," *Proc. Third Int'l. Joint Conf. Neural Networks,* (San Diego, CA, June 17-21, 1990), I-21-I-26.

Stornetta, W.S., and Huberman, B.A. (1987). "An improved three-layer back-propagation algorithm," *Proc. First IEEE Int'l. Conf. Neural Networks,* (San Diego, CA, June 21-24, 1990), II-637-II-644.

Stubbs, D.F. (1990). *Personal communication.*

Tishby, N, Leven, E, and Solla, S.A. (1989). "Consistent inference of probabilities in layered networks: Predictions and generalizations," *Proc. First Int'l. Joint Conf. Neural Networks,* (Washington, D.C., June 18-22, 1989), II-403-II-409.

Tsoi, A.C. (1989). "Multilayer perceptron trained using radial basis functions," *Elec. Letters, 25,* 1296-1297.

Vrckovnik, G., Chung, T., and Carter, C.R. (1990). "Classifying impulse radar waveforms using principal components analysis and neural networks," *Proc. Int'l. Joint Conf. Neural Networks,* (San Diego, CA, June 17-21, 1990), I-69-I-74.

Waibel, A., Sawai, H., Shikano, K. (1989). "Modularity and scaling in large phonemic networks," *IEEE Trans. Acoustics, Speech, and Signal Processing, 37,* 1888-1898.

Werbos, P. (1988). "Generalization of back-propagation with application to a recurrent gas market model," *Neural Networks, 1,* 339-356.

Werbos, P. (1989). "Back-propagation and neurocontrol: A review and prospectus," *Proc. Int'l. Joint. Conf. on Neural Networks,* (Washington, D.C., June 18-22, 1989), I-209-I-216.

Yoon, YO., Brobst, R.W., Bergstresser, P.R., Peterson, L.L. (1989). "A desktop neural network for dermatology diagnosis," *Journal of Neural Network Computing, 1* (Summer), 43-52.

16

ELECTRONIC HARDWARE IMPLEMENTATIONS

Steven G. Morton

16.0 OVERVIEW

This chapter discusses integrated circuit implementations of neural networks. These implementations are divided into three categories: analog, digital, and hybrid. By analog, we mean chips whose signals are continuously varying, and that rely upon the linear behavior of transistors to act as multipliers and adders. By digital, we mean chips whose signals have a predetermined number, typically two, of stable states, and whose processing functions rely upon boolean operations. By hybrid, we mean the combination of analog and digital hardware in a single chip.

Since many neural network chips have already been described in the literature, and new neural network chips are appearing frequently, we will not make any effort to review them all. Instead, we will include a general overview by discussing a typical implementation for each of the three categories of neural network chips, describing the speed, storage capacity, fabrication technology, precision, modularity, etc., of each implementation. In addition, the relative advantages and disadvantages of each implementation will be reviewed. This overview sets the stage for our presentation of a tabular method for quantizing the capabilities of different implementations. This method should prove useful in comparing the capabilities, strengths and weaknesss of the widely varying implementations of both current and future devices.

16.1 ANALOG IMPLEMENTATIONS

Analog hardware implementations of neural networks are electronic circuits composed of a combination of linear and nonlinear devices. Transistors are the primary component, but parasitic capacitors and resistors are also found. All currents and voltages are evaluated over a continuum between two extreme values that are determined by the power supply voltages. The currents and voltages on circuit nodes internal to the circuit

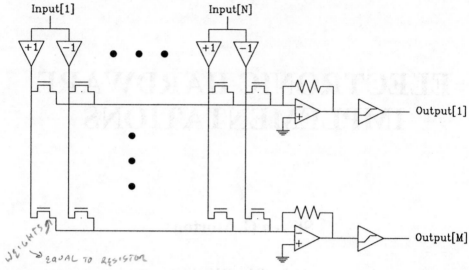

Figure 16.1 Basic analog neural network chip

are continuous functions of time that are generally dedicated to a single set of storage and processing nodes. Extremely small variations in these currents and voltages can thus produce measurable changes in the state of the devices. Such systems are thus continuous, nonlinear, dynamic systems. See Figure 16.1.

16.1.1 Advantages

Our objectives as scientists and engineers are to develop devices that perform useful functions and provide solutions that have benefits over alternative means. The notion that the human brain uses analog components in a massively parallel fashion and that we should therefore build chips that way, too, is beside the point. Birds have long been inspirations for building devices that fly, but mimicking their construction has never led to practical aircraft.

If analog chips are to be successful, then there must be sound economic and engineering reasons for doing so. There are, in fact, several sound reasons for building massively parallel, analog neural network chips.

First, a single transistor can perform a key function of a neural network, namely, multiplication. This operation requires many transistors in a digital implementation. In a MOS (metal oxide semiconductor) transistor, which is the type of transistor used in most integrated circuits, the current from one terminal (the source) is roughly proportional to the product of the voltage from a second terminal (the gate) to the first terminal, and to the voltage from a third terminal (the drain) to the first terminal. Since the precise relationship depends upon exponentials, the smaller the voltages are, the higher quality is the multiplication since a linear approximation to an exponential can be made. Over carefully chosen ranges of operating voltages, the accuracy of the multiplier can be held to several percent. Much more accurate multipliers can be built using bipolar transistors. A bipolar transistor can produce an output current that is proportional to the logarithm of its input. Operation over several orders of magnitude can be provided. Since the sum of two logarithms gives the product of the inputs, the currents from two transistors can be added and then an inverse logarithm computed to give the product. While this bipolar circuit is much more accurate than a MOS circuit, it requires far more transistors and does not provide for another critical element of a neural network chip, namely, the storage of a weight matrix that can be varied electrically. However, a chip fabricated from a combination of MOS and bipolar technologies would provide the best of both worlds, and such technology is becoming available economically.

Second, another key function in a neural network is addition. A simple wire, when held at a fixed potential, typically ground, can sum the currents from many multipliers. In contrast, an adder in a digital chip that handles multiple inputs simultaneously requires many transistors. Naturally, simply passing the sum of currents directly to ground does not accomplish anything, so the wire is fed to the inverting input, called a summing junction, of an amplifier. The output of the amplifier is connected to its inverting input via a resistor. When the sum of currents is flowing toward the input to the amplifier, the output voltage of the amplifier goes sufficiently negative to be able to exactly remove these currents. When the current flows away from the input, the output voltage goes positive. The value of the resistor thus sets the gain of the amplifier by determining the relationship between input current and output voltage.

Third, since there is presumably no switching of states in an analog neural network chip other than to convey changing inputs to the chip, no power is wasted charging and discharging capacitors.

In a digital chip, a single processing element must service multiple storage nodes sequentially. For each circuit node, the amount of power dissipated is proportional to the product of the capacitance and the square of the voltage to which the capacitor is charged. Such capacitors are everywhere in a chip: between wires, between the wires and the chip substrate, and especially at the gate input of each transistor. Although each capacitor within a chip is extremely small, the vast number of them adds up rapidly. Since the voltages in a digital chip must be several times the switching voltage of a transistor in order to give sufficient immunity to noise, substantial amounts of power are dissipated as the switching frequency increases.

Furthermore, a typical digital chip is constructed of CMOS (complemetary MOS), which places two transistors at the output of each gate, one connected to ground to provide a low potential and the other connected to a positive supply (typically five volts) to provide a high potential. Ideally, only one of these two transistors is on at a time. Unfortunately, both are on briefly as a transition is made from one voltage state to the other. This provides a brief, direct path from the positive supply to ground, limited by the current-carrying capabilities of the two transistors. As the switching frequency increases, an ever larger fraction of time is spent with both transistors conducting and thus wasting ever more power.

Fourth, neural network chips require the storage of a weight matrix. If the chip is to implement an adaptive algorithm, then the values of the weights must be electrically variable. MOS chips have the marvelous ability to store an electrical charge where one wants it, not just under every single wire. This charge can set the potential at the gate of a MOS transistor. This transistor can be used as a multiplier, where varying the charge and thus the potential changes the multiplication factor.

A charge can be stored for tens of milliseconds at the otherwise unconnected output of an MOS transistor. This design provides dynamic random access memory chips which are manufactured in vast quantities and are found in every computer. Since the charge is connected to an output of a transistor, and a MOS transistor can both supply and remove current, the charge can be changed quickly from one extreme state to another, in as little as a few nanoseconds. This output can, of course, be connected to the gate of a transistor for sensing, which is often done in so called "dynamic flipflops."

Alternatively, a charge can be stored for very long periods of time, for tens of years or more. This charge is stored at the gate of a MOS transistor to which nothing else is connected. Such a gate is called a "floating gate." Since the electrode at the gate of a MOS transistor is isolated from all other electrodes by a thin layer of silicon dioxide (glass), the impedance is extremely high, so the charge leaks off very slowly even though the amount of capacitance is extremely small.

The problem, then, is how to manipulate the charge since there is no connection to it. A quantum mechanical effect called "tunneling" is used. When the thickness of the oxide is sufficiently thin, much thinner than it would be for a conventional MOS transistor, electrons will, in effect, move through the insulator when the voltage across it is sufficiently high. Since the tunneling current is extremely small, the amount of time required to make the maximum voltage transition can be relatively long — several milliseconds.

In early "EPROM" (electrically programmable read only memory) chips, all of the floating gates would be initialized to a common state by exposing the chip to ionizing radiation in the form of ultraviolet light. The radiation would cause currents to flow in the chip, and these currents would be captured by the floating gates. A high potential would then be imposed across each floating gate, in turn, of those transistors where the opposite state is wanted. More recent techniques allow many floating gates to be initialized electrically at one time.

In a neural network chip, one must be able to both increase or decrease the value of each element of the weight matrix without starting the whole chip over each time. This requires the use of additional transistors for each storage node. By varying the duration of the tunneling current, one can make small changes in the potential on a floating gate and thus make small changes in the multiplication factor provided. One could also vary the potential across the floating gate in order to vary the charging rate, but several volts are required for tunneling to occur at all, and it is much easier to varying the charging time by several orders of magnitude rather than the charging voltage.

Fifth, massive parallelism can be provided in an analog neural network chip. If a memory array has tens of thousands or more storage cells, each controlling a transistor that acts as a multiplier, then vast numbers of multiplications can presumably be performed at once. Each row of transistors can be wired together to sum their currents, providing the fundamental, sum-of-products operation that is so common in neural networks. This provides a matrix-vector multiplier, where an input vector is multiplied by a weight matrix.

If the signal bandwidth is a relatively low 1 MHz and there are 10,000 storage cells and thus as many multipliers, then the equivalent of 20 billion or more multiplications can be performed per second. The precise number depends upon the effective sampling rate, with at least two samples being required per cycle. Since memory chips with millions of bits are common, then one would suppose that more than a trillion multiplications per second per chip are feasible. This is far, far beyond the capability of any digital chip.

Other forms of analog neural network chips besides matrix-vector multipliers can be built as well. For example, linear amplifiers can be arranged in a checkerboard, or mesh, fashion, and connected to their nearest neighbors. Photosensors can be distributed throughout the mesh and be connected to the amplifiers. When an image is shown upon the chip, the chip can detect motion. Such circuits are being pioneered by Carver Mead of Caltech, who is also one of the fathers of modern integrated circuit design methodologies.

Sixth, fault tolerance can supposedly be provided by massively parallel, analog chips. However, there has not been any convincing demonstration that this is true. In the absence of substantial redundancy in the circuit, which can be provided in digital chips as well, any number of single point failures will still render the chip useless. For example, if a wire that connects one of the inputs to a column of multipliers is open or shorted, then all or part of that column of multipliers will malfunction. Many products will thus be in error, causing errors in many sums of products. If a defective product is at a maximally positive or negative value, then it will overshadow many products that are at smaller values and are operating properly.

Seventh, certain forms of non-linear behavior can be provided easily where desired. For example, a unipolar sigmoidal transfer function can be provided by a simple diode and an amplifier. This function requires far more transistors when implemented by a table lookup in a digital circuit.

Eighth, most sensors provide analog outputs. These outputs can be directly connected to the inputs of analog neural network chips. This avoids the use of analog-to-digital convertors as would be required with digital neural network chips.

Figure 16.2 summarizes the advantages and disadvantages of analog neural network chips.

16.1.2   Disadvantages

Unlike digital chips, whose capability is limited to a large extent by how small the transistors can be made, the capability of analog chips is limited by basic device physics. Transistors are continually being made smaller as photolithography improves, but the physics of the devices are not changing unless the devices are extraordinarily small.

First, a major disadvantage of analog neural network chips is the lack of thermal stability. Unlike the human brain that is kept at a constant temperature of 98.6 degrees F (37 degrees C), electronic circuits must operate over wide temperature extremes. The operating range for common, commercial-grade chips is from 0 to 70 degrees C. The operating range for industrial-grade chips is -40 to +85 degrees C, and the range for military-grade chips is -55 to +125 degrees C.

Since the transfer function of a transistor is related to the exponential of the temperature of the device, each 1 degree C change in temperature produces the equivalent of about a 2 mV change in applied voltage. Since there are at most a few

Analog Neural Network Chips

ADVANTAGES	DISADVANTAGES
• High Noise Immunity	• Lack of Thermal Stability
• Speed of Operation	• Noise
• Precise Computation	• Interconnection Problems
• Readily Designed Using Existing Tools	• Limited Accuracy
• Can Include Programmable Components	• Difficult to Test
• Easy to Multiplex / Demultiplex	• Hard to Build Simple Components (resistors & capacitors)
• Can Store Both Fixed and Adaptive Weights	• Lack of Design Tools
	• Non-Uniform Processing
	• Difficult to Mass Produce Chips with Predetermined Weight Matrices

Figure 16.2   Advantages and disadvantages of analog neural network implementations

volts applied to the gate of a MOS transistor, substantial changes in the effective applied voltage can occur, creating large changes in the operation of the device.

This problem is virtually eliminated in analog devices such as analog to digital convertors that have a high gain amplifier and use negative feedback from the output of the amplifier back to the input. Unfortunately, such a circuit requires far more transistors than the single transistor multiplier that on the surface appears so attractive, and such a circuit requires that the nonlinearity be placed in the feedback loop if compensation is to be provided.

In addition, the speed of transistors depends upon their temperature. As MOS transistors are cooled, their gain and speed increases. As bipolar transistors are cooled, their gain and speed decreases. The operation of MOS circuits at liquid nitrogen temperatures has been proposed, but much more modest cooling, such as to 0 degrees C, has produced useful improvements. These speed changes occur in analog as well as digital chips, but present a more severe problem in analog chips where the frequency response of amplifiers, which depends upon the gain of their transistors, must be carefully controlled to provide stable behavior.

Furthermore, there is no way to place a single floating gate in a feedback circuit to compensate for changes in temperature. At best, one can use a pair of floating gate circuits in the hopes that the two will track each other with temperature and any differences will balance out. This technique is widely used. In practice, however, linearity errors better than several percent are hard to achieve.

The second serious problem in analog chips is noise, and there are many sources of it. Since an analog neural network chip is, by definition, responsive to a continuum of input voltages, then every little change in the inputs to the chip will cause a change in the operation of the chip, so long as the outputs are not at their limits. These changes can occur because signals couple into one another through stray capacitance on the circuit board that bears the chips, and within the chip itself. The frequency response of amplifiers in the chip may have to be reduced to minimize the frequencies present on the chip and therefore reduce the coupling.

Floating gates, which are essential for storing analog weight matrices, are also great sources of noise. In a digital chip, where signals have only two distinct states, the potential on the floating gate would be far from the threshold voltage (the voltage that causes a change in the state) of the transistor. Thus, a small additional noise voltage has no effect. This is not the case in the linear circuit of an analog neural network chip.

The substrate of a chip also couples many signals together. Even if the back of the substrate is metalized and grounded, the distance from the rear of the substrate to any transistor is 100 to 1,000 times greater than the distance between any two transistors, and is thus relatively ineffective at isolating them. Early dynamic random access memories particularly suffered from this problem.

Another noise source results if there are any outputs from the analog neural network chip that change rapidly and drive much capacitance. Such capacitance is found in traces on printed circuit boards. The inductance of the bonding wires that provide power to the chip induces voltages across these wires, and these voltages cause apparent changes in the input voltages to the chip. Potentially, the chip can oscillate if a change in output voltage feeds back to the input and exacerbates the change in the output voltage, and so on.

Storing weight matrices on floating gates has another problem. With a single floating gate per matrix element, only matrices without signs can be stored because the transistor that senses the potential on the floating gate only works with one polarity of gate input. As a result, a pair of floating gates and sense transistors is required per matrix element, where one pair multiplies a positive input and the other pair multiplies a negative input. This requires an inverting amplifier on each vector input. Furthermore, the behavior of the sense transistor as a multiplier varies depending upon whether the voltage across it is positive or negative.

Interconnecting analog neural network chips is another problem. If fully connected networks are required and a single wire is dedicated to each input or output, then many wires are required. This is not so much of a problem within a single chip because wires can be very small and close together. However, there are practical limits as to how large integrated circuits can be, generally about 10 mm on a side, so it must be possible to connect multiple chips together to provide more capability than a single chip provides alone.

If the chip has a 256-row by 256-column weight matrix, providing only 65,536 weights, then it needs 256 inputs and 256 outputs, plus power, ground and any control signals. This many inputs and outputs requires a package with over 512 pins, which is very expensive. Since bonding wires, which connect each pin on a package to a signal in a chip, are typically connected to bonding sites, or pads, that are placed around the perimeter of a chip at about .25 mm center-to-center spacing, then the size of the chip would have to be over 32 mm on a side just to provide enough space for all of the wires. An alternative is to distribute bonding pads over the entire chip and to simultaneously solder these many pads to corresponding pads on a package. This is called a "flip chip," but it has been done only on a much smaller scale, and at great expense.

In digital chips, many signals are carried by a single wire in sequence. This is called multiplexing and places demands upon the circuitry driving the wire to charge and discharge any capacitance on the wire quickly. It also requires that circuits that receive the data can correctly capture only the data that pertains to them.

If multiple analog signals are multiplexed together, then they must likewise be sampled and held anywhere that a single signal is desired. In addition, the frequencies in each signal must be constrained, and the sampling frequency must be sufficiently high, that a faithful rendition of each signal is obtained. Furthermore, the circuit (the "multiplexer") that places multiple signals on a single wire must be sufficiently fast, and enough time must be provided between samples, so that sufficient accuracy is provided when the combined data is sampled.

Any multiplexed outputs from a chip are thus charging and discharging the capacitance of wires external to the chip. The capacitance of a wire external to a chip is 100 or more times the capacitance of a wire within a chip. Thus currents must flow rapidly in the power and ground wires of the chip, causing noise to appear in the very inputs to the chip that the chip is trying to capture. One can provide many power and ground inputs to a chip, and try to isolate different portions of a chip from one another, such as input portions from output portions, but these various circuits on a chip must share a common ground connection so that information can be transferred between them.

The accuracy of analog neural network chips is quite limited because of the many sources of error within them. For these chips to be successful, algorithms are required that can give good results without requiring the overall computations to be much better than about one percent. These errors limit the number of inputs that can be combined.

For example, in a matrix-vector multiplier, multiple products are computed and then summed. If the accuracy of each of the products is low, then summing large numbers of products would be pointless since the contributions from small products will be hidden by the errors in large products.

While it is often argued that the human brain does a marvelous job of recognizing patterns and controlling behavior using analog components, and that analog chips should therefore not need much precision either, it is often desirable to build electronic systems that can outperform their human counterparts.

Testing massively parallel analog chips is difficult. In a digital chip, each circuit node has only two states, and nodes can be viewed accurately via other nodes. In an analog chip there is no set of states since voltages are continuous, so it may be necessary to test the linearity of each device by varying its inputs over their operating ranges. This presupposes that the operation of the nodes is independent. This is certainly not true when many currents are being summed into a wire whose resistance causes a voltage to be built up along its length; this voltage adversely affects the operation of the multipliers connected to it. In addition, one does not have direct access to the nodes within the chip, so results can only be viewed via intermediate nodes, yet the intermediate nodes introduce their own errors.

Another problem is that simple components like resistors and capacitors are hard to build on chips. Resistors are needed for setting gains, and capacitors are used for controlling the frequency response, of amplifiers. While discrete resistors are made from carbon, carbon cannot be integrated with chips. Instead, one uses a semiconductor with an ion implant to control its resistance. This produces a transistor with an unconnected gate, hence its behavior depends upon the voltage across it and is nonlinear. Furthermore, while one can control the ratios of resistors fairly well (within 5-10%) by controlling their areas, one cannot control their absolute values well at all. Variations of several-to-one between different chips are common. One can control the ratio of areas even more by increasing the physical size of each resistor, so that small variations are less significant, but this increases the size of the chip.

Capacitor values that are taken for granted with discrete components, such as 100 pF, would take a large amount of area on a chip and are impractical. Typical values are far smaller, such as 0.1 pF. Furthermore, if one builds a capacitor by placing silicon dioxide between two conductors, then one risks the possibility that tiny holes in the oxide will cause the capacitor to short out when the upper layer of metal is deposited on the oxide and flows through the holes.

Another problem with massively parallel analog neural network chips is the lack of design tools. While a regular design can be laid out by defining one module and then replicating it, simulating the behavior of the chip is very difficult. With digital chips, one can simulate entire chips and even systems of chips at the switch level, which models each transistor as either on or off. With analog chips, one must work at the level of the device physics of the transistors, which requires an enormous amount of computation even for a few transistors. The most widely used analog simulator is Spice, which was originally developed at the University of California at Berkeley, but without a supercomputer, one rarely wants to simulate more than about 1,000 transistors because it takes too long even for simple tests. Testing neural network algorithms using Spice is out of the question. Furthermore, there is no way to model interactions between transistors if the governing equations are not known.

Thus if one simulates a neural network algorithm on a computer assuming that a chip that implements it will behave linearly, then there is no guarantee that the results of the simulation will match the results of the actual chip.

In addition, analog devices suffer from non-uniformities in their processing. A transistor in one corner of the chip may have a gain that varies by 10% from the gain of a presumably identical transistor elsewhere on the same chip. Variations between transistors in different wafers can vary by much more, as much as four-to-one. Thus the behavior of a multiplier has an additional scale factor that is poorly controlled. These non-uniformities also exist in digital chips but are not directly apparent except, for example, that the chip runs at a clock frequency that is determined by its slowest part.

Finally, in view of all these facts, if it is important to achieve a particular value of an element of a weight matrix using charge stored on a floating gate, then one must place the floating gate in a feedback loop. One must inject or remove a little charge from the floating gate and then apply a test voltage to the sense transistor to give an indirect measure of the stored value. This sequence is performed again and again until the desired response is achieved. It is not possible to mass produce analog chips with predetermined weight matrices.

16.1.3   Example of an Analog Neural Network Chip

The Intel M64 Electrically Trainable Artificial Neural Network Chip has a storage array of 64 rows by 128 columns of floating gate synapses. Each synapse has two floating gates, one to provide a positive weight and the other to provide a negative weight. Both work together to improve the linearity and temperature stability of the multipliers. Bias circuits are also provided to improve the operation of the multipliers.

The chip accepts 64, bipolar, analog inputs which are fed to the first 64 columns of the weight matrix. It internally produces 64 bipolar sums-of-products that are fed back to the second set of 64 inputs to the weight matrix. Each of the 64 sums-of-products is passed through a threshold circuit, giving 64 binary outputs.

The weights are changed by addressing them one at a time and causing a charge or discharge current to flow. Depending upon the extent of the change desired, anywhere from a few microseconds to a millisecond may be required. Based upon the size of the change required, one can estimate the time required to make most of the change, but the chip must be operated in a feedback loop to complete the adjustment.

According to the specification sheet available in June 1990, the multipliers are linear to several percent, and the accuracy of the overall computation is seven bits. The long term storage accuracy of the weights is about four bits, although extremely small adjustments can be made. Data flows through the chip from input to about in under three microseconds. Using the entire weight matrix, this provides performance of more than 2.4 billion connections per second. The specification sheet is marked, "Experimental," which is unusual in the semiconductor industry. It is customary to mark data sheets as "advance information," "preliminary" or "final."

16.2   DIGITAL NEURAL NETWORK CHIPS

Current digital neural network chips use circuitry that has only two states — on or off. They realize various Boolean functions, such as AND, that enables one to multiply two 1-bit values. The switching of states occurs at discrete points in time that are controlled by a clock signal that is provided to the chip. See Figure 16.3.

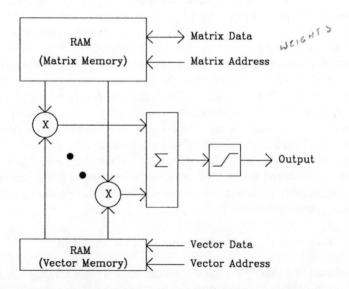

Figure 16.3   Basic digital neural network chip

Memories within these chips may in fact be built from analog circuits to give small size, high speed and low power, but are still switching between only two states.

16.2.1   Advantages

One of the many advantages of digital circuits is their high noise immunity. Since there are only two logic states, the circuits can tolerate noise that is a substantial fraction of the voltage difference between those two states. While this voltage difference depends upon the type of circuitry used, a half of a volt or more of noise can easily be tolerated by CMOS chips that have a supply voltage of five volts.

Another advantage is the speed with which a single operation can be performed. Clock rates of 40 MHz and more are common in CMOS chips. Bipolar chips can have clock rates of several hundred MHz, and gallium arsenide (GaAs) chips can have clock rates of several GHz.

Digital chips can perform computations precisely. Any amount of precision can be provided. The accuracy of the computations is independent of the physical size of the transistors. The results of computer simulations of algorithms will precisely match the behavior of physical hardware if comparable precision is used throughout.

Digital circuits may be rapidly, easily, and reliably designed and with less skill required on the designer's part than is the case for analog circuits. Much more mature design tools are available for digital chips than for analog ones.

It is easy to include programmable components in the digital system design. This allows the operation of the system to be determined by software without changing the hardware configuration.

Multiple signals can easily be multiplexed and demultiplexed, drastically reducing the number of wires required to provide complex network topologies. This allows the complexity of systems to increase by adding more components, rather than being limited by the number of connections from a single device. Many memory technologies are available to store weight matrices and input vectors. If nonvolatility is desired, then identical copies of weight matrices that are never changed can be mass produced very inexpensively using photomasks at the time of chip manufacture. Electrically programmable read only memory can be used if occasional changes are required. If nonvolatility is not required, then dynamic random access memory (DRAM) technology can be used if a large storage capacity, high density of memory cells and fast update rates are desired, so long as the operation of the accompanying logic does not upset the memory. Static random access memory (SRAM) technology can be used if faster speed is desired and density is less important; it has the advantage of being less noise sensitive than DRAM technology.

Figure 16.4 summarizes the advantages and disadvantages of digital neural network implementations.

16.2.2   Disadvantages

A fundamental disadvantage of digital neural network chips compared to analog ones is power dissipation. Since digital chips can operate at high clock frequencies, it is natural to share one processing element over multiple inputs or matrix elements. This

Digital Neural Network Chips

ADVANTAGES	DISADVANTAGES
• High Noise Immunity	• Requires More Transitors to
• Speed of Operation	Implement Fundamental Operations
• Precise Computation	• High Power Dissipation Rates
• Readily Designed Using	• Requires Analog-to-Digital
Existing Tools	Conversions of Most Incoming
• Can Incude Programmable	Sensor Data
Components	
• Easy to Multiplex / Demultiplex	
• Can Store Both Fixed and	
Adaptive Weights	

Figure 16.4   Advantages and disadvantages of digital neural network implementations

sharing requires that the capacitance in many circuit nodes be charged and discharged, consuming power. Since analog chips do not share processing elements, this power is not consumed.

Digital chips require more transistors to implement fundamental operations such as multiplication and addition than analog chips. This allows analog chips to have more processing elements than digital chips, enabling them to compensate for their lower useful speed of individual components. (While the transistors may be very fast, their useful speed is much less since time must be allowed for the results to become stable.) However, care must be taken in the analysis of speed because the precision of the two types of chips is usually radically different. An analog chip with the equivalent of 4-bit inputs and 4-bit weights, and having 64 processors, may only realize seven bits of precision — less than the precision of a single perfect multiplication. A typical digital chip would realize fourteen bits of precision — one hundred times better.

Most sensors are analog, and therefore an analog to digital conversion is required to use a digital chip. Analog chips avoid this conversion UNLESS the sensor is distant from the processing chips, in which case it is more reliable and uses less cabling to digitize the data for transmission.

16.2.3   Example of a Digital Neural Network Chip

The Oxford Computer "64K-by-1b / 64p Intelligent Pattern Recognition Memory (IPRM) Chip" is shown in Figure 16.5. It is a matrix-vector multiplier and a 2-D convolver. It stores one bit of each element of a 1024-row by 64-column weight matrix, and multiple bits of each element of a 64-element input vector. A wider matrix with fewer rows can also be handled. Multiple matrices and input vectors can be stored at

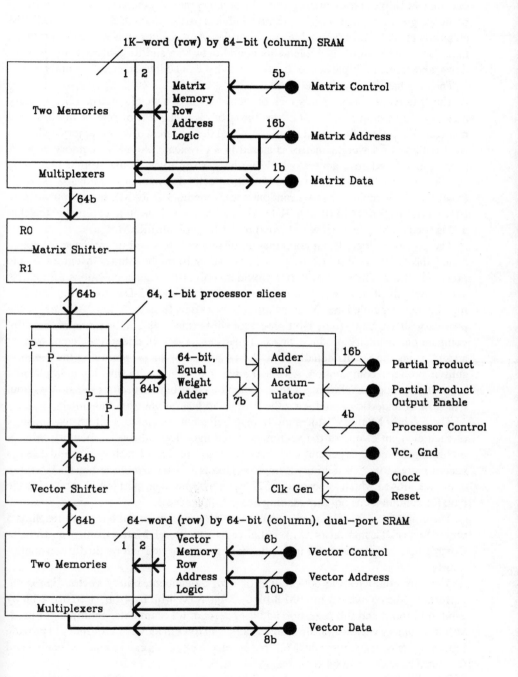

Figure 16.5   Intelligent pattern recognition memory chip. Used with the permission of Oxford Computer, Inc., Oxford, CT.

one time. Multiple chips work together like ordinary memory chips to give any precision of the weight matrix. An additional chip combines partial products from multiple IPRM chips and provides programmable nonlinearities. Any vector precision can be provided under software control. It implements neural networks and 8-by-8 convolution windows, and stores multiple sets of coefficients that can be interchanged instantly.

The chip has three main sections:

The "Matrix Memory" stores 1-bit of each element of a weight matrix. Like ordinary memory chips, chip number N stores bit N of each element of one or more weight matrices. Each row of a weight matrix is stored in a physical row of memory cells, and each column of a weight matrix is stored in a physical column of memory cells. A programmable address generator selects the desired rows of the weight matrix by sequencing through the desired rows of memory cells. The memory is built as a 2K-word (row) by 32-bit (column) static random access memory (SRAM), but it appears as a 64K-word by 1-bit RAM (the "64K-by-1b" in the name of the chip) external to the chip and is used on-chip as a 1K-word (row) by 64-bit (column) RAM.

The processor logic has a combination of two registers and interconnection logic, called the "Matrix Shifter," to pass information from the Matrix Memory to the processor slices. The Matrix Shifter provides straight-through connections for matrix-vector multiplication as well as moving windows for 8-by-8, 2-D convolution. Another pipeline register called the "Vector Shifter" passes data from the Vector Memory to the processor slices. Sixty-four, 1-bit processor slices (the "64p" in the name of the chip) compute either binary products or exact matches. A 64-bit, adder, the "equal-weight adder," sums the results equally from the 64 processor slices, providing the familiar, sum-of-products computation. An adder and accumulator, with shift logic, implement the common add and shift algorithm, to handle eight bits of vector precision without overflow or underflow, producing the Partial Product output from the chip.

The "Vector Memory" stores multiple, 64-element vectors to 8-bit precision. Bit N of each element of one of the vectors is stored in a physical row of memory cells. A programmable address generator sequences through the bits of each vector, and through successive vectors. The memory is implemented as a 64-word (column) by 64-bit (row), dual-port SRAM which appears as a 512-word by 8-bit, dual-port RAM external to the chip for simultaneous on-chip reading and off-chip access.

The chip has a 20 MHz maximum clock rate over commercial temperatures, uses 5 volts and has TTL interfaces. Once initialized via the Reset, Matrix Control and Vector Control signals, matrix-vector multiplications are performed automatically and continuously.

The chip computes the dot-product of two, 64-element, binary vectors in 50 nS, performing the equivalent of 20M 8-bit by 8-bit, 80M 4-bit by 4-bit, or 320M 2-bit by 2-bit, multiplies and full precision adds per second. It accumulates multiple dot-products from binary vectors that are stored in adjacent rows in its Vector Memory to provide input vectors with a gray-scale. Multiple 64-element vectors can be chained end-to-end for matrices wider than 64 columns.

The chip is built in 1-micron CMOS and is packaged in a 132-pin, pin-grid-array. Many pins are unused. Additional memories are on-chip for test purposes but are unused.

A second generation chip will have 256, 1-bit processors and operate at 40 MHz. It will provide 1.28 billion, 8-bit by 8-bit, to 10.24 billion, 1-bit by 8-bit, multiplies and full precision adds in a multi-chip module that contains eight of these chips and a ninth chip of a different type that combines the outputs and has programmable nonlinearities.

16.3 HYBRID NEURAL NETWORK CHIPS

Hybrid neural network chips combine analog and digital techniques to provide the best features of each. For example, one may combine a digital memory with digital, binary multipliers, but perform the sum and threshold functions using analog circuitry. This avoids the storage problems of analog floating gate technology, and avoids the many gates required by a digital circuit to build an adder that sums the many products. Comparing the sum to a threshold provides a binary output, which has a good noise margin, rather than providing an analog output or using an analog to digital convertor to provide a multibit digital output.

16.4 METHOD FOR COMPARING NEURAL NETWORK CHIPS

Using our own mental neural network, we have deduced a set of features that can be used to compare the capabilities of neural network chips. We believe that this set of common metrics will help provide a more thorough comparison of neural network chips.

We have created a table whose columns reflect these features and whose rows describe different chips. We have selected the chips from several vendors and filled in this table using the specifications published by them. We apologize for any errors on our part. Since new chips are being introduced and specifications are improving all the time, we encourage the reader to prepare his or her own table with more recent data.

We have left out price and availability information because they do not reflect fundamental measures of the chips and because they are too volatile. One might argue that price is a useful measure of the difficulty of implementing a device, but where production volumes are still small, it more likely reflects the desire of the chip company to recover a portion of its research and development costs.

Detailed fabrication parameters, such as die size, design methodology, and semiconductor fabrication technology, could also be included, but this information is often difficult to obtain from the vendor. Information on design methodology is relevant if one is trying to prepare a figure of merit for the design, such as operations per square millimeter, for a given feature size. A vendor may choose to use gate array technology, which can be designed quickly and has relatively low nonrecurring engineering costs but gives the largest die size and slowest performance, or standard cell technology which is slower and more expensive to design but gives a smaller die size and better performance, or a full custom design which is very time-consuming and expensive to design, but gives the smallest die size and highest performance.

The criteria we have used are these:

Table 16.1   Comparison of Neural Network Chips

Technology	Inputs	Outputs	Matrix Memory

Intel
M64 Electrically Trainable Artificial Neural Network Chip

analog, 1 um CMOS, full custom design	64 direct analog inputs, signed, 4-bit resolution	64 direct analog outputs after thresholds	64 rows by 128 columns, floating-gates-signed, 4-bit precision long term, more short term; external random access for loading

Oxford Computer
64K-by-1b/64p Intelligent Pattern Recognition Memory Chip

digital, 1 um CMOS, standard cells	8-bit wide path, 64 elements per vector, time-division multiplexed, signed or binary, programmable precision	16-bit wide path, time division-multiplexed, up to 1024 elements, signed	1024 rows by 64 columns;1b per element per chip, signed; dual-port static RAM; reconfigurable width; stores multiple matrices at once; external random access for loading

Vector Memory	Processing Function	Run-time Addressing	Connections per Second

Intel
M64 Electrically Trainable Artificial Neural Network Chip

none	matrix-vector multiply	none - entire weight matrix is used	2+ billion at 7 bits overall

Oxford Computer
64K-by-1b / 64p Intelligent Pattern Recognition Memory Chip

512 bytes dual-port double-buffered; external random access for loading	matrix-vector multiply; sum of products and sum of XORs; 2-D convolution	fully programmable sequencing through matrix and vector memories	20M @ 8b x 8b multiplies, 80M @ 4b x 4b multiplies, 160M @ 2b x 2b, multiplies, full precision sums

Table 16.2   Comparison of Neural Network Chips (continued)

Transfer Function   **Notes**

Intel
M64 Electrically Trainable Artificial Neural Network Chip

selectable, including sigmoid	Size of network is limited to 64 x 128. 64 internal outputs are fed back to input. Multipliers are compensated for improved linearity (several percent) and stability. Loading of weights requires iteration.

Oxford Computer
64K-by-1b / 64p Intelligent Pattern Recognition Memory Chip

linear or truncation	Matrix width is programmable in multiples of 64 columns. Multiple chips handle any matrix precision and size, multiple cycles handle any vector precision and size. Requires external chip to combine outputs from multiple chips and to provide nonlinear transfer functions. Handles multiple layers per chip. Loading of matrix and vector is simultaneous with computation.

- Technology. This is analog, digital, or hybrid, and specifies any particulars about it.
- Inputs. The number of elements in the input, their precision, and whether they are signed or not.
- Outputs. The number of elements in the output, their precision, and whether they are signed or not.
- Matrix memory. The capacity, organization, precision and technology of the memory that stores the weight matrix.
- Vector memory. The capacity, organization, precision and technology of the memory, if any, that stores the input vector.
- Processing function. The mathematical operation performed and whether it is fixed or programmable. Any hard-wired learning algorithms implemented by the chip would be stated here.
- Connections per second. The number of multiply-adds per second, and the precision. The combination of one multiply and one add counts as one connection. Wherever possible, the precision of each multiply, as x-bits times y-bits, and the precision of the final sum-of-products are given separately.
- Run-time addressing. The way the matrix and vector memories (if any) are addressed from within the chip while the chip is performing computations.
- Transfer functions. The way computations are modified, if at all, before being passed to the output of the chip.

Using these criteria, a comparison of the Intel and Oxford Computer chips described herein is given in Table 16.1.

16.5 SUMMARY

We have reviewed in detail the advantages and disadvantages of analog and digital neural network chips from an engineering perspective. We have avoided broad generalities and have addressed specific implementation issues. We have described an analog chip and a digital chip to show how two companies have addressed these issues with their own commercial designs. We have provided a quantatative, tabular framework for comparing existing and future chips.

FURTHER READING IN NEURAL NETWORK HARDWARE IMPLEMENTATION

Akers, L.A., Ferry, D.K., Grondin, R.O., Synthetic Neural Systems in VLSI. In S. Zornester, J.L. Davis, C. Lau (eds.), An Introduction to Neural and Electronic Networks, Academic Press, San Diego, CA, 1990.

Aleksander, I., Thomas, W. V., Bowden, P.A., "Wisard: A radical step forward in image recognition," Sensor Review, Vol. 4, No. 3, 1984, pp. 120-124.

Aleksander, I., (Ed.), Neural Computing Architectures: The Design of Brain-Like Machines, MIT Press, Cambridge, MA, 1989.

Caianiello, E., "Is There a silicon way to Intelligence?," IEEE Micro, December 1989, pp. 75-76.

Coon, D., A. Perera, "New hardware for massive Neural Networks," in Neural Inf. Proc. Systems, AIP, New York, 1988, pp. 201-210.

Cotter, N.K. Smith, M. Gasper, "A pulse width modulation design Approach and programmable logic for artificial Neural Networks," in Proc. 5th MIT Conf. Advanced Research in VLSI, MIT, Cambridge, 1988, pp. 1-15.

Cushman, B., "Matrix crunching with massive parallelism," VLSI Systems Design, December 1989, pp. 18-32.

Defense Advanced Research Projects Agency (DARPA) Neural Network Study, government printing office, Washington, DC, 1988.

Faggin, F., Mead, C., VLSI Implementation of Neural Networks. In S. Zornetzer, J.L. Davis, C. Lau (eds.), An Introduction to Neural and Electronic Networks, Academic Press, San Diego, CA, 1990.

Goser, K., U. Hilleringmann, U. Rueckert, K. Schumacher, "VLSI Technologies for Artificial Neural Networks," IEEE Micro, December 1989, pp. 28-44.

Graf, H.P., Jackel, L.D., "Analog electronic Neural Network circuits," IEEE Circuits and Devices Magazine, Vol. 5, No. 4, July 1989, pp. 44-49.

Graf, H. P., Jackel, L. D., Hubbard, W. E., "VLSI Implementation of a Neural Network Model," Computer, Vol. 21, No. 3, March 1988, pp. 41-51.

Hamilton, A., A. Murray, L. Tarassenko, "Programmable analog pulse-firing Neural Networks," in Advances in Neural Infomation Processing Systems I, Morgan Kaufmann, San Mateo, 1989, pp. 671-677.

Holler, M., Tam, S., Castro, H., Benson, R., "An Electrically Trainable Artificial Neural Network (ETANN) with 10240 Floating Gate Synapses," Proc. International Joint Conf. on Neural Networks, Washington, DC, Vol. II, June 18-22, 1989, pp. 191-196.

Hutchinson, J., Koch, C., Luo, J., Mead, C., "Computing motion using analog and Binary resistive Networks," Computer, Vol. 21, No. 3, March 1988, pp. 52-64.

Intel Corporation, "Specification for an Electrically Trainable Artificial Neural Network (Experimental)," 1990.

Johnson, R. C., "Neural Microcopies Debut," PC AI, January/February 1990, pp. 38-41.

Kampf, F., Koch, P., Roy, K., Sullivan, M., Delalic, Z., DasGupta, S., "Digital implementation of a Neural Network," Electrical Engineering Dept., Temple University, Philadelphia, PA, June 1989.

Kauffman, S.A., Principles of adaptation in complex systems. In D.L. Stein (Ed.), Lectures in the sciences of complexity, Addison-Wesley, Redwood City, CA, 1989.

Koch, C., Resistive Networks for Computer Vision: An Overview. In S. Zornester, J. L. Davis, C. Lau (eds.), An Introduction to Neural and Electronic Networks, Academic Press, San Diego, CA, 1990.

Lyon, R.F., Mead, C., "An analog electronic cochlea," IEEE Trans. Acoustics, Speech, and Signal Processing, Vol. 36, No. 7, July 1988, pp. 1119-1134.

Mead, C. (1989). Analog VLSI and Neural Systems, Addison-Wesley.

Morton, S., "Intelligent Memory Chips: Neural Network Building Blocks," Journal of Neural Network Computing, Fall 1989, pp 39-53.

Morton, S., "Intelligent Memory Chips Provide Ultra-High Performance Digital Signal Processing and Pattern Recognition," Proc. of Govt. Microcircuits and Applications Conference (GOMAC), Nov 1989, pp 345-350.

Morton, S., "'Intelligent' Memory Chips," Defense Computing, November-December 1988, pp. 14-17.

Murray, A.F., "Pulse Arithmetic in VLSI Neural Networks," IEEE Micro, December 1989, pp. 64-74.

National Science Foundation Workshop, Hardware Implementation of Neuron Nets and Synapses, January 14-15, 1988, San Diego, CA.

Newcomb, R.W., "MOS Neuristor Lines," in Constructive Approaches to Mathematical Models, C. Coffman, G. Fix (eds.), Academic Press, 1979.

Rosetto, O., C. Jutten, J. Herault, I. Kreuzer, "Analog VLSI Synaptic Matrices as Building Blocks for Neural Networks," IEEE Micro, December 1989, pp. 56-63.

Strohbehn, K., Andreou, A.G., "A Bit-Serial VLSI receptive field accumulator," IEEE 1989 Custom Integrated Circuits Conference (CICC), San Diego, CA, May 15-18, 1989.

Treleaven, P., M. Pacheco, and M. Pacheco, and M. Vellasco. (December 1989). "VLSI Architectures for Neural Networks," IEEE Micro, pp.8-27.

Van den bout, D.E., Miller III, T.K., "A Digital Architecture employing stochasticism for the simulation of Hopfield Neural Nets," IEEE Trans. Circuits and Systems, Vol. 36, No. 5, May 1989, pp. 732-738.

Van den bout, D.E., Miller III, T.K., "TinMANN: The Integer Markovian Artificial Neural Network," Proc. IEEE International Conference on Neural Networks, 1989, Vol. II, pp. 205-211.

Van den Bout, D.E., Miller III, T.K., "Practical, Non-VLSI Neural Networks," Proc. WNN-AIND 90, Auburn, AL, Feb. 5-6, 1990, p. 185.

Verleysen, M., P. G. A. Jespers, "An Analog VLSI Implementation of Hopfield's Neural Network," IEEE Micro,

Yestrebsky, J., Basehore, P., Reed, J., "Neural Bit-slice Computing Element," reprint, Micro Devices.

17

OPTICAL
NEURO-COMPUTING

Harold H. Szu

17.0 OVERVIEW

In this chapter, optical implementations of neural networks are reviewed, comparisons among electronic and optic implementations are given, and a special emphasis is placed on hybrid, optical electronic implementations. After a brief introduction of history, a review of learning algebra and architecture, an itemized look at current U.S. and Japanese efforts in the area is taken with a complete bibliography. The pros and cons of implementation issues are discussed as well. Finally, a superconductor optical neurocomputer is introduced in the following manner. A super-triode designed for optical sensing has a radiation grid along the major c-axis, orthogonal to the superconducting plane where cathode and anode supercurrents crossflow each other in a highway overpass architecture. By design, the overpass anode current occludes the radiation from the underpass cathode current, with their resistance ratio producing a sensitive, single photon sensing capability.

A neurocomputer, designed to operate at a high speed, has a double layer architecture for feedback control, similar to the adaptive resonance theory for self-clustering pattern classification. A super-triode focal plane array also has many advantages over traditional CCD arrays in the field of imaging. The point-by-point direct and dense read-out supports instantaneous wave front sampling. This quality exhibits a direction finding capability, as well as the possibility of wave-front correction for sharper image formation when coupled in-situ with a neurocomputer.

17.1 HISTORICAL INTRODUCTION OF OPTICAL NEUROCOMPUTING

Neural network research was introduced to the optics community through its frequent recurrence in that field. The first introduction was dramatic. Although the optics community was ready, it had never been exposed to the Proceeding of National

Academy of Sciences, where Hopfield neural networks had appeared [Hopfield,1982 and1984]. Nor had they ever published in Biocybernetics and Cognitive Sciences, where Grossberg neural networks had appeared. Nevertheless, since this sudden appearance occurred, it gained momentum fairly quickly. The optics community soon contributed to the much needed teamwork in neural network research. Eventually, this led to the formation of the International Neural Network Society (INNS) by Stephen Grossberg et al. in 1987. Several active members of the founding governing board members of INNS are from the optics community.

During the last decade, optical computing was dominated by three major schools of thoughts or approaches: (1) The digital optics approach (notably: Adolf Lohmann at Erlangen and Alan Huang at AT&T), (2) the matrix-vector approach (notably: Joe Goodman at Stanford and David Casassant at Carnegie-Mellon), and (3) the holographic approach (notably: John Caulfield then at Aerodyne now at Univ. of Alabama and Sing Lee at UCSD). In the summer of 1984, Nabil Farhat visited JPL and came across the Hopfield neural network. He realized that it was actually an analog/threshold version of what the optical computing field had been doing all along. Technically it operated by "writing by the matrix-vector outer product and reading by the matrix-vector inner product." As he was visiting JPL, he discussed his realization with Demetri Psaltis, then at Caltech. Their collaboration led to the first optical implementation of Hopfield neural networks as a model of associative memory, published in both Applied Optics and Optics Letters in 1985 [Farhat and Psaltis, 1985] [Psaltis and Farhat, 1985].

Concurrently, Robert Hecht-Nielsen, then at TRW/San Diego, was working with Ira Skurnick at DARPA, under the ADAPT program. Their basis was an electronic neural network based on the *Adaptive Resonance Theory* (ART) of Stephen Grossberg and Gail Carpenter. The first DARPA neural network workshop was held in April of 1985 at Santa Barbara, California. There, John Hopfield met John Neff (transfered from AFOSR to DARPA) and some of his key optical computing researchers. (Szu became Neff's Technical Representative (COTR), and optical neurocomputing was separately tasked by Farhat and Psaltis.)

In the next spring, (24-27 March 1986, recommended by John Caulfield), Szu edited a book on "Optical and Hybrid Computing" sponsored by the SPIE Institute as the second book on Advanced Optical Technology. The word "hybrid computing" included not only the digital (algorithms for linear and nonlinear systems) and the analog (application driven devices and system development), but also the distributed (neural networks for computing) and the localized (architectures based on bistable and molecular devices). All the aforementioned neural network researchers (including Teuvo Kohonen at Helsinki, Finland) were invited to attend and to meet all the aforementioned optical computing researchers. International representation included Desmond Smith from Heriot-Watt, Britain, Satoshi Ishihara from MITI, and Toyohiko Yatagai from Tsukuba, Japan. A four day retreat and peer-review of 35 chapters of the book was held at Xerox International Education Center (Leesburg, Virginia) sponsored by SPIE. Additionally, the IEEE Spectrum editor, Trudy Bell, among DoD observers (James Ionson, Tim Coffey, Bob Guethur, Bill Miceli, and others), wrote a detailed account in a special Spectrum issue. Additionally, Grossberg and Carpenter edited a special issue of Applied Optics Volume 26, No. 23, December 1987.

The result of all this was a strengthened and enhanced impact of optics on the neural net community. The annual meeting of the Optical Society of America was held at

Washington, D.C., during that Fall; program chair, Ravi Athale at NRL invited Grossberg to give a talk at a special neural network symposium. Then, Grossberg organized a NSF-sponsored optics and neural workshop at Boston University in the winter. EE Times correspondents Chappell Brown and Colin Johnson both made news accounts (Dec. 1, 1986) about the event, including the design of an optical Cauchy machine. This machine was a speedup version of the Boltzmann Machine presented by Terence Sejnowski at the Leesburg workshop, cf. the accompanying chapter by Szu and Maran. Since then, there has been continued activity in the area including the formation of INNS in March of 1987, the first International Conference of Neural Networks (ICNN) organized by IEEE/San Diego Section (Bart Bosko and Hecht-Nielsen in June of 1987), and the Snowbird Conference (IEEE Information Theory) of neural networks held in November of 1987 by John Denker at AT&T. Once set in motion, neural networks proved to be expansively self-perpetuating: 4000 members enrolled in INNS and subscriptions to the IEEE Transaction reached 5000.

17.2   REVIEW OF LEARNING ALGEBRAS AND ARCHITECTURES

The following summary may help the community adopt a standardization of notations. Neural network computing entails a nonlinear system that satisfies four known principles with the fifth principle yet to be worked out. These are: (1) a nonlinear threshold logic of neurons, (2) a nonlocal distributed memory, (3) a nonstationary neural dynamics, and (4) a nonconvex system optimization, meaning more than one extreme in the energy landscape. The first principle came from the McCulloch-Pitts (M-P) neuron model proposed in 1943. After Rosenblatt had proposed that a random collection of M-P neurons can be intelligent for perception, Minsky and Papert showed that was insufficient even for a simple Exclusive-OR operation, thus proposed alternative approach to the intelligence namely a rule-based AI.

Although these four principles are known separately, they are actually interwoven in a complicated system. They may be approximated respectively by (1) piecewise-linear (namely, binary), (2) piecewise-local (namely, the rank-1 vector outer product), (3) piecewise stationary (namely, iterative revisions), and (4) piecewise convex (namely, local gradient descents). In such controlled approximations, these interwoven complex principles become decoupled and amenable to powerful computer simulations. But even though NI has come a long way in that respect, there still remains the fifth principle to be formulated: non-programmed learning. Several versions of the fifth principle have been proposed by pioneer researchers, but they basically differ only in the degree of supervision; some kind of teacher/programmer clause is still needed. This is the current state of the art in neurocomputing theory.

- Learning Algebras
    Given the physiology of ion transport through a junction, (firing rate of approximately 100 ion transpasses per second, denoted as ui), neurocomputing has limited mathematical options as far as manipulating and extracting the information content. As a matter of fact, known learning algebras are based on

variants of Hebbian product rule of neuron firing rates. The change of the impedance at the synaptic junction between the ith and the jth neurons, ΔW_{ij}, must be proportional to the frequency of the ion passage through the junction. This is not unlike a car engine where tuning up the firing rate is accomplished by the gapping of the spark plug.

- Correlation Learning $\Delta W_{ij} \approx u_i\, u_j$

  Maximum information-exchange rule given a pair of random firing rates.
- Gradient Learning $\Delta W_{ij} \approx (D_i - v_i)\, u_j$

  Error correction by $(\partial E/\partial v_i)$ where $E = (D_i - v_i)^2$ and D_i is the output goal that decides when the change of actual output v_i stops: the delta rule.
- Competitive Learning $\Delta W_{ij} \approx u_i\, (u_j - W_{ij})$

  New change must balance against the old cluster establishment W_{ij}.
- Differential Learning $\Delta W_{ij} \approx (du_i/dt)\, (du_j/dt)$

  Only time rate changes matter, as derived by Taylor series expansion of $u_i(t)$.

- Learning Architectures

  Architectures are important for executing parallel and distributed computing. Almost all neural networks have fixed architectures which belong to one of these categories: one layer of Hopfield's Associative Memory (AM), two layers of Grossberg's Adaptive Resonance Theory (ART), and three layers of Rumelhart's Back Error Propagation, BEP (Figure 17.1) Learning Algorithm-Architecture:
  - In the left-hand column, similar inputs X_i are mapped into similar outputs Z_k in a feature space. Such a (hetero-associative) matrix memory is formed by the vector outer product forming a matrix denoted as $|Z_k\, X_i^T|$, where the superscript T stands for the transpose of the column vector X (indexed with the

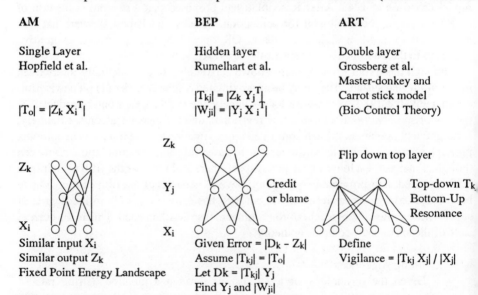

AM	BEP	ART
Single Layer	Hidden layer	Double layer
Hopfield et al.	Rumelhart et al.	Grossberg et al.
		Master-donkey and
	$\lvert T_{kj}\rvert = \lvert Z_k\, Y_j^T\rvert$	Carrot stick model
$\lvert T_o\rvert = \lvert Z_k\, X_i^T\rvert$	$\lvert W_{ji}\rvert = \lvert Y_j\, X_i^T\rvert$	(Bio-Control Theory)

AM: Z_k / X_i — Similar input X_i — Similar output Z_k — Fixed Point Energy Landscape

BEP: Z_k / Y_j / X_i — Credit or blame — Given Error $= \lvert D_k - Z_k\rvert$ — Assume $\lvert T_{kj}\rvert = \lvert T_o\rvert$ — Let $D_k = \lvert T_{kj}\rvert\, Y_j$ — Find Y_j and $\lvert W_{ji}\rvert$

ART: Flip down top layer — Top-down T_k — Bottom-Up Resonance — Define Vigilance $= \lvert T_{kj}\, X_j\rvert\, /\, \lvert X_j\rvert$

Figure 17.1 Learning algorithm-architectures

component i) and the column vector becomes a row vector. Matrix memory is a static version of the Hopfield neural network, because the fixed point coding between the input and the output requires no learning. Fixed point coding can be otherwise expressed as "write-by-outer-product" and "read-by-inner-product" using the matrix-vector operation without iterations.

- In the middle column, where similar inputs can produce surprising outputs, an extra layer is introduced to interpolate these abnormal results by means of supervised training. The difference $|D - Z|$ of the output Z from the desired output D corresponds to error back-propagation by means of a local gradient descent methodology. This system has the potential for generalization. There are several theories about the size of the so-called hidden layer and the ability to do generalization (with fewer neurons than the input nodes) or abstraction (more neurons than the output nodes). The optimum degree of freedom of the internal knowledge representation must match the number of clusters to be classified provided that the orthogonal feature extraction uses the mini-max concept described in the following section. In such a near-orthogonal storage, this rule of thumb seems to be reasonable in assigning credit-or-blame for the output success-or-error.

- When the desired output D is not yet known, the Grossberg model of Adaptive Resonance Theory (ART) becomes handy. It might be thought of as flipping down the unknown output in order to compare the unknown input directly as shown in the righthand column. Thus, the number of interconnections is identical to the BEP model. The master neurons at the top layer have their own top-down wires T_{jk} (shown by dotted lines), while the donkey neurons at the bottom layer have their own bottom-up wires b_{ij}. In order to carry out a follow-the-leader clustering technique, the strategy is that the first arrival is always the leader (the early bird catches the worm), then the ith born leader has the right to reject or to accept any newcomer. If it rejects, the newcomer becomes another leader and forms a new class. However, if the leader accepts the new comer, then the newcomer has the absolute veto power to change the character of its leader by the Hebbian update rule: new $T'_{ik} = T_{ik} x_k$. This cluster rule operates by putting the top layer master feet into the donkey's shoes and checking its bottom-up recommendation of the input (maximum i of the matrix vector product $b_{ij} x_j$). Let $j = i$. Then, calculate the inner product direction cosine $\cos (T_{ik}, x_k)$ which is compared with a constant vigilance parameter r to be set between 0 and 1 (e.g., choose $\rho = 0.71$. If the direction cosine angle $\theta < 45°$, then indeed, $\cos(\theta)$ r and the newcomer is accepted). Otherwise, the newcomer is rejected from the class, and it must be further compared with other classes to find a home within 45° of any leader, or else, to form its own class. Such an update (namely, Hebbian) product rule can extract the common feature of each class. For example, in ART-1 for binary imagery, a zero bit pixel of a newcomer can nullify a non-zero pixel and thin the leader feature. Unfortunately, it can eventually whiten the leader such that the strongest leader is a white picture having all zero pixels. Nevertheless, when the number of samples is open and the number of clusters is unknown, ART is ideally suited for such clustering and simultaneous feature extraction. In fact, the difference between the traditional control theory with negative feedback and the neural network is that both the incentive/carrot and the punishment/stick are used in the biological model having both the excitation and the inhibition exerted at different parts of the self-organized system.

17.3   ASSOCIATIVE MEMORY VS. WIENER FILTER AND SELF-ORGANIZATION-MAP VS. KALMAN FILTERS

The interconnected matrix will use supervised learning, trained by two steps: at the beginning, a top-down design initiates the interconnected weights by a supervised training in a parallel batch mode operation, and then a bottom-up iterative and sequential procedure gives a fine tuning. The two steps are easily coupled together because of the simple geometric interpretation given as follows. The weight is first estimated by the hetero-associative matrix (HAM) memory:

$$T_{ij} = \Sigma \, [U \, V^T]_{ij} \qquad\qquad \text{Eq. 17.1}$$

where the superscript T denotes the transpose of the column vector becoming the row vector in the outer product formula, and the summation symbol indicates a batch mode loading of the HAM module with all possible pairs of $U \, V^T$. Such a supervised initiation is further extended to include a sequential update for fine adjustments.

- Relationship between HAM and SOM

     Now, reading the net matrix T_{ij} horizontally in terms of a set of row vectors M_i^T:

$$T_{ij} = [M_i^T]_j, \qquad\qquad \text{Eq. 17.2}$$

we find that the ith vector M_i takes an input loaded at various j nodes line-by-line directing toward the specific ith output node.

     From such a simple geometric interpretation of HAM, a supervised learning algorithm follows. Giving the new information (indicated by vector U_i to be targeted at the ith action V_i), an iterative procedure for finding the ith vector M_i given the old centroid vector M'_i is the following addition rule of the standard parallelogram constructed by three vectors (M'_i, M_i, U_i):

$$M_i = M'_i + (U_i - M'_i)/2, \qquad\qquad \text{Eq. 17.3}$$

     We will now show that such a update rule of HAM is quite general, related to the linear (learning) vector quantization, Kohonen Self-Organization Map (SOM) [Kohonen, 1987], and Kalman filter as follows. Note that SOM is when the factor 1/2 of Eq. 17.3 becomes an adjustable constant α, and a pair of two such equations like Eq. 17.3 are used for adjusting two neighborhood classes.

- Relationship between SOM and Kalman Filter

     If M_i were a random variable with sample value m_i, the averaged value over N trials and errors may be denoted by the N-subscript angular bracket $\langle M_i \rangle_N$,

$$<M_i>_N = (\Sigma_n m_i^{(n)})/N, \qquad\qquad\qquad \text{Eq. 17.4}$$

Then, by definition,

$$<M_i>_{N+1} = (\Sigma_n m_i^{(n)})/(N+1) \qquad\qquad \text{Eq. 17.5}$$

This can be split into old data (using the definition) and new datum:

$$<M_i>_{N+1} = (N+1-1/N+1)<M_i>_N + m_i^{(N+1)}/(N+1) \qquad \text{Eq. 17.6}$$

We have inserted +1-1 in righthand side in order to rewrite the result in terms of the important concept of "updating the difference only"

$$<M_i>_{N+1} = <M_i>_N + K (m_i^{(N+1)} - <M_i>_N) \qquad \text{Eq. 17.7}$$

where K, known as Kalman gain [Szu, Caulfield, Werbos 1990], is given by

$$K = 1/(N+1) \qquad\qquad\qquad \text{Eq. 17.8}$$

in the uniform average case. The recursion stops when the error/difference vector becomes orthogonal to the input vector.

- Relationship between HAM and Wiener Filter: Then the sequential-update Kalman filter becomes identical to the following batched-mode-update Wiener filter T_{ij} defined by

$$O'_i = S_j T_{ij} I_j \qquad\qquad\qquad \text{Eq. 17.9}$$

Let image I_k be multiplied on both sides, and then averaged, by means of $<I_j I_k> = |I|^2 \delta_{jk}$. Then, Wiener filter is deduced as the averaged HAM

$$T_{ik} = <O_i I_k>/|I|^2 \qquad\qquad\qquad \text{Eq. 17.10}$$

where the estimated object O'_i is replaced with the actual object O_i, because the important concept that the difference/error $(O'_i - O_i)$ becomes orthogonal to the input image I_k as proved by the variation of the least square error function $\partial E/\partial T_{ik} = (O'_i - O_i)I_k = 0$ where $E = |O' - O|^2$. Note the formal similarity between Eq. 17.10 and Eq. 17.1. Also note the similarity between Eq. 17.3 and Eq. 17.7 as the vector constituents of those tensors. If the new input is $U_i = m_i^{(N+1)}$, the new centroid is $M_i = <M_i>_{N+1}$ and the old centroid is $M'_i = <M_i>_N$. Thus, we conclude that not only control and neurocontrol are similar, but these filters are similar to associative memory HAM to Wiener, and SOM to Kalman. A large class of such recursive algorithms known as the linear (learning) vector quantization method or the Kohonen self-organization map [Kohonen, 1986] are similar because of the important concept of "update the difference only" useful for the stability and convergence.

An adaptive novelty filter has been developed by Szu and Messner [1986], and implemented optically by Anderson and Erie [1987].

17.4   OPTICAL IMPLEMENTATIONS OF NEURAL NETWORKS

- *One-Layer Hopfield-like (Associative memory)*: Farhat and Psaltis [1985] first implemented this using LEDs, PDs, and electronic circuits for threshold logic and feedback iteration. Athale, Szu, and Friedlander later [1986] developed a sightly different implementation using LED's interlaced with PD strips and used a controlled nonlinearity in the correlation domain.
- *One-Layer Kohonen-like (Self Organization Map (SOM))*
- *Two-Layer Groosberg-like (ART)*:
  - *BAM*: Adaptive Birection Associative Memory (BAM) [Kosko, 1987] utilizes a top-down Hetero-Associative Memory (HAM) and a bottom-up HAM efficiently via a single rectangular matrix and its transpose to operate on both input long vector and the output short vector, in an alternatively reverse role. This has been optically implemented by [Guest and Tekolste, 1987].
  - *Counter-propagation networks*: Hecht-Nielsen [1987] proposed a hybrid between SOM and ART.
  - *Super-triode ART*: Following the analog version of the Adaptive Resonance Theory, or ART-2 [Carpenter and Grossberg, 1987], Szu has utilized the two-dimensional nature of the High Tc superconductor ceramic material at the liquid nitrogen temperature to design a Super-Triode [U.S. Patent, 1990]. The top-layer is an electricity-mediated photo-refractive crystal SLM sandwiched with CGM. The electrical currents are wired from the bottom layer serving as the bottom up interconnections. The bottom layer consists of NxM super-triodes located in a checkerboard fashion for the input supercurrent of N ports and the output supercurrent of M ports. Then, NxM light sources image on a one-to-one basis through lenslets onto the NxM supertriods. This serves as a top-down optical interconnection. Details will be described later in the optical electronic implementation.
- *Three-Layer Rumelhart-like (Backprop):* The hidden-layer backward error propagation model has been developed by Wagner and Psaltis [1987] and recently implemented by Ishikawa et al. at University of Tokyo and Hamamatsu [1990] and Itoh et al. at NTT [1990].
- *Multiple layers:* Associative memory for optical character recognition was designed by Szu [1987]. Later, a programmable holographic version was implemented by Owechko and Soffer [1988].
- *3/2-D:* Holography with a fractal dimensionality in neural networks was reviewed by Psaltis et al. [1990]. Since the crystal volume has three dimensions, and the image plan has two dimensions, then the coupling between two unequal dimensions suggests the fractal dimension of three halves.
- *4-D:* Four dimensional interconnections for two dimensional images were designed using lenslets by Jenkins at USC and using holographic elements by Caulfield at UAH.
- *Higher Order networks:* Lee and Maxwell [1987] summarized optical implementations of higher order networks. While the disadvantage is implementing the

tensor interconnections, the advantage is the ability to simultaneously load the input correlation for in variance.

- *Stochastic/Wavefront Neural networks:*
  - optical Boltzmann Machine was developed by Farhat [1987].
  - one dimensional Cauchy version by Scheff and Szu [1987].
  - Resonator wave mixing has been considered by Owechko [1987] and in terms of coherent optical eigenstate by Anderson [1986].

17.5   COMPARISON BETWEEN ELECTRONIC AND OPTIC IMPLEMENTATIONS OF NEURAL NETWORKS

Traditional electronic chip implementations are designed with a two-dimensional configuration. The planar technology is preferred because of an equal proximity to the electrical ground level. The size of these planar chips has been miniaturized progressively. As a result the density of the components has been increasing by a factor of two almost every (other) year. Until recently, chip fabrication technology had not been challenged. The N2 interconnection wires among N-neuroprocessors for neural network global interconnection becomes a "spaghetti on a plate" phenomenon.

Furthermore, each electronic neuroprocessor needs the gain mechanism and the adaptive weight learning/adjustment. The trade-off consideration between the complexity of network architecture/communication and the size/capability of each neuroprocessor led to the state of art fabrication of about one thousand neurons or less (with less than million interconnections) built per module. After Tank and Hopfield, Japan recently reported a neurochip which can solve the Travelling Salesman Problem (TSP) over 100 cities (with 100! combinatorial explosion) by means of a Hopfield-like neural network in less than a few seconds [IJCNN-San Diego, 1990].

Optical-neurocomputing utilizes the standard tricks useful for optical computing, and invents a few new ones, e.g., spatio-temporal multiplexing. It employs garden variety tools and develops a few more, e.g., those needed for stochastic optics for Boltzmann Machine. All together, these are: (1) interconnection communication: diffraction optics, holographic element, Bragg cell, lenslet, and fiber optics, (2) nonlinear interaction: Spatial Light Modulator (SLM) [Fisher and Lee, 1987], Light Emitting Diode (LED), Photon Detector (PD), and Phase Conjugate Mirror (multiple wave nonlinear mixing with gains) [Yeh et al., 1987], (3) static storage media: Computer Generated Hologram (CGH) and photographic film mask, and (4) dynamic storage media: photo-refractive crystals [Fainman, and Lee, 1987], ring laser [Anderson, 1986], and write-read erasable optical disk [Psaltis, 1989].

Optics is parallel, real time, fast, and most importantly, wireless in communication. Nevertheless, optical implementation of neurocomputing is limited by two main factors. One, by neuroprocessor-size or two, by the dynamic range resolution ranges (see below). At the present time, no more than several hundred neurons that remain distinguishable throughout the iteration processing have been built.

1. **Neuron-Size Limited:** Due to the positive number of photons, a minimum number of about one hundred photons will be needed for carrying the information

of visibility. Thus, in order to enjoy wireless interconnections, we must pay the price of electrical interface through underlying circuitry for thresholds and light from LEDs and PDs. Emulation of neurons with LEDs requires two PDs for each, one for the positive interconnections and the other for the negative interconnections (light is always positive in nature).

2. Limited Dynamic Range: Employing a lenslet interconnection, Jenkins et al. at USC has cleverly designed a four dimensional interconnection for two dimensional images. Since each pixel of the image may be counted as a neuron, such a network apparently has tens of thousands of neurons. But the intrinsically low resolution (about 2 bits) of the iteration processor renders the result meaningless. Due to this resolution difficulty, the result after several cycles through the system becomes unrecognizable. With some expansive gain mechanism for the signal-to-noise enhancement, an improvement has been proposed by Caulfield et al. using lenslets and active holographic elements.

As yet, both techniques can not built up to several hundred distinguishable neurons per module in a fixed neuroarchitecture.

17.6   HYBRID NEUROCOMPUTING

Neurocomputing seems to be most promising in the hybrid approach. In its original meaning, hybrid neurocomputing was based on the intrinsic property of linear super-position of light and the nonlinear interaction of electricity. Unfortunately, the coupling between electricity and light is weak. The conversion between the two requires devices such as an analog-to-digital (A/D) converter, Light Emitting Diode (LED), Photon Detector (PD), and Electron Multiplier Tubes (EMT), e.g., a Hamamatzu micro-channel plate (MCP). Almost all Japanese efforts in neural networks are in this hybrid area.

- *Hitachi* company in Japan has been applying neural networks to expert systems which can assist investments in security markets; announced as early as September 20, 1988.
- *Mitsubishi* Electric Corporation had developed an optical neural computer which has 36 neurons and 1024 nodes which can recognize A, E, J, etc., in a form of a 6x6 matrix in February, 1988.
- *Product Science Laboratory*, an agency of industrial science and technology, under the Ministry of International Trade and Industry (MITI) had applied the spatial optical modulation tube made by Hamamatsu Photonics to recognize three letters expressed by 4x4 matrices.
- *Fujitsu* developed a neural network to be able to appreciate artistic nature of a picture and a music. NEC had produced a personal neural computer based on personal computer PC9800 with a board for neural networks. It can recognize alphabets in terms of 11x9 matrices.
- *Toshiba* has developed a hybrid of traditional and neural computers which can overcome the computional time-disadvantage of Hopfield model.

- *Nippon Telephone and Telegram (NTT)* modified the backward error propagation model to reduce the time for learning voice recognition.
- *Advanced Technology Research International* (ATR) has four laboratories in communication systems, automatic translations, audiovisual systems, and optical electric wave communication, has been working on Japanese letter 'Kanji' recognition, image processing, voice recognition, and robotic control.
- *Nagano* at Housel University, Japan, has developed a forced learning algorithm to replace backprop for applying to the two-to-three dimensionality recognition problem of human visual system.
- *Electronic Technology Synthetic Laboratory*, an agency of industrial science and technology under MITI, has investigated neural network pattern recognition and biocomputers.
- *NHK Broadcast Technology Laboratory* developed a selective attention subsystem to augment the classical cognition and neocognitron designed early by Fukushima [1987].

17.7 APPLICATION TO PATTERN RECOGNITION AND IMAGE PROCESSING

See the chapter in this text on Boltzmann machines for details using simulated annealing to extract features for pattern recognition.

The image space is 2-D, but the search space can be 1-D, provided that a space-filling scanning technique is adopted for mapping 2-D imagery space to 1-D search space while preserving the local neighborhood relationships [Szu and Scheff 1990]. In principle, the space-filling can be done to any desired degree of resolution meaningful to the original dynamic range and image pixel resolution.

A video sequence has been taken of an object seen through atmospheric turbulence, downward through a wavy surface of turbulence simulated with a specific time-correlation scale. A shift-and-add technique was used in the second look to produce a sharp template of the submerged object. The second look was a regional re-summation that was redone piecewise with respect to the identical set of imagery that produced, in the first pass, a blurred template. The second had correct statistics of image pieces, by means of a straightforward pointing-and-tracking summation of many frames (about 16 distorted fields) according to the centroid of the whole frame [Szu and Blodgett 1982]. This effect demonstrated the need of a smart sensor concept. The eye, for example, can see a weak star during an "instance of good seeing" [Szu et al. 1980] through the turbulent sky. On the other hand, the indiscriminate, dumb telescope camera can only produce a blurred picture of the weak star in an over-exposed picture by whole-frame summation. Such a picture is produced based on the straightforward pointing-and-tracking gimbal, without any adaptive phase for turbulence medium phase correction.

Recently, a sequence of distorted imagery that consists of a training set of 15 samples of hand-written characters (each 4 by 4 pixels) was used to demonstrate the ability to generalize: to recognize a new class of letter [Szu and Scheff 1989]. This was done by means of critical feature extraction using the "mini-max concept." By itself, it discovered a new class of five more hand-written characters by analyzing the intra-interclass clustering property on the self-constructed feature space. The Gram-Schmidt orthogonal feature extraction was based on the associative memory employing the Fixed-Point Cycle Two Theorem [Szu, Scheff 1989]. Such a procedure of parallel Gram-Schmidt constrained orthogonalization could be exceedingly useful for covert communications constrained by call signs and known scrambling instructions. This is so for feature extraction by means of straightforward projection will not obliterate critical portions of the signal. However, any practical construction of a large set of orthogonal feature vectors could be subject to a real time processing bottleneck. In this paper, the Cauchy Simulated Annealing technique is proposed to alleviate the bottleneck problem.

It was introduced by Szu that such an analytical Taylor series expansion (see Figure 17.2) yields, from the top-down design, all those higher order interconnections [Giles and Maxwell, 1987], provided that a Peano-Hilbert N-curve Vi is adopted along which the local derivatives, e.g., $T_{ij} = \partial^2 E/\partial V_i \partial V_j$ exist everywhere [Szu and Scheff 1990].

17.8   THE SUPERCONDUCTING MECHANISM

Single electrons in copper wires move irregularly and incoherently during electrical conduction creating a lot of wasteful heat (not unlike bachelor courtship). However, pairs of electrons in Cooper wires move in synch (they get married and settle down). One electron stream is perfectly cancelled by another electron counter-streaming with the opposite spin angular momenta. This detail-balanced pairing limits severely the roaming phase space P on the Fermi surface resulting in a much smaller entropy production: S = Log (P). Consequently, much less heat is produced: Q = ST. Of course, zero heat is due to quantum phenomenon. Those Boson-like electron-pairs having zero

$$\text{Energy} = \sum \frac{1}{|\text{Interclass Distance}|} + \sum |\text{Intraclass Distance}|$$

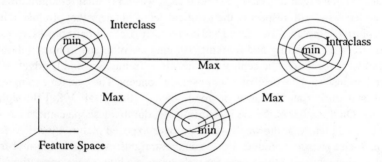

Taylor Expansion to Derive multiple layer interconnections

$$\text{Energy} = -\frac{1}{2} \sum_{i,j} T_{i,j} \, V_i \, V_j - \frac{1}{6} \sum_{i,j,k} T_{i,j,k} \, V_i \, V_j \, V_k - \dots$$

Figure 17.2   Top-down design of hard-wired neural networks mini-max principle

angular momentum enjoy, at low temperature, the Bose-Einstein condensation at the ground state, i.e., the only one state implying P=1 and consequently S = Log (P) = 0.

A ceramic material cooled with liquid nitrogen can conduct electricity without resistance, i.e., a supercurrent. This supercurrent is two dimensional consisting of pairs of electron holes streaming in synch along the plane parallel to the orthorhombic crystalline minor axes composed of copper and oxygen atoms. Ceramic superconductors do not completely expel electromagnetic fields from their interiors, therefore permitting radiation interactions.

Almost all ceramic high-temperature superconductors, except two produced recently in January and March of 1989 [3], have a common peculiarity. The electric current is actually carried by pairs of electron holes, rather than electrons. These holes are empty spaces of the crystal material caused by the doping of Yttrium in BaCuO crystal (Yttrium has fewer electrons than the Barium it replaces). As electrons from other crystals move backward to fill these holes, the holes appear to propagate forwards. The actual forward path through the crystal is along the buckled copper-oxygen planes sandwiching the Yttrium atom, which Wu et al. [2] used to replace Lanthanum in February of 1987.

Under external pressure alone, Chu was able to raise the temperature of superconductance of the LaBaCuO molecule to 40°K (Kelvin) in December of 1986 [4] (previously 30°K created first by Muller and Bednorz [1] in April of 1986). This inspired the ingenious concept of applying molecular pressure to replace external pressure (suppressing lattice thermal agitations while raising the critical temperature, T_c). By substituting Yttrium for Lanthanum (which is located in the same rare earth group of the periodical table as is Lanthanum but with a smaller atomic radius), Chu et al created internal pressure generated through two adjacent Barium atoms (sandwiching the smaller Yttrium atom) which buckled the copper-oxygen planes necessary to raise the T_c to 95°K without the need of external pressure.

With a permanent magnet, pellets of cooled superconductor are loosely anchored by penetrating magnetic field lines in magnetic suspension experiments. This anchoring may be a result of flux pinning at irregularities and inhomogenealities along the crystal axis and/or impurities at the grain boundary. This sluggish lateral motion is characteristic of type II superconductors, as opposed to the free sliding motion of type I superconductors in magnetic levitation experiments. Although such a type-II superconductor may not yet be suited for high-field applications, it is idea for designing low-field radiation sensors.

17.9   THE SUPER-TRIODE

A super-triode, which is a superconducting triode consisting of anode cathode and grid, is a molecular optical electronic device. The design solely depends on the two-dimensional superconducting nature operating at liquid nitrogen temperatures. A super-triode designed for optical sensing has a radiation grid along the major crystal c-axis, orthogonal to the superconducting plane (a-b axes), where cathode and anode supercurrents crossflow each other in a highway overpass architecture. A superlattice structure is indicated for the enhancement of interaction. With occlusion of the radiation from the underpass cathode current by the overpass anode current, the resistance ratio enables sensitive, single photon sensing capability.

Type II superconductor characteristics, i.e., flux creep, leads to the Meissner effect with imperfect field-expulsion. This material seems to be a normal conductor along the c-axis, about the length of 11.71A° while it is superconducting along the minor a and b axes about the size of 3.83A° and 3.88A°, respectively. The direction of the c-axis, serving as a radiation penetration window, is used for grid control of the super-triode anode current. Therefore, by the Lenz induction law, a truly three-dimensional Type-I superconductor that takes zero resistance to produce a supercurrent to shield off external field penetration in any direction, cannot be useful for the super-triode radiation grid.

The external radiation can indirectly impede supercurrents by enhancing or entangling the magnetic field concentration. Thus, the synergism of a specific radiation when coupled with an external magnetic field can control the flux creeping effect and enhance the light-supercurrent interaction. For example, one can fabricate a 1-2-3 compound superlattice of specific structures matching the optimum radiation interaction area that has a diffraction-limited spot size of the impinging radiation (ranging from microns to Angstroms).

The super-triode measures the external magnetic field and the impinging radiation in terms of the ratio of resistances along two crosspasses. Such a measurement by the resistance ratio becomes sensitive to a single photon because: (1) the overpass anode current has physically occluded the radiation from the underpass cathode current, (2) there is no shot noise for spatially distributed Cooper pairs, and (3) the Johnson/thermal noise at liquid nitrogen temperatures is negligible. A single photon can break a Cooper pair thus creating quasiparticles with electrical resistance. Consequently, the super-triode can take advantage of the resistance ratio measurement using the technique of dividing the quasiparticle resistance by the zero resistance, elucidating one of the most sensitive processing techniques. The term switch, not used in the sense of Josephson tunnelling, is used in the effect of relative current changes in anode and cathode crosspasses.

17.10   THE SUPER-TRIODE NEUROCOMPUTER

A super-triode neurocomputer is fast and compact because it is built on molecular electronics. The bottom layer is kept at liquid nitrogen temperature, and is made of super-triodes in a checker-board lattice array. The radiation grids, pointing upward to the top layer of Spatial Light Modulators (SLM) [Fisher and Lee, 1987], are imaged one-to-one from a super-triode to an aperture of a SLM lenslet. The lower layer, consisting of super-triodes located at the intersection points, computes thousands of output currents flowing from north to south by switching thousands of input currents which flow from west to east. This upper layer SLM has two parts consisting of both fixed and adaptable elements. The fixed elements, CGH films or CD, store application associative memories of millions of interconnectional values among thousands of neurons. These values are 1-1 loaded onto millions of super-triodes by means of a uniform radiation source shining through the SLM layer. The adaptive portion of the SLM is a photo-refractive electro-optical crystal having millisecond response time for changing the long term memory coded in films. This crystal is electrically addressed by the output from the bottom layer of super-triodes having picosecond transit time in a simple negative feedback manner. According to the adaptive resonance theory, the present double layer architecture employing bottom-up electrical and top-down optical interconnections, can be further augmented with a direction cosine vigilance module for the implementation of self-clustering pattern classifications [6].

17.11 WAVE-FRONT IMAGING TELESCOPE WITH A FOCAL PLANE ARRAY OF SUPER-TRIODES

Super-triodes have many advantages over traditional CCD arrays in the field of imaging. Due to the fine crystal grain boundaries, a super-triode has a negligible ratio of filler element to active sensing area per pixel, as opposed to CCD's whose filler is considerably larger due to the underlying electronic circuitry per pixel definition. As a result of such undersampling, current telescope imaging cannot easily resolve wavefront arrival directions. Additionally, super-triodes have no need for heat-dissipation mechanisms, allowing dense pixel packaging. Finally, in contrast to the planar CCD readout, the utilization of non-planar readouts allows an array of super-triodes to have all of its pixel elements read simultaneously. This avoidance of time-multiplexing circuitry, meets the quick time resolution requirements necessary for wavefront detection. This ability has many implications as follows.

For imaging device applications, a dense array of the super-triodes is located on a Focal Plane Array (FPA) sensor with each c-axis pointing outwards to receive imaging. It differs from the CCD FPA in four main ways: (1) its single pixel filler-to-active ratio is nearly zero, (2) the pixels are densely packaged due to the elimination of heat dissipation-related limitations, (3) there is no underlying circuitry normally needed for photon detector and time-multiplexing, and (4) the point-by-point direct and dense read-out permits instantaneous sensing of arriving wave fronts. Thus, the super-triode FPA exhibits direction finding capability as well as the possibility of wave-front correction for sharper imaging.

17.12 SPACE-BORNE IN-SITU SMART SENSING WITH NEUROCOMPUTING

A Mega-Cray neurocomputer can execute 1015 associative memory matrix vector operations per second. This capability can be contained within a one by one by one foot cubic package comprising less than a pound of payload weight. Additionally, by locating it in the constant shadow of a solar panel, superconductive temperature can be maintained by a simple electrical cooler. The inputs to the space-borne neurocomputer are the supercurrents from each super-triode pixel located in the FPA of the telescope. Meanwhile, only the neural network computation for the novelty detection needs to be down-linked to the ground station.

Wavefront sensing capability, when coupled with an in-situ neurocomputer would allow "smart" novelty exploration of deep space. The super-triode FPA would allow space-borne ultraviolet, visible, or infrared telescopes to pinpoint arriving radiation directions of objects such as brown dwarfs, or possibly even water or chlorophyll spectra from other planetworks. These possibilities might encourage us to develop future space deployment; there remains much to be accomplished.

17.13 CONCLUSION

New generations of material sciences may support new implementation technologies of neurocomputing, be it electronic, optic, or optical electronic. We are optimistic about the future of the optical neurocomputer.

In the earlier superconductors, the binding force among Cooper pairs is photon-mediated, i.e., a solid state equivalence to the liquid state negative Bernoulli pressure (not unlike the doubling of passing speeds of two oppositely travelling cars on a highway). However, the binding force, in the order of millivolts, is known to be limited to 40oK because 40oK times the Boltzmann Constant KB equals the Plank Constant h times the lattice Debye cutoff frequency w. Thus, a natural question is "why do Cooper pairs of holes in high-temperature ceramic superconductors remain bounded?" The theoretical challenge becomes two-fold, to explain both Cooper pairs of holes and Cooper pairs of electrons when using an exactly opposite choice of doping ceramic material consisting of excess electrons.

In emerging sciences, new discoveries and new devices may help us to sharpen the theory.

Since Testardi initiated the study of light-superconductor interactions in 1970, nonequilibrium superconductivity was born. Thus, we expect that the photon interaction of quasiparticles under a scanning external field may unlock the secret of superconducting mechanisms that might be useful for the creation of stable, recyclable, and malleable superconductors at even higher temperatures.

BIBLIOGRAPHY

Abu-Mostafa, Y. F., and Psaltis, D. (California Institute of Technology) "Optical Neural Computers." In: Scientific American, March, 1987, pp. 66-73.

Anderson, D. Z., and Erie, M. C. (JILA, Univ. Colorado), "Resonantor Memories and Optical Novelty Filters." Opt. Eng. Vol. 26, pp. 434-439, 1987.

Anderson, D. Z., and Lininger, D. M. (JILA, Univ. Colorado), "Dynamic Optical Interconnections: Volume Holograms as Optical Two-Port Operators." Appl. Opt., Vol. 26, No. 23, pp. 5031-5038, 1 Dec. 1987.

Anderson, D. Z. (JILA, Univ. Colorado), "Coherent Optical Eigenstate Memory." Opt. Letters, Vol.11, pp. 56-8, 1986.

Athale, R. A., Szu, H. H., and Friedlander, C. B., (Naval Research Lab, Washington, D.C., BDM Corp. McLean, VA), "Optical Implementation of Associative Memory With Controlled Nonlineararity in the Correlation Domain." Opt. Letters, Vol.11, P. 482-484, July 1986.

Batacan, P. "Can Physics Make Optics Compute?" Computers in Physics, Mar/Apr. 1988, pp. 9-15.

Brady, D., Hsu, K., Psaltis, D. (California Institute of Technology), "Multiple Exposed Photorefractive Holograms with Maximal Diffraction Efficiency." Optical Computing 1989 Technical Digest Series, Vol. 9, pp. 429-32, 1989, (27 Feb.-1 March 1989, Salt Lake City, UT, Opt. Soc. America).

Barnard, E., Casasent, D. (Carnegie Mellon Univ.), "New Optical Neural System Architectures and Applications." Proc. SPIE - Int. Soc. Opt. Eng. (USA) vol.963, pp. 537-545, 1989 (Optical Computing 88, 29 Aug. 2 Sept., 1988, Toulon, France).

Boyd, G. D. (AT&T Bell Lab., Holmdel, NJ), "Optically Excited Synapse for Neural Networks." Appl. Opt. (USA) Vol. 26, No. 14, pp. 2712-2719, July 15, 1987.

Botha, E., Barnard, E., Casasent, D. (Carnegie-Mellon Univ), "Optical Neural Networks For Image Analysis: Imaging Spectroscopy and Production Systems." IEEE International Conference on Neural Networks, pp. 541-546, Vol. 1, 1988 (24-27 July 1988, San Diego, CA).

Carpenter, G., and Grossberg, S. (Boston Univ.), "Self-Organization of Stable Category Recogniton Codes For Analog Input Patterns." Appl. Opt., Vol. 26, No. 23, pp. 4919-4930, Dec. 1, 1987.

Caulfield, H. J., Kinser, J., Rogers, S. K. (Alabama Univ., Huntsville), "Optical Neural Networks." Proc. IEEE (USA) Vol. 77, no. 10, pp. 1573-83, Oct. 1989.

Caulfield, H. J. (Univsity of Alabama, Huntsville), "Optical Neural Networks." Proc. SPIE — Int. Soc. Opt. Eng. (USA) Vol. 960, pp. 242-252, 1989 (Real-Time Signal Processing for Industrial Applications, 27-28 June 1988, Dearborn, MI).

Caulfield, H. J. (University of Alabama, Huntsville), "Optical Processing of Optical Correlation Plane Data." Proc. SPIE — Int. Soc. Opt. Eng. (USA) Vol. 1053, pp. 93-95, 1989 (Optical Pattern Recognition, 7-18 Jan. 1989, Los Angeles, CA).

Caulfield, H. J. (University of Alabama, Huntsville), "Applications of Neural Networks to the Manufacturing Environment." Proc. SPIE — Int. Soc. Opt. Eng. (USA) Vol. 954, pp. 464-467, 1989 (Optical Testing and Metrology II, 27-30 June 1988, Dearborn, MI).

Caulfield, H. J.(University of Alabama, Huntsville), "Continuous Time Neural Networks." Proc. SPIE — Int. Soc. Opt. Eng. (USA) Vol. 882, pp. 43-46, 1988.

Chevallier, R. C., Sirat, G. Y., Heggarty, K. J., and Maruani, A. D. (Dept. Images, Ecole Nat. Superieure des Telecommun., Paris, France), "Frequency Multiplexed Raster Scheme of an Optical Neural Network: Shift Invariant Recognition." Proc. SPIE — Int. Soc. Opt. Eng. (USA) Vol. 963, pp. 522-526, 1989 (Optical Computing 88, 29 Aug. 2, Sept. 1988, Toulon, France).

Collins, D. R., Sampsell, J. B., Hornbeck, L. J., Florence, J. M., Penz, P. A., and Gately, M. T. (Texas Instrum., Dallas), "Deformable Mirror Device Spatial Light Modulators and Their Applicability to Optical Neural Networks." Appl. Opt. (USA) Vol. 28, No. 22, pp. 4900-7, 15 Nov. 1989.

Fainman, Y. and Lee, S. H., "Applications of Photorefractive Crystals to Optical Signal Processing." In: "Optical and Hybrid Computing." H. Szu Ed. SPIE Vol. 634, 1987 (24-27 March 1986, Leesburg Va.).

Farhat, N. H., "Optoelectronic Analogs of Self-Programming Neural networks: Architecture and Methodologies For Implementing Fast Stochastic Learning by Simulated Annealing." Appl. Opt., Vol. 26, No. 23, pp. 5093-5103, Dec. 1, 1987.

Farhat, N. B., Psaltis, D., Prata, A., and Paek, E., (University of Pennsylvania, and California Institute of Technology), "Optical Implementation of Hopfield Model." Appl. Opt. Vol. 24, pp. 1469-1475 (1985).

Farhat, N. H., and Psaltis, D., "Optical Implementation of Associative Memory Based on mModels of Neural Networks." Optical Signal Processing, J.L. Horner, Ed. (Academic, New York 1987), pp. 129-162.

Fisher, A. D., Lippincott, W. L., and Lee, J. N. (Naval Research Lab, Wash. DC), "Optical Implementations of Associative Networks With Versatile Adaptive Learning Capabilities." Appl. Opt. Vol. 26, pp. 5039-5054 (1987).

Fisher, A. D., and Lee, J. N. (Naval Research Lab, Wash. DC), "Current Status of Two-Dimensional Spatial Light Modulator Technology." H. Szu Ed. SPIE Vol. 634, 1987, (24-27 March 1986, Leesburg Va).

Fukushima, K. (NHK Japan), "Neural Network Model For Selective Attention in Visual Pattern Recognition and Associative Recall." Appl. Opt. Vol. 22226, No. 23, pp. 4985, Dec. 1, 1987.

Ghosh, A. and Tien-Sek Chiun (Iowa Univ., IA), "Almost Necessary Conditions For Optical Neural Networks." Appl. Opt. (USA) Vol. 27, No. 24, pp. 5002-5004, 15 Dec. 1988.

Giles, L. and Maxwell, T. (AFOSR and NRL), "Learning, Invariance, and Generalization in High-Order Neural Networks." Appl. Opt. Vol. 26, No. 23, pp. 4972-4978, Dec. 1 1987.

Gmitro, A. F. and Gindi, G. R. (Dept. of Diagnostic Radiol., Yale Univ.), "Optical Neurocomputer For Implementation of the Marr-Poggio Stereo Algorithm." Caudill, M. and Butler, C. (Editors) IEEE First International Conference on Neural Networks pp. 599-606 Vol. 3, 1987 (21-24 June 1987, San Diego, CA).

Guest, C. C. and Tekolste, R. (UCSD), "Designs and Devices for Optical Bidirectional Associative Memories." Appl. Opt. Vol. 26, No, 23, pp. 5055-5060, Dec. 1 1987.

Gustafson, S. C. and Little, G. R. (Dayton Univ. Res. Inst., OH), "Optical Neural Classification of Binary Patterns." Proc. SPIE — Int. Soc. Opt. Eng. (USA) Vol. 882, pp. 83-89, 1988.

Hara, T., Ooi, Y., Suzuki, Y. and Wu, M. H. (Hamamatsu Photonics KK, Japan), "Transfer Characteristics of the Microchannel Spatial Light Modulator." Appl. Opt., Vol. 28, No. 22, pp. 4781-4786, Nov. 15 1989.

Hara, K., Kojima, K., Mitsunga, K. and Kyuma, K. (Mitsubishi Electr. Corp., Hyogo, Japan), "Differential Optical Switching at Subnanowatt Input Power." IEEE Photonics Technol. Lett. (USA) Vol. 1, No. 11, pp. 370-2, Nov. 1989.

Hecht-Nielsen, R. (HNC), "Counterpropagation Networks." Appl. Opt. Vol. 26, No. 23, pp. 4979-4984, Dec. 1 1987.

Hopfield, J. J., "Neurons With Graded Response Have Collective Computional Properties Like Those of Two-State Neurons." Proc. Nat'l Acad. Sci. USA, Vol. 81, pp. 3088-3092, May 1984.

Hopfield, J. J., "Neuron Networks and Physical Systems With Emergent Collective Computational Abilities." Proc. Nat'l Acad. Sci. USA, Vol. 79, pp. 2554-2558, April 1982.

Hopfield, J. J. and Tank, D. W., "Neural Computation of Decisions in Optimization Problems." Bio. Cyb., 52, pp. 141-152, 1985.

Ishikawa, M., Mukohzaka, N., Toyoda, H., and Suzuki, Y., (Univ. Tokyo, Hamamatsu Photonics), "Optical Associatron: A Simple Model for Optical Associative Memory." Appl. Opt. Vol. 28, pp. 291-301, 1989.

Ishikawa, M., Mukohzaka, N., Toyoda, H. and Suzuki, Y., (Univ. Tokyo, Hamamatsu Photonics), "Experimental Studies on Learning Capabilities of Optical Associative Memory." Appl. Opt. Vol. 29, No. 2, pp. 289-295, Jan. 10 1990.

Ishikawa, M., Mukohzaka, N., Toyoda, H. and Suzuki, Y., (Univ. Tokyo, Hamamatsu Photonics), "Experimental Studies on Adaptive Optical Associative Memory." Proc. Soc. Photo-Opt. Instrum. Eng. Vol. 963, pp. 527-536, 1988.

Itoh, F., Kitayama and Tamura, Y., (NTT), "Optical Outer-Product Learning In a Neural Network Using Optical Stimulatable Phosphor." Opt. Letters, Vol. 15, No. 15, pp. 860-862, Aug. 1 1990.

Ittycheriah, A. P., Walkup, J. F., Krile, T. F. and Lim, S. L. (TI, and Texas Tech, Lubbock), "Outer Product Processor Using Polarization Encoding." Appl. Opt. Vol. 29, No. 2, pp. 275-283, Jan. 10 1990

Johnson, K. M. and Moddel, G. (Colorado Univ., Boulder), "Motivations For Using Ferroelectric Liquid Crystal Spatial Light Modulators in Neurocomputing." Appl. Opt. (USA) Vol. 28, No. 22, pp. 4888-99, Nov. 15 1989.

Kohonen, T. (Helsinki Univ.), "Representation of Sensory Information in Self-Organizing Feature Maps, and Relation of These Maps to Distributed Memory Networks." In: "Optical and Hybrid Computing." H. Szu Ed. SPIE Vol. 634, 1987, (24-27 March 1986, Leesburg Va) or Kohonen, T., "Adaptive, Associative, and Self-Organizing Functions in Neural Computing." Appl. Opt., Vol. 26, No. 23, pp. 4910-4918, Dec. 1 1987.

Kosko, B. (USC), "Adaptive Bidirection Associative Memories." Appl. Opt., Vol. 26, No. 23, pp. 4947-4960, Dec. 1 1987.

Kubota, K., Tashiro, Y., Kasahara, K. and Kawai, S. (Optical-Electron. Res. Labs., NEC Corp., Kawasaki, Japan), "Optical Crossbar Interconnection Using Vertical-to-Surface Transmission Electro-Photonic Devices (VSTEP)." Proc. SPIE - Int. Soc. Opt. Eng. (USA) Vol. 963, pp. 255-259, 1989 (Optical Computing 88, 29 Aug. 2, Sept. 1988, Toulon, France).

Kyuma, K., "Optical Neural Networks." Solid State Phys. (Japan) Vol. 24, No. 11, pp. 990-996, Nov. 1989.

Lemmon, M. and Kumar, B.V. K.V. (Carnegie Mellon Univ., Pittsburgh, PA), "Competitively Inhibited Optical Neural Networks Using Two-Step Holographic Materials." Optical Computing 1989 Technical Digest Series, Volume 9, pp. 36-9, 1989 (Feb. 27, March 1 1989, Salt Lake City, UT, Opt. Soc. America).

Li, Y. and Eichmann, G. (Dept. of Electr. Eng., City Coll., City Univ. of New York), "Parallel N2 Weighted Reconfigurable Networks for Optical Neuron and Chip-to-Chip Interconnections." Opt. Commun. (Netherlands) Vol. 67, No. 4, pp. 251-255, July 15 1988.

Lin, S. H., Kim, J.H., Katz, J. and Psaltis, D. (Jet Propulsion Lab., California Institute of Technology, Pasadena, CA), "Integration of High-Gain Double Heterojunction GaAs Bipolar Transistors With a LED For Optical Neural Network Application." (IEEE; Cornell Univ) Proceedings. IEEE/Cornell Conference on Advanced Concepts in High Speed Semiconductor Devices and Circuits, pp. 344-52, 1989 (Aug. 7-9 1989, Ithaca, NY, IEEE).

Lu, T., Choi, K., Wu, S., Xu, X. and Yu, F.T.S. (Penn. State Univ., University Park, PA), "Optical Disk Based Neural Network." Appl. Opt., Vol. 28, No. 22, pp. 4722-4724, Nov. 15 1989.

Lu, T., Wu, S., Xu, X. and Yu, F.T.S. (Penn. State Univ., University Park, PA), "Two-Dimensional Programmable Optical Neural Network." Appl. Opt. Vol. 28, No. 22, pp. 4908-13, Nov. 15 1989.

Lu, T., Wu, S., Xu, X. and Yu, F.T.S., (Penn. State Univ., University, Park, PA), "An Inter-Pattern Association Neural Network Model and its Optical Implementation." Optical Computing 1989 Technical Digest Series, Volume 9 (papers in summary form only received) pp. 437-40, 1989 (Feb. 27–March 1 1989 Salt Lake City, UT, Opt. Soc. America).

Lu, T., Wu, S., Xu, X. and Yu, F.T.S. (Penn. State Univ.), "Optical Implementation of Programmable Neural Networks." Proc. SPIE — Int. Soc. Opt. Eng. Vol. 1053, pp. 30-39, 1989 (Optical Pattern Recognition, Jan. 17-18 1989, Los Angeles, CA).

Lu, T., Wu, S., Xu, X. and Yu, F.T.S. (Penn. State Univ. University Park, PA) "Optical Neural Network Using a High Resolution Video Monitor." Proc. SPIE — Int. Soc. Opt. Eng. Vol. 1134, pp. 32-39, 1989 (Optical Pattern Recognition II, 26-27 April 1989, Paris France).

Lu, T., Xu, X., Wu, S. and Yu, F.T.S., (Penn. State Univ., University Park, PA), "Neural Network Model Using Interpattern Association." Appl. Opt. Vol. 29, No. 2, pp. 284-88 (Jan. 10 1990).

Lu, T., Xu, X., Wu, S. and Yu, F.T.S. (Penn. State Univ., University Park) (IEEE), "A Neural Network Model Using Inter-Pattern Association (IPA)." IJCNN: International Joint Conference on Neural Networks, pp. 596 Vol. 2, 1989 (18-22 June 1989, Washington, DC, IEEE TAB Neural Network Committee).

Lee, L. S., Stoll, H.M. and Tackitt, M.C. (Northrop Res. and Technol. Center, Palos Verdes Peninsula, CA), "Continuous-Time Optical Neural Network Associative Memory." Opt. Lett. (USA) Vol. 14, No. 3, pp. 162-164, Feb. 1 1989.

McAulay, A. D. (Wright State Univ., Dayton, OH), "Optical Neural Network For Engineering Design." Proceedings of the IEEE 1988 National Aerospace and Electronics Conference, pp. 1302-1306, Vol. 4, 1988 (23-27 May 1988, Dayton, OH).

Meng, R. L. Kinalidis, C., Sun, Y. Y., Gao, L., Tso, Y. K., Hor, P. H. and Chu C. W., "Manufacture of Bulk Superconducting YBCuO by Continuous Process." Nature, Vol. 345, pp. 326-328, May 24, 1990 (M-K Wu, et al., C.W. Chu, Phys. Rev. Lett., March 1987).

Mezard, M. (Lab. de Phys. Theor., Ecole Normale Superieure, Paris, France), "The Space of Interactions in Neural Networks: Gardner's Computation With the Cavity Method." J. Phys. A, Math. Gen. (UK) Vol. 22, No. 12, pp. 2181-2190, June 21 1989.

Midwinter, J. E. (Univ. Coll., London, UK), "Digital Optics, Smart Interconnections or Optical Logic?" (Phys. Technol. (UK) Vol. 19, No. 3, pp. 101-108, May 1988.

Muller, A., Bednorz, G., Z. fur Physik 1986, April. Science Year, 1990, The World Book. Chicago, p. 299.

Noehte, S., Manner, R., Hausmann, M., Horner, H., Cremer, C. (Phys. Inst., Heidelberg Univ., West Germany), "Classification of Normal and Aberrant Chromosomes By an Optical Neural Network in Flow Cytometry." Optical Computing 1989 Technical Digest Series, Volume 9., pp. 14-17, 1989 (Feb. 27 March 1 1989, Salt Lake City, UT, Opt. Soc. America).

Owechko, Y. (Hughes Res. Lab., Malibu, CA), "Self-Pumped Optical Neural Networks." Optical Computing 1989 Technical Digest Series, Volume 9. pp. 44-7, 1989 (Feb. 27–March 1 1989 Location: Salt Lake City, UT, Opt. Soc. America).

Owechko, Y. (Hughes Res. Lab., Malibu, CA), "Nonlinear Holographic Associative Memories." IEEE J. Quantum Electron. (USA) Vol. 25, No. 3, pp. 619-634, March 1989.

Owechko, Y. and Soffer, B.H. (Hughes Res. Lab., Malibu, CA), "Programmable Multilayer Optical Neural Networks With Asymmetric Interconnection Weights." IEEE International Conference on Neural Networks, pp. 385-393, Vol. 2, 1988, July 24-27 1988, San Diego, CA).

Peterson, C. and Redfield, S. (Microelectron. and Comput. Technol. Corp., Austin), "Adaptive Learning With Hidden Units Using a Single Photorefractive Crystal." Proc. SPIE — Int. Soc. Opt. Eng. (USA) Vol. 963, pp. 485-496, 1989 (Optical Computing 88, Aug. 29 Sept. 2 1988, Toulon, France).

Psaltis, D., Brady, D., Gu, X. G. and Lin, S., "Holography in Artificial Neural Networks." Nature, Vol. 343, pp. 325-330, Jan. 25 1990.

Psaltis, D. and Farhat, N, (California Institute of Technology, and Univ. Penn.), "Optical Information Processing Based on Associative Memory Model of Neural networks With Threshold and Feedback." Opt. Letters, Vol. 10, pp. 98-100, 1985.

Psaltis, D., Park, C.H. and Hong, J. (California Institute of Technology), "Highter Order Associative Memories and Their Optical Implementation." Neural Networks, Vol. 1 pp. 149-163, 1988.

Psaltis, D., Brady, D. and Hsu, K. (California Institute of Technology), "Learning in Optical Neural Networks." In: Parallel Processing in Neural Systems and Computers pp. 543-548, 1990, March 9-21 1990, Dusseldorf, West Germany, North-Holland, Amsterdam.

Eckmiller, R., Hartmann, G. and Hauske, G. (Editors) Robert Bosch; IBM; Philips; Siemens; et al.

Psaltis, D., Brady, D. and Wagner, K, (California Institute of Technology), "Adaptive Optical Networks Using Photorefractive Crystal. " Appl. Opt. Vol. 27, pp. 1752-1759 (1988).

Rogers, S.K. and Kabrisky, M. (USAF Inst. of Technol., Wright-Patterson AFB, OH), "1988 AFIT Neural Network Research." Proceedings of the IEEE 1989 National Aerospace and Electronics Conference NAECON 1989, pp. 688-94 Vol. 2, 1989 (May 22-26 1989 Dayton, OH).

Sage, J.P. (Lincoln Lab., MIT, Lexington, MA) "Electronic vs. Optical Implementations of Neural Networks." (Opt. Soc. America; Air Force Office Sci. Res.; Defence Adv. Res.Projects Agency; et al) Optical Computing 1989 Technical Digest Series,

Volume 9., pp. 5-8, 1989 (Feb. 27 - March 1 1989, Salt Lake City, UT, Opt. Soc. America).

Scheff, K. and Szu, H., "1-D Optical Cauchy Machine Infinite Film Spectrum Search." IEEE Int. Conf. Neural Networks, p. III 673, San Diego 1987.

Shariv, I. and Friesem, A.A. (Dept. of Electron., Wizmann Inst. of Sci., Rehovot, Israel), "All-Optical Neural Network With Inhibitory Neurons." Opt. Lett. (USA) Vol. 14, No. 10, pp. 485-487, May 15 1989.

Slinger, C. (EM1 Div., R. Signals and Radar Estab., Great Malvern, (UK), "Fidelity In Weighted, Volume Interconnects (Optical Neural Networks)." IEE Colloquium on 'Optical Connection and Switching Networks for Communication and Computing' (Digest No. 076, May 14 1990, London, UK), p. 9, Jan. 5 1990, IEE, London, UK.

Wu, S., Lu T., Xu X. and Yu, F.T.S. (Penn. State Univ., University Park, PA) "An Adaptive Optical Neural Network Using a High Resolution Video Monitor." Microw. Opt. Technol. Lett. (USA) Vol. 2, No. 7, pp. 252-257, July 1989.

Sirat, G.Y., Maruani, A.D. and Chevallier, R.C. (Telecom Paris, France), "Frequency Multiplexed Raster Neural Networks. 1. Theory." Appl. Opt. (USA) Vol. 28, No. 7, pp. 1429-1435, April 1 1989.

Stormon, C.D. (New York State Center for Advanced Technol. in Comput. Appl. and Software Eng., Syracuse Univ., NY) "Optical Neural Network Models Applied to Logic Program Execution." Proc. SPIE — Int. Soc. Opt. Eng. (USA) Vol. 882, pp. 23-29, 1988.

Stoll, H.M. and Lee, L.S. (Northrop Res. and Technol. Center, Palos Verdes Peninsula, CA), "A Continuous-Time Optical Neural Network." IEEE International Conference on Neural Networks, pp. 373-384, Vol. 2, 1988 (July 24-27 1988, San Diego, CA).

Szu, H., (Naval Research Lab (NRL)), "Globally Connected Network Models for Computing Using Fine-Grained Processing Elements." Proceedings International Conf. on laser — 1985, Ed. C. P. Wang, (Soc. Opt. and Quant. Elect. STS Press, P.O. Box 177, Mclean VA) pp. 92-97 (In: Opt. proc and Dev. Chaired by J. Neff). (Las Vegas, Dec. 2-6 1985).

Szu, H., (NRL), "Three Layers of Vector Outer Product Neural Networks for Optical Character Recognition." In: "Optical and Hybrid Computing." H. Szu, Ed. SPIE Vol. 634, 1987 (March 24-27 1986, Leesburg Va.).

Szu, H. (NRL), "Superconducting Optical Switch." U.S. Patent No. 4,904,882, (7 Claims, 3 Drawing Sheets), February 27, 1990.

Szu, H. (NRL), "Superconductors: Friend or Foe to Optical Computing?" Optical Computing NewsLetters, Vol. 1, No. 2, January 1988 in: Laser Focus p. 59, 1988.

Szu, H. (NRL), "High Tc Superconductor Optoelectronic Switch for High Density Sensing and Neurocomputing." In: "Molecular Electronics." Editor: F. Hong, IEEE Biomedical Eng. Soc. Meeting, Philad. Penn. Nov. 2-4, 1990.

Szu, H., Caulfield, J. and Werbos, P. (NSF), "Convergence Theorems of Neurocontrol Systems." Trans. IEEE Neural Networks, submitted 1990.

Szu, H. and Blodgett, J "Self-Reference spatio-temporal Image-Restoration Technique." J. Opt. Soc. Am., Vol. 72, pp. 1666-1669, 1982.

Szu, H. and Messner, R., "Adaptive Invariant Novelty Filters." Proc. IEEE, V. 74, p. 519, 1986.

Szu, H. and Scheff, K., "Gram-Schmidt Orthogonalization Neural networks for Optical Character Recognition." Int Joint Conference on Neural Networks, Vol. I, pp. 547-555, Washington D.C., June 18-22 1989 apperared in the J. Neural Network Computing 1989.

Szu, H., "Fast Simulated Annealing." In: "Neural Networks for Computing." AIP Conf. Vol. 15, pp. 420-425, Edited by J. Denker, Snow Bird U.T., 1987.

Szu, H. and Hartly, R., "Fast Simulated Annealing." Phys. Letters A 122, pp. 157-163, Jun 8, 1987.

Szu, H. and Hartly, R., "Non-Convex Optimization by Fast Simulated Annealing." Proc. IEEE, V. 75, pp. 1538-1541, 1987.

Takefuji, Y. and Szu, H., "Parallel Distributed Cauchy Machine." Int. Joint Conf. Neural Networks-89, p. I-529, Washington D.C. June 18-22 1989.

Szu, H., "Reconfigurable Neural networks by Energy Convergence Learning Principle Based on Extended McCulloch and Pitts Neurons and Synapses." Int. Joint Conf. Neural Networks 1989, p. I-485, Washington D.C. June 18-22 1989.

Szu, H. and Scheff, K., "Simulated Annealing Feature Extraction from Occluded and Cluttered Objects." Int. Joint Conf. Neural Networks 1990, p. II-76, Washington D.C. Jan. 15-18 1990.

Testardi, L. R., Phys. Rev. B4, 2189, 1971.

Yao, K.C. (Lab. des Systemes Photoniques, Univ. Louis Pasteur, Strasbourg, France), "Perspective of a Neural Optical Solution of the Traveling Salesman Optimization Problem." Proc. SPIE — Int. Soc. Opt. Eng. Vol. 1134, pp. 17-25, 1989 (Optical Pattern Recognition II, April 26-27 1989, Paris, France).

Yeh, P. and Chiou, A.E.T. (Rockwell at Thousand Oaks), "Optical Matrix-Vector Multiplication Through Four-Wave Mixing in Photorefractive Media." Opt. Letters, Vol. 12, No. 2, pp. 138-140, Feb. 1987.

Yu, F.T.S., Lu, T., Yang, X. and Gregory, D.A. (Penn State Univ, and Redstone Army M. C.), "Optical Neural Network With Pocket-Sized Liquid Crystal Television." Opt. Letters, Vol. 15, No. 15, pp. 863-865, Dec. 1 1990.

Wagner, K. and Psaltis, D. (California Institute of Technology), "Multilayer Optical Learning Networks." Proc. SPIE — Int. Soc. Opt. Eng. Vol. 752, pp. 86-97, 1987 (Digital Optical Computing, 3 Jan. 14 1987, Los Angeles, CA).

Wagner, K. and Psaltis, D. (California Institute of Technology), "Multilayer Optical Learning Networks." Appl. Opt. Vol. 26, No. 23, pp. 5061-5076, Dec. 1 1987.

Special Collective References:

"Neural Networks." Applied Optics, Editors: S. Grossberg, G. Carpenter, Vol. 26, No. 23, Dec. 1 1987.

SPIE Book, "Optical and Hybrid Computing." H. Szu Ed. SPIE Vol. 634, 1987, pp. 450, (March 24-27 1986, Leesburg Va.).

IEEE Spectrum special issue Optical computing, Article by Trudy Bell, 1986.

Peter Batacan, "Can Physics Make Optics Computing?" Computers in Physics, Mar/Apr 1988, pp. 9-15.

18

NEURAL NETWORKS FOR SPATIO-TEMPORAL PATTERN RECOGNITION

Alianna J. Maren

18.0 OVERVIEW

The key issue in using neural networks for pattern recognition with temporal or spatio-temporal data is that there needs to be some means of recognizing and storing the *temporal* nature of the pattern. That is, clear ways to represent temporal relationships and processes in network structures, dynamics, and (for real-time learning systems) in learning must be found. From a neurally based viewpoint, A. Harry Klopf and James Morgan state that "real-time considerations are fundamental to natural intelligence" [Klopf and Morgan, 1989].

We devote a whole chapter to spatio-temporal pattern recognition because so many real-world applications require these techniques. It is very easy to apply most of the well-known neural networks (such as discussed in Parts II and III) to the recognition of spatial patterns, or *snapshot* patterns. It is the patterns which draw from the considerations of previous history, context, or any of the other subtleties which crop up in the more challenging applications, which require subtle adaptations of network design. An exemplar of this type of application is speech recognition. (For excellent reviews, see [Lippmann, 1989] and [Smythe, 1989].) There are many similar applications. This chapter deals with neural network approaches suitable for these class of application as a whole.

There are multiple ways in which neural networks can represent temporal information. These are to:

- Create a spatial representation of temporal data
- Put time-delays into the neurons or connections
- Use back-connections (recurrent connections) so that a network can create a temporal signal sequence,

- Use neurons with activations which sum inputs over time
- Use both short-term and long-term *synaptic* connections,
- Use frequency-encoding for presenting data to nets and for network operations
- Use combinations of the above

18.1   CREATING SPATIAL ANALOGUES OF TEMPORAL PATTERNS

Many methods rely on transforming temporal data into some type of spatial, or at least static, representation. This is probably the easiest strategy to use, and several researchers have opted for this choice. The basic idea is to represent a sequence of incoming temporal data simultaneously in the input layer of the network. The network used is typically a back-propagating Perceptron, and it performs classification on the input data. When a new data element is accessed, the oldest data element in the input layer is bumped out, the remaining ones are shifted to the next position in line, and the new data element is inserted at the tail end of the stored data stream.

If the data is sampled at a rate at which the network can do this processing (including the pattern recognition step once a set of data is in the net), then the clocking of the network can be set equal to the data access rate. Sometimes, the data sampling rate is much faster than the network can work, or the information is spread over a large temporal interval with many data points. Also, sometimes the raw data is transformed into a new representation space (e.g., power spectrum), and the resulting spectrum is no longer clearly linked to each individual data element. In these cases, network designers have to arbitrarily select time windows. These may be overlapping or non-overlapping.

Researchers who have taken this approach include Simpson na Deich, et al. [1988], Gorman and Sejnowski [1988a], Bengio et al. [1989] and Goldberg and Pearlmutter [1989]. The work by Goldberg and Pearlmutter [1989] exemplifies the simplist version of this approach, and they obtain fairly decent results when applying this method to trajectory learning for a robotic arm. (Best results are root mean square error divided by standard deviation, or RMSS, of .027 on training data and .10 on test data.)

Another effective way of constructing a spatial representation of temporally-occurring information is to create a frequency spectrum (or power spectrum) of the incoming information. This can be represented as sampling from a two-dimensional space, where one coordinate is frequency or power and the other is amplitude. This approach has been used by Roitblat et al. [1989]. Bengio et al. [1989] used a similar two-dimensional grid to represent frequency versus time for speech detection. Gorman and Sejnowski [1988 (a) & (b)] and Khotanzad et al. [1989] have used a similar approach for passive sonar interpretation. Gevins and Morgan [1988] have achieved impressive results using a system of multiple, interacting feedforward neural networks to achieve automatic noise removal and waveform detection/analysis of multichannel evoked potential response measurements.

Tom and Tenorio [1989] use a somewhat more complex version of the same approach. For speech analysis, they extract each *word* from the temporal data and segment it into 32 frames. If the temporal interval during which the word was spoken was very short, then those frames may overlap. 256 data points are sampled to provide input data for each frame. Each *frame* contains not raw data, but rather four features extracted from the 256 data points, and four coefficients from a linear predictive coding of that set of points. The overall input into the neural network is thus 256 data elements, which are 32 frames of 8 features each.

Lippmann [1989] reviews a number of other workers in the speech recognition area who have used similar approaches. He describes two different classification networks used with this type of *static* representation of temporal information. One of them is the multilayer, back-propagating Perceptron. This is the method of choice of all researchers previously cited in this subsection. The other method is a hierarchical network that computes kernal functions. This network (developed by Huang and Lippmann [1988]) gives similar performance results to the back-propagating Perceptron, but is much faster to train. It uses a combination of a Kohonen-like bottom layer which performs learning vector quantization, followed by a Perceptron-like top layer which performs classification.

Klimasauskas [1989] shows how a *spatial* representation of temporal information may be used in a back-propagating multilayer Perceptron to do noise filtering. He demonstrates that the same method can be used to perform dimensionality reduction, which in this case is very similar to the task done by the adaptive data sampler when it outputs the frequency and amplitude decomposition of the incoming data.

We can build systems of multiple, interconnected networks to combine these capabilities. An example would be to use a neural net for noise filtering followed by a spatio-temporal pattern recognizer, as Simpson and Deich [1988] have proposed.

This type of data presentation may also be used in more complex networks, such as the ones developed by Smythe [1989]. These networks are discussed in the following subsections appropriate to their major architectural features.

In summary, the approach of using a static representation of temporal information is perhaps the easiest and most well-known of the neural network methods for temporal data pattern recognition. The advantage of this approach is that it is relatively straightforward to implement. Networks can be trained to recognize certain patterns. Preliminary *featurization* of raw data may facilitate classification of temporally periodic data.

There are also several disadvantages with this approach, as discussed by Elman [1990]. First, this approach requires some interface or buffering with the real world. How should a system know how often to examine the buffers? Second, this approach (and that of the time-delay network, discussed in the next section), imposes a rigid limit on the duration of patterns. This fixed limit is not necessarily appropriate to the incoming stimulus. Third, this approach does not distinguish between relative temporal position and absolute temporal position. Finally, all the known architectures for using this form of data representation are all "standard" (e.g., back-propagation) pattern classification networks. These networks do not handle novel inputs well.

18.2   NEURAL NETWORKS WITH TIME DELAYS

The previous subsection dealt with how we could make a spatial representation of temporal data. In contrast, this and the next few subsections deal with how we can explicitly handle the temporal aspect of incoming data. There are three major approaches: use of time delays, use of recurrent connections, and use of temporal integration in the nodes. Many networks involve combinations of these three.

In a time-delay network, information taken at one moment in time is *shifted right* down a series of nodes, and new information is inserted in the leftmost node. (See Figure 18.1.) The number of nodes determines the number of time intervals over which information is kept. Each network operation is still done in a single network cycle; there is no explicit integration of temporal information, nor does the information stored in any of the rightmost nodes degrade with time. When the information in the rightmost node(s) is bumped out of the network altogether, its effect simply disappears.

Tam and Perkel [1989] have developed a neural network which can recognize sequences of spike trains by modifying a back-propagating Perceptron to have multiple hidden layers. All but the first hidden layer are on time-delay shifts from a previous hidden layer. (See Figure 18.1.)

A simple implementation of a time delay network is incorporated into the ring network developed by Tabor and Deich [in review]. Their architecture is well-suited for classifying temporally periodic data, and they have applied it to classification of acoustic ship signatures. In their *ring* approach, they have adapted Hecht-Nielsen's spatio-temporal pattern recognizer [Hecht-Nielsen, 1987]. The processing element in the ring structure is a *Dot Product Neuron with Latches* (DPLN) that hold time and activation values. Using the DPLN allows for incoming patterns to be effectively compared with stored temporal patterns even when they are *out of phase* with the stored pattern. Analysis includes use of fuzzy primitives.

Lippmann's review article [1989] cites a number of neural networks which use time delays and achieve good results for speech classification. Unnikrishnan, Hopfield, and Tank [1988] and Tank and Hopfield [1987] have developed a time-delay network which has variable length delay lines designed to disperse impulsive inputs such that longer delays result in more dispersion. Impulsive inputs to these delay lines are formed by

A Time-Delay Network

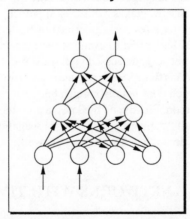

Figure 18.1   A time-delay network.

enhancing spectral peaks in the outputs of 32 bandpass filters. Outputs of delay lines are multiplied by weights and summed to form separate matched filters for each word.

Lang, Waibel, and Hinton [1990] have developed a time-delay neural network which distinguishes among the difficult class of "E" words (B, D, V, and E) with up to 94 percent accuracy on training data and 90.9 percent accuracy on test data. This can be compared with human performance at 94 percent and the performance of the IBM recognition system of 80 percent and 89 percent performance of a specialized Hidden Markov Model classifier which included a back-propagation network. Hampshire and Waibel [1990] have further developed this work, using a system containing three similar networks, each trained using different objective functions. One objective function is the typical mean-squared error (which is the basis of the back-propagation training rule), another is the cross-entropy method developed by Hinton [1987], and a third is their newly defined *classification figure of merit* (CFM). This last approach uses the ideal activation outputs only to identify which node is providing the most optimal response to a pattern. The CFM seeks to maximize the difference between that node's output and all the other nodes. By using a simple arbitration mechanism to resolve conflicts betweent the decisions of the three networks, they have been able to achieve error rates consistently below 2 percent on distinguishing between B, D, and G sounds produced by different speakers. All of this work has been done using time-delay networks.

Bottou et al. [1990] have recently reported using a time-delay neural network to achieve results on speaker-independent isolated digit recognition which are highly comparable to those of a Dynamic Time Warping (DTW) system. Their best results were 99.21 percent on training patterns and 99 percent (99 out of 100) correct identification of a test set. Typical results on their final network architecture were never less than 98 percent correct on training data and 94 percent on test data. The DTW achieved 99 percent performance, similar to their best-performing network. They note that the network operates faster in real-time than DTW, and that has better potential to handle poorly-segmented data.

Lippmann [1989] cites other work in which researchers compared learning vector quantization networks with time delays with the results of multilayer Perceptrons with time delays. The error rates were about the same (1.7 percent) on identification of similar-sounding consonants.

Smythe [1989] has developed a network which has two stages: a preliminary stage with time delays implemented by lateral connections, and a later stage for syllable classification.

Some interesting things happen when neural networks simply have time delays in their connectivity. For example, an architecture which leads to convergence to a steady state when there is no time delay might produce sustained oscillations or even chaotic behavior when even small time delays are introduced [Marcus and Westervelt, 1989].

18.3   STORING AND GENERATING TEMPORAL PATTERNS VIA RECURRENT CONNECTIONS

One of the most well-known ways of making a neural network store temporal information is to introduce certain types of recurrent connections or back-connections into a set of neurons. Grossberg [1969, 1970] did this when he developed the Avalanche network, and several others have used this idea. Typically, these back-connected architectures will store a pattern which they can produce when stimulated. However, avalanche-type networks are not able to classify incoming patterns, compare an incoming pattern with their own internally stored pattern, adaptively change their stored patterns, or create new patterns which they store. However, if we will decide to build a network which will compare an incoming temporal signal against a stored one, then these types of networks can be used to generate the model temporal signal pattern. Also, these networks may have some use in pacing certain cyclic operations.

It is possible to use recurrent or back-connections in other ways. (See, e.g., Rumelhart et al. [1986] for description of some early work on recurrent networks.) We can create a back-propagating Perceptron or Boltzmann machine in which there are two types of input. One is the input obtained from the current incoming data. The other is the output just produced by the network, which is fed back to the input level to represent the *state* of the network at the preceding moment in time. Figure 18.2 illustrates this type of recurrently connected feedforward network. Lippmann [1989] cites a number of researchers who have chosen this route. While these networks seem to have performed well on certain types of speech classification, they have also required a great deal of training time.

Jordan [1989, and references contained therein] has used recurrently-connected back-propagation Perceptrons for guiding robotic motion and for speech. Elman [1990], in a detailed study, has explored the use of recurrent networks for representing structure in time. He has concluded that a time-varying error-signal can be used as a clue to temporal structure, in that temporal sequences are not always uniformly structured. He also found that representation of time — and memory — is highly task-dependent. Internal representations mix the demands of the task with the demands imposed by carrying out the task over time. This means that there is no separate, identifiable

A Recurrently-Connected Network

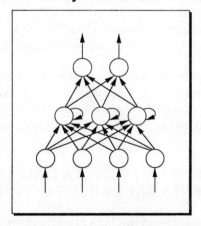

Figure 18.2   An illustration of a recurrently-connected network.

representation of time. There is simply representation of input in its context. He has also found that it is possible to represent complex structures using recurrent networks. These include, for example, categorical relationships and hierarchical relationships such as type/token distinctions. Because representations are structured, relations between representations are preserved.

Servan-Schreiber et al. [1988] have explored using the recurrent network described by Elman to predict successive elements in a sequence. They find that a network trained with samples from a finite-state grammar can become a perfect finite-state recognizer for that grammar. Their work shows that long distance sequential contingencies can be encoded by the network even if only subtle statistical properties of the embedded strings depend on the early information. Brown [1990] has used a system of recurrent networks to explore auditory-verbal short-term memory.

Almeida [1987] and Pineda [1987a, 1987b, 1988] have both developed special types of recurrent back-propagation networks. They each treat the network as a dynamical system in which the behavior of the network obeys coupled differential (for Pineda) or difference (for Almeida) equations without making distinctions between input and output. Almeida notes that the network can converge to more than one stable state given an input vector, so that its final output vector depends on the previous state. This leads to sequential behavior. Pineda's network bears some resemblence to an asymmetric Hopfield network and the Lapedes and Farber [1986] master-slave network as well as to a back-propagation network. Deprit [1989] has implemented their network on a Connection Machine (TM).

18.4   USING NEURONS WITH TIME-VARYING ACTIVATIONS AND SUMMING INFORMATION OVER TIME INTERVALS

Another way in which neural networks recognize time-varying information is to modify the individual neurons so that they are able to sum time-varying (analog) data over time, instead of being limited to single presentations of data at each cycle of operation. Even without extension to analog capabilities, artificial neurons can be created which keep some residue of previous signals, and allow slow decay of historical information.

If the neurons have recurrent connections, as described in the previous subsection, the feedback of information acts as a form of creating hysteresis in each neuron. In a time-varying situation, this helps to perpetuate an induced state beyond the original stimulus which created that state [Norrod et al., 1989; Yanai and Sawada, 1990]. Alternatively, the individual neuron may have a time-dependent activation function in which the activation decays over time [Horn and Usher, 1989; Uchiyama et al., 1989]. Cohen et al. [1987] have used such an approach in their masking field network and applied it to speech analysis and synthesis.

Wang and King [1988] have used neurons which integrate information over time to create three different types of networks. These networks can achieve frequency ordering, frequency filtering, and temporal-sequence memory.

Sung and Priebe [1988] have developed a neural network in which the input nodes undergo temporal decay, and the output nodes are Gaussian classifiers which effectively sum information over time. The architecture has some similarity to a simple two-layer Perceptron, except that the output nodes sum their activation over time. The input layer has specific nodes devoted to specific, identifiable input values. This network incorporates some interesting ideas, but there are issues which still have to be worked out (such as a handy way to represent incoming data — the method just described seems a bit ad hoc).

18.5 NEURAL NETS WHICH HAVE SHORT-TERM AND LONG-TERM MEMORIES

Another option is to change the connection weights between neurons to allow for different types of information transfer: short-term and long-term. To be useful, a neural network should, ultimately, handle time-varying information and be able to store and distinguish between *instantaneous* information, short-term memory, and long-term memory.

Kleinfeld and Sompolinsky [1989] have developed an associative network model which can generate different types of temporal patterns: either linear sequences or cycles of patterns. Their network has a unique feature: each pairwise connection between two neurons is actually encoded as two connections. One has a short response time, and the other a much longer one. The time scale for the short response time neurons determines the time needed for the network to settle into one of its embedded states. After being in that state for a while, the information being transmitted by the connections with longer response time causes the network to shift into a new state. This process continues, and the physical link-up of connections between neurons determines whether the temporal pattern of states produced is a simple linear pattern or cycles among a set number of states.

Amit [1988, 1989] has similarly described a Hopfield-like network with connections such as used by Kleinfeld and Sompolinsky. The network which they describe mathematically should be able to recall different types of sequences, and is also able to count the observation of multiples reoccurences of the same type of pattern, as in counting chimes.

Grzywacz and Poggio [1990] have shown how temporally sustained inhibition facilitates motion detection in the visual cortex. This is an example of the powerful role of inhibition in biological systems, with strong implications for neural network design for vision and related applications.

18.6 FREQUENCY CODING IN NEURAL NETWORKS

Dayhoff [1990, 1988, 1987], Dress [1987] and others have suggested that it could be useful to develop neural networks which respond to frequency-encoded or pulse-coded information rather than single presentations of *static* information. They point out that this is the method of choice of biological systems, and that our current reliance on single presentations of data is an artifact due to our use of sequential machines. Dayhoff states that pulse patterns, in particular, have the additional advantage of high information content which can be encoded into the pulse structure. Pulse-coded information can be decoded and represented using a topographic map [Tam, 1990]. This is an advance over similar topographic map representation of sequences described by Tolat and Peterson [1989]. Goerke et al. [1990] have developed a two-module network for generating different sequences of pulse codes. Such a network can be used to drive actuators for robotics applications. Beerhold et al. [1990] have developed a hardware implementation of a network for processing pulse-coded information.

18.7 NETWORKS WITH COMBINATIONS OF DIFFERENT TEMPORAL CAPABILITIES

A final approach might be to combine previously mentioned various options.

Ryckebusch, Bower, and Mead [1989] have developed networks which generate *oscillating*, or *burst patterns*, of activation. These analog VLSI networks are back-connected, and integrate current over time. Like their biological counterparts, each neuron in this network can also output a train of pulses. Temporal delays of different lengths are also incorporated into his networks. They have shown how different network configurations can crudely model the observed activity in several well-known biological Central Pattern Generators (CPGs).

Another very interesting and useful piece of work has been done by Rossen and Anderson [1989]. They have investigated different representational formats for inputting temporal information into a speech recognition system. Their network is a concatenation of a back-propagation network and a recurrent connection in the output layer leading to a Brain-State-in-a-Box type of autoassociative output among categorization possibilities.

They tested several different types of inputs, in different combinations. Their inputs were based upon overlapping time-slices of frequency log-amplitude data. They used 10 neurons to represent each smoothed frequency log-amplitude. Neuron activation would be proportional to a Gaussian function of its distance (in terms of amplitude) from the amplitude of the the incoming signal at that frequency. They recorded information from a number of frequencies, and produced a total of 256 real frequency log-amplitudes within a specified range. They used this preprocessed signal to produce several different input representations. This input would be gathered in ten different time-slices, and thus ten presentations would be made to the neural network.

The input representations they investigated included smoothed spectral log-amplitude, cepstral representation (Fourier transform of the preprocessed signal), time difference spectral representation, frequency discriminators and modulation detectors, and spectral slope representation. These input representations were combined in various forms and used as input to the neural network. The combination which worked best produced an error rate of about 2.6 percent for a one-speaker system. This was a combination of frequency amplitudes in both 1-D and 2-D representations, along with 2-D representations of 2-frame time differences and frequency discriminators.

Their system was trained to reject noise and to notice when superfluous categorizations were made. Both the hidden neurons and the recurrent connections each contributed about 5 percent improvement to system performance. The strength of their system seems to lie in the combination of different architectures, good training strategy, and comprehensive, multitype input representation. Their system will not respond well to recognizing novel input, but some ideas from their work could be used in a system that has such a capability.

18.8 SUMMARY

There are several methods for using neural networks for spatio-temporal pattern recognition. The simplest, and most common, is to spatially encode the temporal information, as can be done by using Fourier transform preprocessing. Time-delay networks are also a valid choice. In these networks, certain input neurons receive their input from neurons storing information from the previous clock cycle, rather than from an external source. Time delay networks have shown great promise in such applications as speech recognition, and are highly competitive with existing technologies. However, both spatial encoding and time-delay approaches have some inherent problems with scaling, in terms of how much (or how long) of a temporal pattern can be encoded. Recurrent networks are a useful and interesting possibility, and have been applied to speech recognition and robotic control tasks. There is a background of theoretical research in recurrent networks (e.g., studies of stability), and they are likely to be developed more over the coming years. Other alternatives, such as frequency (or pulse) coding of information, neurons with leaky integrator or temporal summing properties, and networks containing connection weights with different time constants, are all under development but are less well established in the applications domain.

REFERENCES

Almeida, L.B. (1987). "A learning rule for asynchronous perceptrons with feedback in a combinatorial environment," *Proc. First Int'l. Conf. Neural Networks,* San Diego, CA; June 21-24, 1987, II-609–II-618.

Amit, D.J. (1989). *Modeling Brain Function,* Cambridge Univ. Press, Cambridge, England.

Amit, D.J. (1988). "Neural networks counting chimes," Proc. Nat'l. Acad. Sci. USA, *85,* 2141-2145.

Beerhold, J.R., Jansen, M., and Eckmiller, R. (1990). "Pulse-processing neural net hardware with selectable topology and adaptive weights and delays," *Proc. Third Int'l. Joint Conf. Neural Networks,* San Diego, CA, June 17-21, 1990, II-569–II-574.

Bengio, Y., Cardin, R., de Mori, R., and Merlo, E. (1989). "Programmable Execution of Multi-Layered Networks for Automatic Speech Recognition," *Communications of the ACM, 32,* 195-199.

Bottou, L., Fogelman Soulie, F., Blanchet, P., and Lienard, J.S. (1990). "Speaker-independent isolated digit recognition: Multilayer Perceptrons vs. dynamic time warping," *Neural Network, 3,* 453-465.

Brown, G.D.A. (1990). "Short-term memory capacity limitations in recurrent speech production and perception networks," *Proc. Second Int'l. Joint Conf. Neural Networks*, Washington, D.C., Jan. 15-19, 1990, I-43-I-46.

Cohen, M.A., Grossberg, S., and Stork, D. (1987). "Recent developments in a neural model of real-time speech analysis and synthesis," *Proc. IEEE First Int'l. Conf. Neural Networks*, San Diego, CA; June 21-24, 1987, IV-443-IV-453.

Dayhoff, J. (1990). "Regularity properties in pulse transmission networks," *Proc. Third Int'l. Joint Conf. Neural Networks*, San Diego, CA; June 17-21, 1990, III-621-III-626.

Dayhoff, J.E. (1988). "Temporal structure in neural networks with impulse train connections," *Proc. IEEE Second Int'l. Conf. Neural Networks*, San Diego, CA; July 24-27, 1988, II-33-II-45.

Dayhoff, J.E. (1987). "Detection of favored patterns in the temporal structure of nerve cell connections," *Proc. IEEE First Int'l. Conf. Neural Networks*, San Diego, CA; June 21-24, 1987, III-63-III-77.

Deprit, E. (1989). "Implementing recurrent back-propagation on the Connection Machine," *Neural Networks, 2*, 295-314.

Dress, W.B. (1987). "Frequency-coded artificial neural networks: An approach to self-organizing sytems," *Proc. IEEE First Int'l. Conf. Neural Networks*, San Diego, CA; June 21-24, 1987, II-47-II-54.

Elman, J.L. (1990). "Finding structure in time," *Cognitive Science, 14*, 179-211.

Gevins, A.S., and Morgan, N.H. (1988). "Applications of neural-network (NN) signal processing in brain research," *IEEE Trans. Acoustics, Speech, and Signal Processing, 36*, 1152-1161.

Goerke, N., Schone, M., Kreimeier, B., and Eckmiller, R. (1990). "A network with pulse processing neurons for generation of arbitrary temporal sequences," *Proc. Third Int'l. Joint Conf. Neural Networks*, San Diego, CA, June 17-21, 1990, III-315-III-320.

Goldberg, K.Y., and Pearlmutter, B.A., (1989). "Using Backpropagation with Temporal Windows to Learn the Dynamics of the CMU Direct-Drive Arm II," in D.S. Touretzky (Ed.), *Advances in Neural Information Processing Systems I*, Morgan Kaufmann, San Mateo, CA, 356-363.

Gorman, R.P., and Sejnowski, T.J. (1986) "Learned classification of sonar targets using a massively parallel network," *IEEE Trans. Acoustics, Speech, and Signal Processing, 36*, 1135-1140.

Gorman, R.P., and Sejnowski, T.J. (1988a). "Analysis of hidden units in a layered network trained to classify sonar targets," *Neural Networks, 1*, 75-89.

Grossberg, S. (1969). "Some networks that can learn, remember, and reproduce any number of complicated space-time patterns, I," *J. Math. and Mechanics, 49*, 53-91.

Grossberg, S. (1970). "Some networks that can learn, remember, and reproduce any number of complicated space-time patterns, II," *Studies in Applied Math., 49*, 135-166.

Grzywacz, N.M., and Poggio, T. (1990). "Computation of motion by real neurons," in S.F. Zornetzer, J.L. Davis, and C. Lau (Eds.), *An Introduction to Neural and Electronic Networks*, Academic, San Diego, 379-403.

Hampshire, J.B. II, and Waibel, A.H. (1990). "A novel objective function for improved phoneme recognition using time-delay neural networks," *IEEE Trans. Neural Networks, 1*, 216-228.

Hecht-Nielsen, R. (1987). "Nearest matched filter classification of spatio-temporal patterns," *Applied Optics, 26*, 1892-1899.

Hinton, G.E. (1987). *Connectionist Learning Proceedures*, Carnegie-Mellon Univ. Tech Report CMU-CS-87-115 (Version 2), Dec. 1987.

Horn, D., and Usher, M. (1989). "Motion in the space of memory patterns," *Proc. First Int'l. Joint Conf. Neural Networks,* Washington, D.C., June 18-22, 1989, I-61–I-66.

Huang, W.Y., and Lippmann, R.P. (1988). "Neural net and traditional classifiers," in D. Z. Anderson (Ed.), *Neural Information Processsing Systems,* American Institute of Physics, New York, 387-396.

Jordan, M.I. (1989). "Generic constraints on underspecified target trajectories," *Proc. First Int'l. Joint Conf. Neural Networks,* Washington, D.C., June 18-22, 1989, I-217–I-225.

Khotanzad, A., Lu, J.H., and Srinath, M.D. (1989). "Target detection using a neural network based passive sonar system," *Proc. Int'l. Joint Conf on Neural Networks,* Washington, D.C., June 18-22, 1989, I-335–I-340.

Kleinfeld, D., and Sompolinsky, H. (1989). "Associative network models for central pattern generators," in C. Koch and I. Segev (Eds.), *Methods in Neuronal Modeling* MIT, Cambridge, MA, Chapter 7, 195-246.

Klimasauskas, C. (1989) "Neural nets and noise filtering," *Dr. Dobb's Journal* (January), 32-48,ff.

Klopf, A.H., and Morgan, J.S. (1989). "The role of time in natural intelligence: Implications for neural network and artificial intelligence research," *Proc. First Int'l. Joint Conf. Neural Networks,* San Diego, CA, June 18-22, 1989, II-97–II-100.

Lang, K.J., Waibel, A.H., and Hinton, G.E. (1990). "A time-delay neural network architecture for isolated word recognition," *Neural Networks, 3*, 23-43.

Lapedes, A., and Farber, R. (1986). "Programming a massively parallel, computation universal system: Static behavior," in J.S. Denker (Ed.), *Neural Networks for Computing* (AIP Conference Proceedings 153, *Proc. of Neural Networks for Computing Conf. at Snowbird, UT, Apr. 13-16, 1986),* 283-298.

Lippmann, R.P., "Review of Neural Networks for Speech Recognition," *Neural Computation, 1* (1989), 1-38.

Marcus, C.M., and Westervelt, R.M., "Dynamics of analog neural networks with time delay," in D.S. Touretzky, *Advances in Neural Information Processing Systems I,* San Mateo, CA: Morgan Kaufmann, 1989), 568-576.

Norrod, F.E., O'Neill, M.D., and Gat, E. (1989). "Feedback-induced sequentiality in neural networks," *Proc. First Int'l. Joint Conf. Neural Networks,* San Diego, CA; June 21-24, 1989, II-251–II-258.

Pineda, F.J. (1987a). "Generalization of back-propagation to recurrent neural networks," *Phys. Rev. Letters, 59*, 2229-2232.

Pineda, F.J. (1987b). "Generalization of back-propagation to recurrent and high-order neural networks," *Proc. IEEE Conf. Neural Inf. Proc. Systems.* IEEE: New York.

Pineda, F.J. (1988). "Dynamics and architecture for neural computation," *J. of Complexity, 4*, 216-245.

Roitblat, H.L., Moore, P.W.B., Nachtigall, P.E., Penner, R.H., and Au, W.W.L (1989). "Dolphin echlocation: Identification of returning echoes using a counterpropagation network," *Proc. Int'l Joint. Conf. Neural Networks,* Washington, D.C., June 18-22, 1989, I-295–I-299.

Rossen, M.L., and Anderson, J.A. (1989). "Representational issues in a neural network model of syllable recognition," *Proc. Int'l. Joint Conf. on Neural Networks,* Washington, D.C., June 18-22, 1989, I-19–I-25.

Rumelhart, D.E., Hinton, G.E., and Williams, R.J. (1986). "Learning internal representations by error propagation," in D.E. Rumelhart and J.L. McClelland (Eds.), *Parallel Distributed Processing: Explorations in the Microstructure of Cognition, 1,* MIT Press: Cambridge, MA, 318-362.

Ryckbusch, S., Bower, J.M., and Mead, C. (1989). "Modeling small oscillating biological networks in analog VLSI," in D.S. Touretzky (Ed.), *Advances in Neural Information Processing Systems* (San Mateo, CA: Morgan Kaufmann), 384-393.

Servan-Schreiber, D., Cleeremans, A., and McClelland, J.L. (1988). *Encoding Sequential Structure in Simple Recurrent Networks,* Carnegie-Mellon University Technical Report CMU-CS-88-183.

Shamma, S. (1989). "Spatial and Temporal Processing in Central Auditory Networks," in C. Koch and I. Segev (Eds.), *Methods in Neuronal Modeling* (Cambridge, MA: MIT, 1989), 247-289.

Simpson, P.K., and Deich, R.O. (1989). "Neural networks, fuzzy logic, and acoustic pattern generation," *Proc. of the AAAIC 88* (Dayton, Ohio, September).

Smythe, E.J. (1989). "Temporal representations in a connectionist speech system," in D.S. Touretzky (Ed.), *Advances in Neural Information Processing Systems I,* Morgan Kaufman, San Mateo, CA, 240-247.

Sung, C.-H., and Priebe, C.E. (1988). "Temporal Pattern Recognition," *Proc. IEEE Int'l. Conf. on Neural Networks,* San Diego, CA; July 24-27, 1988, I-689–I-696.

Tabor, W.R., and Deich, R.O., "A Comparison of Feedforward Neural Networks: Fuzzy Operators and Acoustic Ship Signatures," *Int'l. Journal of Intelligent Systems* (in review).

Tam, D.C. (1990). "Decoding of firing intervals in a temporal-coded spike train using a topographically-mapped neural network," *Proc. Third Int'l. Joint Conf. Neural Networks,* San Diego, CA, June 17-21, 1990, III-627–III-632.

Tam, D.C., and Perkel, D.H. (1989). "A model for temporal correlation of biological neuronal spike trains," *Proc. Int'l. Joint Conf. on Neural Networks,* Washington, D.C., June 18-22, 1989, I-781–I-786.

Tank, D., and Hopfield, J.J. (1987). "Concentrating information in time: Analog neural networks with applications to speech recognition problems," *Proc. IEEE First Int'l. Conf. Neural Networks,* San Diego, CA; June, 1987.

Tolat, V.V., and Peterson, A.M. (1989). "A self-organizing neural network for classifying sequences," *Proc. First Int'l. Joint Conf. Neural Networks,* Washington, D.C., June 18-22, 1989, II-561–II-568.

Tom, M.D., and Tenorio, M.F. (1989). "A spatio-temporal pattern recognition approach to word recognition," *Proc. Int'l. Joint Conf. on Neural Networks,* Washington, D.C., June 18-22, 1989, I-351–I-355.

Uchiyama, T., Shimohara, K., and Tokunaga, Y. (1989). "A modified leaky integrator network for temporal pattern processing," *Proc. First Int'l. Joint Conf. Neural Networks,* Washington, D.C., June 18-22, 1989, I-469–I-475.

Unnikrishnan, K.P, Hopfield, J.J., and Tank, D.W. (1988). "Learning time-delayed connections in a speech recognition circuit," *Proc. Neural Networks for Computing Conference,* Snowbird, UT, 1988.

Wang, D., and King, I.K. (1988). "Three Neural Models Which Process Temporal Information," presented at the 1988 IEEE International Conference on Neural Networks. Copy of the paper available from the authors at Dept. of Computer Science, University of Southern Cal., Los Angeles, CA, 90089-0782, (213)-743-2747.

Williams, R.J., and Zipser, D. (1989). "A learning algorithm for continually running fully recurrent neural networks," *Neural Computation, 1* (Summer), 270-280.

Yanai, H., and Sawada, Y. (1990). "Associative memory network composed of neurons with hysteretic property," *Neural Networks, 3,* 223-228.

19

NEURAL NETWORKS FOR MEDICAL DIAGNOSIS

Dan Jones, M.D.

19.0 OVERVIEW

Despite high hopes and two decades of work, symbolic diagnostic systems have had minimal impact on the practice of medicine. Perhaps neural networks offer new hope for allowing computers to assist in the challenging and costly process of medical diagnosis. Likely market niches for diagnostic neural networks, pragmatic design considerations, and existing neural network systems for medical diagnosis are discussed. Special problems inherent in the development of a large, general-purpose diagnostic network are also discussed. Such a system presents the ultimate challenge to designers of diagnostic networks.

19.1 INTRODUCTION

Neural networks are highly fault-tolerant, and when properly trained, are capable of finding near-optimum solutions from limited or degraded information. If they can allow these elusive qualities of human reasoning to be combined with the compulsive thoroughness, precise logic, and perfect memory of computers, then neural networks can make a significant contribution to medical diagnosis.

Medical data generally has several properties that make it resistant to conventional techniques of computer analysis:

- *Subjectivity.* Does the patient appear pale? Is your pain constant or crampy?
- *Imprecision.* Simple blood pressure measurement, for example, is imprecise due to variability in the patient's arm size, posture, and emotional state; the type of equipment used; and the technique of the person taking the measurement.

- *High noise content.* Symptoms and signs are often encountered that are normal variants, or incidental findings that bear no relationship to the patient's primary problem.
- *Incompleteness.* A medical data base is never complete due to limitations of time, equipment, money, and other resources.

Although the above properties have made medical diagnosis resistant to conventional techniques of computer analysis, they are the very properties most amenable to neural network solutions. As we will see, neural networks have already demonstrated competence in circumscribed areas of medical diagnosis. But much work remains to be done before neural networks become commonplace in the clinic.

19.2 PROSPECTS FOR NEURAL NETWORKS IN MEDICINE

Several strong recent trends in medicine are promoting a more receptive climate for computerized diagnostic aids.

Physicians are under ever-increasing pressure from third-party payers to standardize and optimize diagnosis and treatment. Both Medicare and insurance companies are routinely refusing to reimburse for diagnostic tests, treatments, and hospitalizations that cannot be justified according to standard criteria. The federal government and other third-party payers are clamoring for uniform application of standardized diagnostic and therapeutic criteria. Although physicians are very good at intuiting what is required for a particular patient, they clearly will need computer assistance to cope with the coming onslaught of rules and regulations.

In general, physician-patient relationships are more business-like, and less stable and committed than in the past. Contributing factors include more frequent patient relocation; the increased specialization of physicians, causing patients to see more physicians while spending less time with each; and technological advances causing patients to spend more time with technicians and machines, and less time with their physicians. This deterioration in the physician-patient relationship, combined with the general societal trend toward increasing litigation, has resulted in an explosion in malpractice suits. This hazardous litigation climate compels physicians to overdiagnose and over-treat in a defensive manner, increasing costs and decreasing efficiency. Note that this pressure is in direct conflict with the above-mentioned pressure from third-party payers. These conflicting pressures dispose physicians to embrace intelligent systems that can increase the standardization and efficiency of their care, while decreasing oversights and litigation risks.

Most hospitals now maintain peer review committees. Physicians know that their professional activities are subject to detailed review by their peers at any time. This is clearly another incentive to standardize care, and to be able to justify decisions according to standard criteria.

The explosion of medical knowledge, and the development of new diagnostic and therapeutic modalities continues unabated. Physicians are under constant pressure to keep up. With computerized diagnostic assistance, periodic software updates can assure the physician that the most current diagnostic and therapeutic knowledge is at his fingertips. Working with such a system will help familiarize physicians with new diagnostic and therapeutic developments.

For many years, medical costs have increased at a rate substantially greater than inflation. This is a source of increasing concern by the government, businesses, third-party payers, and the public at large. By increasing the standardization and efficiency of care, computerized diagnostic aids may be able to reduce unnecessary costs, while providing justification for those costs that are essential. Such systems might also reduce costs per patient by facilitating patient flow, allowing more efficient use of nursing resources, and allowing each physician to see and diagnose more patients per hour.

19.3 POTENTIAL NICHES FOR NEURAL NETWORK DIAGNOSTIC AIDS

According to some estimates, by the year 2000, neural network technology will account for half the total revenues of the computer and robotics markets (Obermeier and Barron, 1989). If neural network diagnostic aids can be smoothly integrated into existing medical settings, they are likely to find commercial applications in the following areas:

- Hospital emergency rooms. During peak hours, many emergency departments are overloaded and understaffed. Patients often wait for hours to be seen. An effective neural network diagnostic system would allow a triage nurse to quickly and efficiently derive a list of candidate diagnoses, and to order appropriate lab tests and x-rays. In many cases, the physician would need only to review and verify the network's recommendations. This would allow the doctor to make much more efficient use of his time. Also, since computers are not subject to problems of fatigue and oversight, diagnostic errors (and the subsequent morbidity, mortality, and litigation) would be minimized.
- Doctors' offices and clinics. All the above-mentioned considerations apply here, as well.
- Medical schools. By showing students how diagnostic probabilities vary with the accumulation of data, diagnostic neural networks can provide excellent teaching aids.
- Telephone medical assistance. Oftentimes, patients would like to consult a physician purely to obtain information (e.g., to ask whether a particular symptom could be a sign of serious disease, necessitating an examination). With a general-purpose diagnostic system, a nurse could take a patient's symptoms over the telephone, and give immediate feedback regarding the most likely diagnoses, how soon medical attention should be obtained, and what tests are likely to be required to sort out the diagnostic possibilities. Such a system would allow telephone triage of patients to the most appropriate medical facility, reducing misuse and overuse of medical resources.
- Home diagnostic software. A simplified version of a neural network diagnostic system, using layman's terminology, could be marketed for home computers. Such software could help consumers be more medically knowledgeable, and to make more efficient use of the health care system.

19.4   FACTORS AFFECTING PHYSICIAN ACCEPTANCE

Many software products have been developed for physicians, and considerable research has been done to determine factors affecting physician acceptance of computer aids. Physicians have traditionally been very receptive to new technology when it improves patient care. Nonetheless, certain special considerations apply if a diagnostic product is to impact this market.

In *What Makes Doctors Use Computers?: Discussion Paper* [Young, 1984] the author states, "Though diagnostic systems of all types function better than the average practitioner, their use is still very limited. If performance is enhanced, the likely reason is that these systems replace rather than support the doctor." He concludes, "The greater the change, the less likely the system will be accepted..." A computer aid should integrate smoothly into the existing way of doing things, requiring minimal change, but giving immediate benefit.

In a study designed to determine the needs, expectations, and performance demands of clinicians (Teach and Shortliffe, 1981), 81 percent of physicians expressed acceptance of computer diagnostic aids. The authors state, "Applications that were presented as aids to clinical practice were more readily accepted than those that involved the automation of clinical activities traditionally performed by physicians themselves. The distinction between a clinical aid and a replacement seems to be important....Physicians seem to prefer the concept of a system that functions as much like a human consultant as possible."

In summary, the authors suggest: strive to minimize changes to current clinical practices, and to enhance physician capabilities, rather than infringing on them or replacing them; concentrate on enhancing the interactive capabilities of the system (the system should offer understandable and persuasive explanations, should exhibit "common sense," should be easy to use, and should be easy to update); recognize that 100 percent accuracy is neither achievable nor expected.

In *Needs Assessment for Computer-Based Medical Decision Support Systems*, Berner and Brooks (1988) summarize the requirements for a successful system as follows:

- It should provide information related to specific patient problems and should provide explanations of its reasoning.
- It should have a patient registry feature.
- It should not duplicate or replace the physician as decision maker and should minimize changes to current clinical practices.
- Data in the knowledge base should come from reliable literature and should be easily updated.
- Data in the knowledge base should meet the needs of different user groups.
- The system should be easy to learn and use.

19.5   DIAGNOSTIC NETWORK DESIGN CONSIDERATIONS

Certain problems, although not unique to networks for medical diagnosis, deserve special mention. In particular, the problems of unknown inputs, large and sparse input patterns, and explanation of network results will be discussed.

19.5.1   Unknown Inputs

Medical diagnosis is best conceived of as a process, where one begins with a bare minimum of data (generally a single complaint or problem). The challenge then is to proceed from the initial complaint to a near-definite diagnosis by the most economical route (i.e., by adding the least amount of additional data to the input pattern). Each additional datum is expensive, requiring at minimum additional patient and physician time (for historical or physical exam data), and often dozens or hundreds of dollars for specialized tests. Due to these economic factors, the degree of diagnostic certainty required must often be balanced against the varying risk of misdiagnosis and the variable potential gain from accurate diagnosis. Therefore, any useful diagnostic network must be able to give reasonable outputs when some or most of the potential inputs are unknown.

It is important to point out that, at least in the case of back-propagation networks, there is no reason to assume that the network will respond appropriately to unknown values in the input pattern unless it is explicitly taught how to do so. In other words, the training set must include unknowns. The problem then becomes how best to represent unknown values in the network inputs. For binary inputs, a commonly used technique is to let 1=yes or present, 0=no or absent, and 0.5=unknown (Bounds and Lloyd, 1988; Mulsant and Servan-Shreiber, 1988). For continuous inputs, the problem is more complex. One obvious solution is to use some point on the input range, say zero, to represent the unknown condition. The known values can then be scaled to another part of the input range, say 0.25 to 1.0, leaving a gap (from zero to 0.25) that is not used. This is done to prevent the network having to learn a discontinuity at the junction of the known and unknown values. I have performed experiments with this approach and have found that it works, but training tends to be excessively slow.

The second major approach is to use more than one input for each datum. Hripcsak (1989) has experimented with this method, essentially using two inputs for each datum, one for the known state and one for the unknown. He even allowed proportional activations of these two inputs to reflect variable certainty regarding the input datum. He was able to obtain statistically valid outputs using this method but, again, training was slow even for small networks.

19.5.2   Large Sparse Input Patterns

The problem of large sparse input patterns relates to the preceding discussion. Most existing networks for medical diagnosis are relatively small and deal with a specialized diagnostic domain. But any general-purpose medical diagnostic system will need to accept several thousand input items. For example, DXplain (a non-network diagnostic aid offered via modem by the American Medical Association) accepts about 4700 input terms. I am not aware of any existing diagnostic neural networks with input arrays approaching this size. There seems to be a general consensus that networks of such size are not practical, considering the limitations of current hardware. But given the continuing rapid advances in hardware, such networks will probably become practical within a few years. In the meantime, perhaps advantage can be taken of the sparseness of the large input array required for general medical diagnosis.

Although a general medical diagnostic network would need to accept several thousand different input items, only a small fraction of these (a few dozen to a hundred) are likely to be known for any given patient. A major hurdle to creating a general-purpose neural network for medical diagnosis seems to be finding a way to take advantage of the potential economies inherent in the sparseness of this large input array. One approach is to use a modular system. Stubbs [1990] has suggested using a separate modular network for each major presenting symptom. He feels that a system consisting of about 250 small networks could correctly diagnose a high percentage of presenting cases. A conventional program would be used to organize and orchestrate the large number of modular networks.

19.5.3 Explaining Network Results

Any expert system should be able to explain its results. This is especially important for a system that will be used to guide decisions affecting human health and mortality (see section 19.4, above). Researchers have found this to be a necessary ingredient for physician acceptance of a diagnostic aid.

Unfortunately, explanatory facility has not been a strong feature of neural network diagnostic systems explored to date. Since neural networks do not operate according to verbal rules, deriving verbal explanations for their outputs is problematic. Some progress has been made, however. Yoon, et al. (1989) use a graphical display showing the changing likelihoods of various diseases as additional input data are entered. This provides a nonverbal but quantitative and comprehensible "explanation" of how the outputs are affected by the inputs. Saito and Nakano (1988) have experimented with methods for extracting rules and relation factors from a diagnostic network.

19.6   EXISTING NEURAL NETWORKS FOR MEDICAL DIAGNOSIS

The research thus far on neural network diagnostic systems has been limited but promising. In a recent review of neural networks in medicine, Stubbs (1990) offered several conclusions, including: neural networks have proven ability to diagnose circumscribed areas; accuracy is competitive with other approaches; inputs can be either discrete or continuous; and back-propagation works well as a training procedure. The following examples include only a minority of those published to date, but give a good overview of what has been accomplished. Saito and Nakano [1988] trained a back-propagation network to diagnose 23 causes of headache, on the basis of 216 possible answers to a set of questions. After being trained on 300 patients, this system achieved an accuracy of 67 percent at diagnosing at least one of a patient's problems, compared to 70 percent for a more highly developed symbolic system. Their system used 216 input units, 72 hidden units in a single hidden layer, and 23 output units. They also experimented with extraction of rules and relation factors from the network.

Bounds and Lloyd [1988] used 100 cases to train a network to recognize 4 general causes of low back pain. They used a back-prop system with 50 inputs (1=yes, 0=no, and 0.5=unknown) and 2 or 4 output units. They obtained best performance using only

2 output units to code the 4 possible diagnoses. Using 100 training cases and 100 test cases, they found 1 hidden layer to work as well as 2. They found performance to be only marginally better with 30 hidden units (80 percent correct) than with 0 hidden units (77 percent) or 50 hidden units (78 percent). This implies that their data contained little information not accessible through simple linear regression. The performance of this system was compared with that of 3 groups of physicians: neurosurgeons, orthopedists, and general practitioners. The network performance exceeded that of all 3 groups of physicians.

YoungOhc Yoon, et al. [1989], used a neural network, called Desknet, for the diagnosis of papulosquamous skin diseases. They used a back-propagation network with 96 inputs, 20 units in a single hidden layer, and 10 outputs. Desknet diagnoses 10 skin diseases on the basis of the 96 input findings. The system achieved a 70 percent success rate when tested with 99 real patients, after training on 250 hypothetical cases. Their use of hypothetical training cases is interesting, since obtaining adequate numbers of real training cases is often problematic. Techniques for generating explanations from the network are discussed.

Mulsant and Servan-Shreiber [1988] claimed an accuracy of 61 percent with a novel network that diagnoses 7 causes of dementia on the basis of 80 inputs. Their network uses 58 finding input units encoding 22 clinical attributes, and 22 "confidence" input units. They used 2 hidden layers of 10 and 7 units. The output layer includes 7 units for the 7 diagnoses, and 22 "question units" for the 22 clinical attributes; these units indicate what data should be entered next. This network is interesting because it gives ongoing advice regarding the optimum sequence of data entry.

A group in Milan, Italy [Apolloni et al., 1990], developed a large back-propagation network for the diagnosis of epilepsy. They used 724 binary inputs and 31 binary outputs representing 31 possible diagnoses. Using a single hidden layer they found best performance with 50 hidden nodes, the standard sigmoidal activation function, a learning rate of 0.1, momentum 0.1, initial weights ranging from -0.3 to +0.3, and training to a maximum error of 0.3. This network obtained "plausible generalizations" on 87 percent of test cases. After training, the authors eliminated all inputs which had no output weights above a criterion, to produce a much smaller network with only 74 inputs. This pruned network achieved an accuracy of 80 percent, comparing favorably with the much larger version. This work was performed using only 134 training and 22 test cases.

Calvert and Price [1989] used a back-propagation network with 44 inputs, 10 to 30 hidden units, and 25 outputs to predict the site of a brain lesion (as revealed by computed tomography) from the patients neurological findings. Using 90 cases for training and 40 for testing, they obtained good generalization and reasonable predictions for the test cases.

19.7   EXISTING NEURAL NETWORKS FOR PROGNOSIS AND TREATMENT

Blumenfeld [1989, 1990] has used a Simple Recurrent Network (SRN, first proposed by Elman, 1988) to track the blood glucose of a diabetic patient, proposing an appropriate insulin drip dose on an ongoing basis. This network is trained using back-propagation, but employs feedback from the hidden and output layers. Inputs are the blood glucose level and the current insulin dose; the single output node gives the proposed new insulin dose. Information regarding previous states is retained via recurrent connections. Impressive network performance was obtained using artificial training and testing data. Potential problems with this network, and the possibility of clinical application are briefly discussed.

Hutton, et al. [1989a] used a back-propagation network with 8 inputs for known risk factors, and one output, to predict the likelihood of diabetes developing in Pima Indians. They related the improved performance of a 3-layer network, relative to a 2-layer network, to the presence of secondary structure in the data (structure not detectable by linear regression). In a follow-up paper (Hutton, et al., 1989b) the authors provide additional analysis of the secondary data structure as it relates to the network. They discuss a common phenomenon: although longer training and more hidden nodes invariably improve performance on the training set, performance on the test set reaches a maximum and then decreases again. They suggest that this phenomenon is due to the network learning spurious aspects of the higher order structure of the training set. (Presumably, learning of higher order structure occurs more slowly.)

Stubbs [1989] used a back-propagation network with 5 inputs, 8-12 hidden units, and 3 outputs to predict the outcome (as recovered, dead, or disabled) of patients 6 months after head injury. This system was able to predict death and recovery with near-100 percent accuracy. Its accuracy in predicting disability was about 80 percent. 240 patients were used for training and 80 for testing.

19.8 SUMMARY

Although numerous papers have demonstrated their feasibility, diagnostic neural networks have yet to impact the practice of medicine. Socioeconomic trends are promoting a receptive climate for computerized diagnostic aids, but successful products will need to be carefully tailored to physician needs and existing clinical practices.

Multilayer feedforward networks, trained using back-propagation, give good diagnostic performance in circumscribed areas. Large networks for multisystem diagnosis have not yet been demonstrated. Special problems include representation of unknown inputs, how to derive explanations for the network's results, and how to handle the large sparse input array required for a multisystem diagnostic network.

REFERENCES

Apolloni, B., Avanzini, G., Cesa-Bianchi, N., and Ronchini, G., "Diagnosis of Epilepsy via Backpropagation," Proceedings IJCNN-90-Washington, D.C., Vol. II, pp 571-574.

Berner, E.S., and Brooks, C.M., (1988). "Needs Assessment for Computer-Based Medical Decision Support Systems," Proceedings of the Twelfth Annual Symposium on Computer Applications in Medical Care, pp. 232-236.

Blumenfeld, B., (1989). "A Connectionist Approach To The Recognition of Trends in Time Ordered Medical Parameters," Proceedings, Thirteenth Annual Symposium on Computer Applications in Medical Care, pp 288-294.

Blumenfeld, B., (1990). "A Connectionist Approach To The Processing Of Time Dependent Medical Parameters," Proceedings IJCNN-90-Washington, D.C., Vol. II, pp 575-578.

Bounds, David G., and Lloyd, Paul J., "A Multi Layer Perceptron Network for the Diagnosis of Low Back Pain," Proceedings of the IEEE International Conference on Neural Networks, Vol. 2, 1988, pp. 481-489.

Calvert, P.C., and Price, T.R., (1989). "Neurologic Localization: Learning, Knowledge Representation, and Generalization by a Neural Network Model," Proceedings, Thirteenth Annual Symposium on Computer Applications in Medical Care, pp 283-287.

Elman, JL, "Finding Structure in Time," CRL Technical Report 9901. Center for Research in Language, Univ. of California, San Diego, 1988.

Hripcsak, G., (1989). "Toward Data Analysis Using Connectionist Modules," Proceedings, Thirteenth Annual Symposium on Computer Applications in Medical Care, pp 271-275.

Hutton, L.V., Sigillito, V.M., and Johannes, R.S., (1989a). "Prediction of Diabetes in Pima Indians Using a Neural Network," presented at the Symposium on Biomedical Applications of Neural Networks, April 22, Johns Hopkins Applied Physics Laboratory, Laurel, MD.

Hutton, L.V., Sigillito, V.M., and Johannes, R.S., (1989b). "An Interaction Between Auxiliary Knowledge and Hidden Nodes on Time to Convergence," Proceedings, Thirteenth Annual Symposium on Computer Applications in Medical Care, pp 302-306.

Mulsant, Benoit H., and Servan-Shreiber, Emile, (1988) "A Connectionist Approach to the Diagnosis of Dementia," Proceedings of the Twelfth Annual Symposium on Computer Applications in Medical Care, pp. 245-250.

Obermeier, Klaus K. and Barron, Janet J., "Time to Get Fired Up," Byte, August, 1989, pp 217-224.

Young, D.W., "What Makes Doctors Use Computers?: Discussion Paper," Journal of the Royal Society of Medicine, Aug. 1984; Vol. 77, pp 663-667.

Saito, K., and Nakano, R., (1988). Medical Diagnostic Expert System Based on PDP Model, Proceedings of the IEEE International Conference on Neural Networks, Vol. 1, pp 255-262.

Stubbs, D., (1990). "Multiple Neural Network Approaches to Clinical Expert Systems," presented at SPIE, Orlando, Fla., April.

Stubbs, D., (1989). "Three Applications of Neurocomputing in Biomedical Research," Proceedings, IJCNN-89, Vol. II, p 609.

Teach, Randy L. and Shortliffe Edward H., "An Analysis of Physician Attitudes Regarding Computer-Based Clinical Consultation Systems," Computers and Biomedical Research, 14, 542-558 (1981).

Yoon, YoungOhc, Brobst, Robert W., Bergstresser, Paul R., and Peterson, Lynn L., "A Desktop Neural Network for Dermatology Diagnosis," Journal of Neural Network Computing, Summer 1989, Vol 1, #1, pp. 43-52.

20

NEURAL NETWORKS FOR SONAR SIGNAL PROCESSING

Patrick K. Simpson

20.0 OVERVIEW

Sonar signals are acoustic energy (sound waves) that propagate through the water. Sonar signals are used in both military and commercial areas (see Figure 20.1). In military systems, sonar is used for underwater sensing, navigation, communication, and surveillance. Sonar is used commercially for depth finding and fish finding. Sonar systems come in two broad, but clearly delineated, categories: active sonar systems and passive sonar systems. Active sonar systems utilize a combined transducer/receiver system to emit and receive sonar signals. The time delay and distortion of the returning signals provides useful information about the shape of the object that the sonar signals were reflected from (much like in radar) and the distance between the transmitting source and the reflecting surface. The key areas where active sonar systems are being applied include side scan sonar image processing, depth finders, fish finders, and underwater communication systems. Passive sonar systems are primarily used for monitoring the underwater environment. Most of the passive sonar systems are concerned with underwater surveillance and they have a distinctive military flavor to them.

Neural networks have been applied in a wide range of areas (cf. Simpson, 1990), but they seem to have some of the greatest potential in sonar signal processing. There are extreme demands made on any sonar signal processing system. Sonar signals are typically buried in a large amount of ocean noise *Signal-to-Noise Ratios* (SNRs) of minus 15 decibels (32 times more noise than signal) are not uncommon. Sonar signals are severely deformed as they propagate through the water. The path and speed of the acoustic energy in the sonar signal is affected by water salinity, depth and temperature, as well as the bathymetry of the immediate area. Sonar signals tend to operate in a very wide range of frequencies. For a sonar system to be effective it must have the ability to

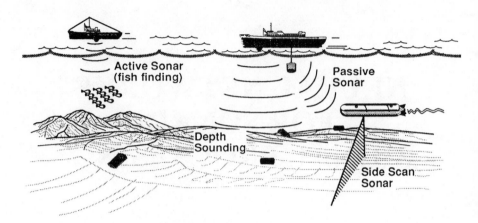

Figure 20.1 Sonar signal processing examples

mask out noise, circumvent multipath interference, and handle broadband signals in real-time. Neural networks seem ideally suited for such problems.

The applicability of neural networks in this area has not gone unnoticed. The *Defense Advanced Research Projects Agency* (DARPA) currently has a program that is funding several researchers to compare the performance of various neural network techniques against "classical" methods of classifying sonar signals (ONR, 1989). The Office of Naval Research (ONR) has several areas where neural networks are being applied as well (ONR, 1989). In addition to the government funded military efforts, there are internal research and development programs being conducted in several other companies, and there are several government research laboratories studying the use of neural networks for various aspects of sonar signal processing.

20.1 INTRODUCTION

Although neural networks are applicable to beam-forming, noise cancellation and feature extraction, they are primarily used in the area of sonar signal detection/classification. In this chapter we will take a macroscopic view of all of these areas of sonar signal processing with neural networks by outlining a generic multistage sonar signal processing system and then discussing the neural networks that can be used at each stage of the system. This chapter will not go into great detail about the myriad of issues related to sonar signal processing, rather it will emphasize those particular portions of sonar signal processing where neural networks have direct applicability. For a more detailed discussion of sonar signal processing and the classical techniques for processing sonar data, please refer to Oppenheim (1978) and NAVSHIPS (1965).

20.2 SONAR SIGNAL PROCESSING SYSTEMS

Figure 20.2 illustrates a generic sonar signal processing system that is broken into four stages: Noise Cancellation, Feature Extraction, Detection/Classification, and Post-processing/Display. Neural networks can be applied to the first three stages as well as to

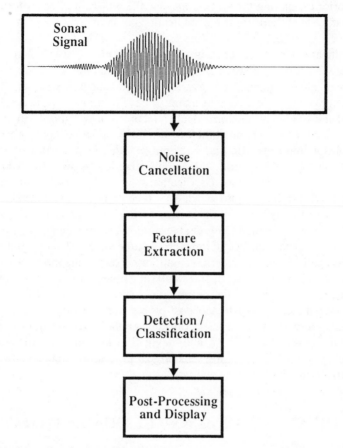

Figure 20.2   Sonar signal processing systems

some earlier signal collection efforts. A similar system designed specifically for localization of sonar signal sources has been described by Mattison, Work & Korbin (1989). Also, similar systems have been discussed within the framework of aerial acoustic emissions (Simpson & Deich, 1989) and side scan sonar object classification (Castellano & Gray, 1990).

Most of the signal processing equipment available today is digital. Hence, the flow of information through the sonar signal processing system begins with the digitization of the sonar signal. The process of receiving and digitizing a sonar signal depends upon the frequency range of the signal, the reliability of the sensors, and the ability of the processing system to handle the data received. The results of the digitization process is a time-series $X = x(1), x(2), ..., x(t), ..., x(N)$ that is created by sampling the continuous time sonar signal at regularly spaced intervals. Often the time-series is created from an array of several hydrophones that have *beam-formed* all the signals into one signal that is passed along to the later stages of the sonar signal processing system (see Figure 20.3). Some systems digitize the signals prior to beam-forming and others after. In our sonar signal processing system we will assume that the beam-forming has been completed prior to entering the system. Despite the omission of beam-forming in our generic sonar system, there are some applications of neural networks in this area and they will be discussed below.

After digitization the time-series is passed to the noise cancellation stage of the system where the non-correlated portions (noise) of the time-series are removed and the portions of interest (signal) are passed along for feature extraction. The feature extraction stage of the system is where a time-slice of the time-series (a window of time between $x(t)$ and $x(t+N)$) is transformed into a set of features. Typical feature extraction consists of taking the Fourier transform to determine the amount of power in pre-defined frequency bins within a specific frequency range. Other feature extraction techniques are also available (see the section on Feature Extraction below). The resulting set of features is passed to the detection/classification stage where it is determined if there is anything of interest and, if so, into which class it belongs. Note that the detection/classification stage of this system will not work effectively if it makes its decisions based completely upon only one time-slice. On the contrary, sonar signal processing requires the classification of the sequence of transformed time-slices. This recognition process, referred to as spatio-temporal pattern recognition, is one of the most difficult because of frequency-warping (caused by doppler shift) and time-warping (caused by an erratic source). Finally, the results of the detection/classification stage of the sonar signal processing system are passed to the post-processing and display stage where the information is presented to the system operator or the next system that will use the results of this system (such as an auto-navigation system onboard an unmanned undersea vehicle). Each stage of the sonar signal processing system shown in Figure 20.2 is described in the following sections.

20.3   BEAM-FORMING AND BEARING ESTIMATION

In some sonar signal processing systems there is an array of hydrophones that are used to receive sonar signals (see Figure 20.3). The process of synchronizing the wave-front across the array of hydrophones is called *beam-forming*. By using several hydrophones in an array and beam-forming the signal it is possible to reduce the noise and localize the direction, and possibly the distance, of the source of the signal.

Neural networks have only recently been described in the area of beam-forming. Speidel (ONR, 1989; pg. 96) has been using Kohonen Spatial Feature Maps to beam-form and cancel interference. In a related problem, neural networks have been applied to bearing estimation (or source localization): determining the direction of the source that is radiating the sonar signal. Accurately estimating the bearing of a signal can improve the beam-forming effort and aide in tracking the source. Following the technique of combinatorial optimization introduced by Tank and Hopfield (1986), the bearing estimation problem has been mapped into a Lyapunov energy function by Jha, Chapman and Durrani (1988) and Rastogi, Gupta and Kumeresan (1987). The bearing estimation energy function is, in turn, mapped into a set of weights and inputs for a neural network. In the later work by Jha, Chapman and Durrani there have been several alterations made to the earlier work of Rastogi, Gupta and Kumeresan that have improved the algorithm's performance.

Figure 20.3   Example of a sonar array

20.4   NOISE CANCELLATION

The process of noise cancellation is vitally important to the success of any sonar signal processing system. Noise cancellation is the process of separating the noise from the signal in the time-series. Mathematically restated, the t'th sample of the time-series, x(t), can be decomposed into signal, s(t), and noise, n(t); x(t) = s(t) + n(t). Noise cancellation systems attempt to estimate n(t) and subtract it from x(t) leaving only s(t). There are two neural network techniques that have been proposed to perform noise cancellation: *ADALINE Finite Impulse Response* (FIR) filters and *Back-propagation FIR* filters.

20.4.1   ADALINE FIR Filter

Although there have been various forms of noise-cancellation techniques available since the early 1900s, the first adaptive method of noise cancellation was developed by Bernard Widrow and his associates in the mid-1970s (Widrow, et al., 1975). The adaptive noise cancellation system, shown in Figure 20.4, attempts to predict the value of x(t) given the last p samples of the time-series. The difference between the predicted value of x(t), x'(t), and the actual value, x(t), is the error that is used to adjust the weights of the two-layer ADALINE. The ADALINE neural network has its weights being constantly updated, resulting in a linear prediction of the next sample based upon the last p-many samples. The operation of this network results in the separation of the linearly correlated portions of the data from the uncorrellated portions of the data. The neural network shown in Figure 20.4 is often seen in signal processing under the name of filter or transverse filter. Because of the similarity with the FIR filter, we will refer to this structure as the ADALINE FIR filter. For a more detailed discussion of this noise cancellation process, refer to Widrow and Stearns (1985).

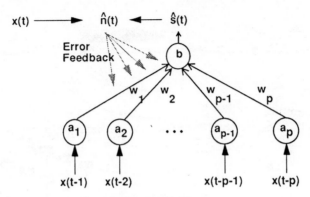

Figure 20.4   Topology of the ADALINE FIR Filter

20.4.2   Back-propagation FIR Filter

Recently it has been possible to achieve better performance than that provided by the ADALINE FIR filter by adding a hidden layer to the ADALINE and employing the back-propagation algorithm (Werbos, 1974; Rumelhart and McClelland, 1986) in place of the *Least Mean Square* (LMS) learning law used by the ADALINE. The resulting neural network, shown in Figure 20.5, is a generalized FIR filter, hence we shall call this extended network the back-propagation FIR filter. The added capability of the back-propagation FIR filter comes from its ability to capture nonlinear correlations. This technique, described by Klimisauskas (1989), Fu (1990), and Hecht-Nielsen (1990), has shown great promise. Hecht-Nielsen (1990) has reported preliminary experiments that have shown an improvement by a factor of 2 to 10 over the ADALINE version. The added power of nonlinear capability comes at the cost of slower adaptation rates. If the time-series contains signals that are fairly consistent and do not abruptly change (i.e., tonals buried in noise) then this noise cancellation technique is very powerful.

20.5   FEATURE EXTRACTION

The goal of feature extraction is to derive as much information as possible from each time-slice of the time-series. Typically, the Fourier coefficients were used as the sole set of features. As the need for more proficient classification with less data has emerged, other feature extraction techniques such as the Wigner-Ville transform (Nuttall, 1988), Gabor transform (Gabor, 1946; Friedlander and Porat, 1989), and wavelets (Combes, Grossman and Tchamitchian, 1989) are being explored. Feature extraction techniques are primarily performed using algorithms developed in signal processing. In addition, advances in hardware have made it possible for most feature extraction methods to be implemented in real-time. Nonetheless, there are still areas where neural networks have the potential to play a role.

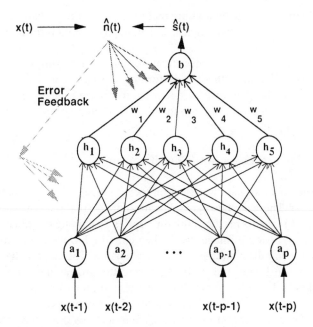

Figure 20.5   Back-propagation FIR Filter

20.5.1   Minimum Mean-Square Error (MMSE) Estimation

Many feature extraction techniques are all based upon unconstrained *minimum mean-square error* (MMSE) estimation. MMSE is one of the most fundamental operations in signal processing — especially in feature extraction techniques like the discrete Fourier transform, Maximum Entropy Method, Gabor transforms, etc. In a largely overlooked 1988 paper, Altes described a method of determining the coefficients for any arbitrary set of basis functions using Tank-Hopfield neural circuits (Tank and Hopfield, 1986). Altes derived a Lyapunov energy function that solved for the coefficients of any MMSE problem and then carefully back-engineered the equation into a set of neural network weights and input terms that would produce transformation coefficients in a parallel network. Altes was able to extend his basic result to weighted MMSE estimation as well. In addition to transformations, an example of how this formalism could be used for noise cancellation was included.

20.6   DETECTION AND CLASSIFICATION

Detection refers to determining that there is some activity in a time-series that is worthy of further investigation. Classification is the process of trying to identify exactly what was detected. Several sonar signal processing systems attempt to perform classification prior to noise cancellation and feature extraction to reduce that amount of processing that is done. In this sonar signal processing system we are lumping the detection and classification problems together by arguing that proper noise cancellation and feature extraction will remove any spurious signals and all the remaining information is ready for classification.

Having reduced the detection and classification problem into simply classification, we now focus on two different methods of performing this task. The first method works by creating a single spatial pattern that represents the entire time-series. The second method works by incorporating time into the classification process — a truly spatio-temporal (space-time) pattern classification process.

20.6.1 Spatial Pattern Detection/Classification

The majority of the sonar signal classification techniques employ the spatial pattern classification method. There are two general forms of the spatial pattern. The first form (shown in Figure 20.6) transforms the entire time-series containing the sonar signal into a *1-dimensional* (1-D) spectral image that is then classified. Several neural networks have been used to classify the resulting 1-D spectral images. This approach is nice from the perspective that the longer the sampling period the greater the accuracy of the transformation (this is because of the inherent noise suppression quality of longer sample periods). Unfortunately this approach also has the side-effect of neglecting important temporal cues that might exist during the transitions from one portion of the signal to another. To capture some of this valuable information, it is possible to create a 2-D spatial pattern (shown in Figure 20.7) by taking successive time-slices of the time-series, creating a spectral image of each time-slice, and concatenating them all together to create a time-frequency image. The 2-D image is now amenable to image classification techniques. In the following sections are several examples of how neural networks have been applied using the spatial pattern classification approach.

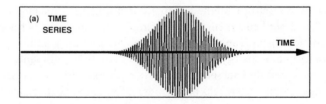

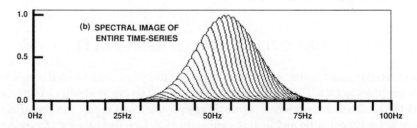

Figure 20.6 1-D Spatial Pattern Created from Entire Time-Series

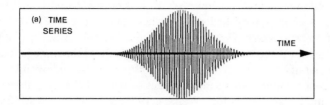

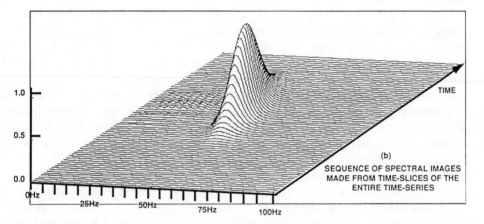

Figure 20.7   Spatial Pattern Created from Time-Slices of the Entire Time-Series

20.6.1.1.   Passive Sonar Signal Detection/Classification   Passive sonar signal detection/classification is the processing of quietly listening to the sonar signals that are emitted from underwater sources and classify the result. Passive sonar signal detection/classification is primarily used for ocean surveillance from arrays of hydrophones. These hydrophones can be towed in arrays behind either a submarine or a surface ship, they can be affixed to the bottom of the ocean in a specific geometric arrangement, and they can be located along the hull of undersea vehicles and submarines. Applications follow:

Submarine Detection with the RCE Network. Raytheon Corporation used the *Reduced Coulomb Energy* (RCE) network (Reilly, Cooper and Elbaum, 1982) to classify sonar signals into one of two classes: submarines and non-submarines (Nestor, 1989). The training set consisted of 108 1-D spatial patterns corresponding to submarines and 20 1-D spatial patterns that were non-submarine. The 1-D spatial patterns had 80 dimensions. The RCE network was able to attain 100 percent correct classification of the submarines and 95 percent correct classification of the non-submarines when given an equally large test set of sonar signals.

Submarine Detection with the Probabilistic Neural Network. Lockheed Palo Alto Research Center has used the *Probabilistic Neural Network* (Specht, 1990) to classify 1-D spatial patterns into submarine/non-submarine classes (Specht, 1989). Lockheed sampled the sonar signal for 2.5 seconds and performed long term integration on a series of Fourier transform spectral images to get a single spatial pattern. Using a proprietary feature extraction technique, a feature vector was created from the spatial pattern that was used for classification by the Probabilistic Neural Network (PNN). The training set consisted of 169 surface ship spatial patterns and 328 submarine spatial patterns. The test set consisted of 4390 surface ship patterns and 3269 submarine patterns. The PNN was able to classify the surface ships with 100 percent accuracy and the submarines with 70 percent accuracy. In an independent study by Lockheed, the same data was classified using back-propagation (Werbos, 1974; Rumelhart and McClelland, 1986) with a fall-off in performance and an immense increase in training time (Specht, 1989).

Submarine Detection with Back-propagation and Kohonen Feature Maps. Results using back-propagation for the same submarine/non-submarine classification problem described above were reported by Planning Systems Inc. (Planning Systems, 1989) and seem to confirm that the accuracy of a back-propagation network is good when tested on the training set, but falls off a bit when using novel test data. To improve the performance of the networks there was an attempt made to combine the Kohonen Topology Preserving Map with the back-propagation network. The Kohonen feature map was used to preprocess and separate the signal into a representation that was then fed to the back-propagation network for classification. This use of two neural networks in a hierarchical fashion was able to improve the performance slightly.

Scale Space Transformation Classification with back-propagation. Rice University has studied the use of a new feature extraction technique that is combined with back-propagation (Maccato and de Figueirido, 1990). The feature extraction technique is a scale space transformation that creates a 2-D spectral image from a time-slice (Witkin, 1984). The 2-D image is represented as a set of syntactic trees (Shaw and de Figueredo, 1990) that can subsequently be described as a binary vector. Finally, the resulting binary vector is passed to the back-propagation network for classification. There were no results reported for this classification technique, but it does illustrate another possible representation scheme that creates a spatial pattern for classification. In addition, because the data is binary, it would be possible to apply the binary adaptive resonance theory (ART1) neural network (Carpenter and Grossberg, 1987a), the Sparse Distributed Memory (Kanerva, 1988), or any one of several other binary pattern classifiers (cf. Simpson, 1990).

Submarine Detection and Tracking with the Willshaw Associative Memory and Back-propagation. Southern Methodist University has created a passive sonar target detection system that utilizes two neural networks that work in tandem (Khotanzad, Lu and Srinath, 1990). The goal of this sonar signal processing system is to find "tracks" in a 2-D spectral image. The tracks are moving objects that are emitting a constant frequency signal that is affected by doppler shift. The experiments were addressing two concerns: (1) the removal of noise from a 2-D spectral image, and (2) the classification of noise reduced sonar tracks. The first portion of the sonar system utilizes a Willshaw autoassociative memory (Willshaw, Buneman and Longuet-Higgins, 1969) that has been trained to recognize specific types of tracks in the 2-D image. When a noisy 2-D image is presented to the Willshaw network it produces a noise removed version. This noise reduction process is followed by a thresholding to produce a binary 2-D image that is classified with the back-propagation neural network. The system was demonstrated by adding Gaussian noise to a set of six 2-D spectral images. The amount of added noise ranged from 9 dB SNR (8 times more signal than noise) to -15 dB SNR (32 times more noise than signal). With only one exception at -15dB, all the 2-D spectral images had enough of the noise removed to allow proper classification. This example illustrates how the combination of two or more neural networks can produce a solution that is more effective than one neural network by itself.

Passive Sonar Source Localization with Several Different Neural Networks. Passive sonar source localization is the process of pinpointing where a classified sonar signal was emitted relative to the receiver. The localization process can be used to find and follow objects that move under the water such as fish and submarines. Computer Sciences Corporation has proposed a system that merges conventional signal processing with neural networks for the purpose of localizing passive sonar signals (Mattison, Work and Korbin, 1989). The processing system described has been broken into three stages of processing: (1) detection, (2) classification and identification, and (3) tracking and localization. Note that in this system the localization is done after classification, where in the earlier bearing estimation application (section 20.1.) the localization was done prior to classification. In this framework neural networks are used for signal detection and signal classification and identification. The neural networks proposed for the signal detection portion of the system are analog pattern classifiers; specifically the analog adaptive resonance theory (ART2) neural network (Carpenter and Grossberg, 1987b) and the Kohonen Self-Organizing Feature Map (Kohonen, 1984). The neural networks that are proposed for classification are back-propagation (Werbos, 1974; Rumelhart and McClelland, 1986) and the probabilistic neural network (Specht, 1990). There were no classification and localization results reported for this system, but it re-emphasizes the use of multiple neural networks in one system.

20.6.1.2. Active Sonar Target Classification   Active sonar target classification is performed by emitting a signal from a transducer and collecting the sonar signal return. By recording the time elapsed between sending and receiving the signal, the distance to the target can be determined and by determining how the signal was deformed during the reflection off the surface that returned the signal, it is possible to determine the shape, size and, sometimes, the density of the object. The subtleties in the differences in these sonar returns and the difficulty in determining exactly what it is in the signal that is determining all these features makes the adaptive nature of neural networks very appealing. By training a neural network to distinguish between the various type of returns, a major difficulty in developing robust active sonar target classification systems can be eliminated. Applications follow:

Target Classification with Back-propagation. One of the first applications of the back-propagation neural network was the classification of sonar targets from active sonar signal returns (Gorman and Sejnowski, 1988a and 1988b). The sonar targets were of two types; a rock and a metal mine, both of similar shape and size. The two targets had sonar returns that were recorded from several aspect angles from close range. From a data set of 1200 returns, 111 mine returns and 97 rock returns were selected for training. The SNR of these patterns was between 4 dB and 15 dB. Each sonar return was processed into a 1-D spectral image that contained 60 components. Each spectral image was used as input to the back-propagation network and the output was the class of the input (either rock or mine). The number of hidden units used in this experiment ranged from 0 to 24. The results of the neural network ranged from 89.4 percent to 99.8

percent correct classification on the training set (depending on the number of hidden units used). In comparison, the same data was used to train a two-layer ADALINE neural network and the classification accuracy dropped by 10 percent. Hence, the importance of the hidden units is demonstrated for this type of classification problem. When the back-propagation network was tested on data that was not included in the training set, the classification accuracy was between 77.1 percent and 84.5 percent. The two-layer ADALINE performance on this second data set was, again, poorer than the three-layer back-propagation network, demonstrating a drop of 5.7 percent in classification accuracy. In a final comparison a conventional nearest neighbor classification technique was employed, resulting in 82.7 percent correct classification on the second data set. The nearest neighbor method provided incorrect responses 1.7 percent more of the time. When human sonar operators were tested, the best performance achieved was 88 percent. Back-propagation provided 97 percent correct classification on the same data set. This set of experiments is one of the most extensive to date. An analysis of both the hidden unit values and the weights was performed on the best performing back-propagation network in addition to the comparisons with the ADALINE's and the human's performance.

Target Classification Comparison between a Dolphin and the Counterpropagation Network. In a similar active sonar target identification task, Naval Ocean Systems Center in Hawaii and the University of Hawaii performed a study that compared the performance of the dolphin with that of the counterpropagation network (Roitblat, et al, 1989) for active sonar target classification. Four targets were used for the classification experiments: an open PVC tube, a closed PVC tube, a solid aluminum cone, and a water-filled stainless steel ball. The training set was created by recording sonar returns from simulated dolphin clicks in a quiet pool. The pool returns were processed into 1-D spectral images with 20 components that were used to train the counterpropagation network. The counterpropagation network (Hecht-Nielsen, 1987b) provided 100 percent accuracy on this data set. Sonar returns from a dolphin were then collected from the same objects in the ocean. The counterpropagation network had 96.7 percent classification accuracy, while the dolphin had 94.5 percent classification accuracy on the same sonar returns. Incredibly, the counterpropagation network outperformed the dolphin at this problem! Admittedly this problem is rather small and confined, but it does illustrate the use of another neural network for sonar signal classification that performs as well as back-propagation did in the previously described experiments by Gorman and Sejnowski (1988a and 1988b) and it demonstrates that neural networks can potentially perform as well as sea mammals.

Target Classification and Fish Speciation with Back-propagation. Yet another active sonar target classification experiment was conducted by a team of researchers at Ontario Hydro Research Division (Ramani, et al., 1989). This set of experiments attempted to classify 7 different types of targets; 3 of the objects were non-fish (a ping-pong ball, bubbles, and a leaf) and 4 of the objects were fish (walleye, rainbow trout, brown bullhead, and sturgeon). The data was collected from several different aspect angles at relatively close range (6-9 meters). The ramifications of this type of experiment are tremendous for the fisheries. If a neural network can effectively determine the species of fish from active sonar returns, this could virtually revolutionize the fisheries. The sonar signals were received and transformed into a 192 point input pattern. There were

48 sonar signals from each of the nine different targets. The neural network that was used to classify the signals was a three-layer back-propagation network with 192 inputs, between 10 and 30 hidden units, and 9 outputs (one for each target). The training data was 38 of the 48 sonar signals from each target; the remaining 10 signals were used for testing. Using this initial configuration, the neural network was able to classify the objects correctly 91 percent of the time. A human was only able to attain 75 percent accuracy with the same data set. The network with 25 hidden units performed the best with equal performance at 20 and 30 hidden units. In a second experiment the problem was broken into three stages with a neural network dedicated to each stage. Stage one classified the object as either fish or debris. Stage two processed those returns that were determined to be fish and classified the fish into their appropriate species. The third stage was used to determine the angle of the fish relative to the transducer. Although the accuracy of the later stages of this system were dependant on the earlier stage's accuracy, by breaking the classification job into smaller chores (again, we see a hierarchical decomposition of the problem), the system's accuracy improved by about 2 percent, yielding 93 percent correct classifications. The number of hidden units used in the best performing three-stage system was 20, with reduced performance seen in systems with 15 and 25 hidden units. In a final experiment, a threshold was placed on the output node values that were used for classification. If an output class did not provide an activation value above the threshold, then the input was classified as a "not sure." Using this approach of discarding those classifications that did not provide a strong response, the accuracy of the three-stage classification system increased to 96 percent. Approximately 10 percent of the response were classified as "not sure" responses.

20.6.2   Spatio-Temporal Pattern Detection/Classification

The majority of the sonar signal detection and classification systems try to eliminate time from the classification process by creating an image (either 1-D or 2-D) that has been created from the portion of the time-series of interest. Although this approach has been very successful, it still has one inherent limitation: time and frequency are easily warped. If a classifier is creating 2-D images from very small time-slices, then small variations in the signal will create large changes in the image and, hence, make the classification much more difficult. If a 1-D image is created from very small time-slices, it is likely that there will not be enough information for correct classification. By incorporating the temporal nature of the occurrence of successive time-slices, i.e., keeping time in the recognition process, it is possible to attain correct classification despite considerable deformation in the signals. There have been few neural networks developed that can handle this type of spatio-temporal pattern recognition. Below we will describe one such approach and then discuss how some techniques that have been introduced for other spatio-temporal pattern recognition problems (such as speech and language) can be applied in the sonar domain. Applications follow:

Passive Sonar Signal Classification with an Extended Grossberg Avalanche. In 1969 Grossberg introduced a neural mechanism called the avalanche that was capable of learning and recognizing sequences of spatial patterns (Grossberg, 1969 and 1970). In 1987 Hecht-Nielsen refined Grossberg's avalanche and made clear the connection between Grossberg's avalanche and the matched filter (Hecht-Nielsen, 1987a). Recently, Hecht-Nielsen (1990) has further improved this neural network. We will refer to the resulting network as the *Avalanche Matched Filter* (AMF). The AMF works by encoding each spatial pattern of a sequence as a weight vector that abuts a processing element and then placing temporal connections between these processing elements to reinforce the temporal order that the data was encoded. Each spatio-temporal pattern is encoded in its own network. During recall a sequence is fed to each of the networks and the best responding network is determined to be the closest match and will be determined to be the class where the signal belongs.

Using the AMF and some fuzzy logic post-processing, General Dynamics Electronics Division had demonstrated the recognition of different types of ships (Taber, Seigel and Deich, 1988). In a similar set of experiments by the same group with the same network, Orca (killer whales) were geographically classified according to their calls (Taber, et al., 1988). In a later set of experiments by General Dynamics Electric Boat Division, an adaptive version of the AMF was used to recognize side scan sonar images by processing each scan as an individual pattern and allowing the temporal nature of the scans to be used to classify the object. The preliminary experiments showed that large objects could be classified with high accuracy, but the smaller objects were not. In addition, the various aspect angles that were not included in the training set, but were used during testing, created problems during recognition. Despite these early problems, the adaptive nature of the algorithm would allow training to be done during use; hence, recognition accuracy would improve over time.

Other Spatio-temporal Pattern Classifiers. There are several spatio-temporal pattern classifiers that have been presented within the framework of speech and language processing that can be brought to bear on the sonar signal classification problem. Specifically, the Viterbi Network (Lippmann and Gold, 1987), the temporal neocognitron (Ito and Fukushima, 1990), the bipolar spatio-temporal pattern recognizer (Homma, Atlas and Marks, 1988), and the temporal adaptive resonance circuit (Nigrin, 1990). In addition to these networks, there have been several temporal extensions to back-propagation that have been proposed, most notably those by Waibel, et al., (1987), Elman (1988), and Williams and Zipser (1988).

20.7 SUMMARY

Sonar signal processing applications will continue to emerge as more and more researchers begin to learn about neural networks. This migration is inevitable because of the alignment of sonar signal processing problems with neural network capabilities. There is still much work that needs to be done before the applications presented here will have scaled up to a size amenable to a real-world environment, but that day will come.

REFERENCES

Altes, R. (1988). Unconstrained minimum mean-square error parameter estimation with Hopfield networks, *Proceedings of the IEEE International Conference on Neural Networks: Vol. II* (pp 541-548). San Diego: IEEE Press.

Carpenter, G. and Grossberg, S. (1987a). A massively parallel architecture for a self-organizing neural pattern recognition machine, *Computer Vision, Graphics, and Image Understanding, Vol. 37,* pp. 54-115.

Carpenter, G. and Grossberg, S. (1987b). ART2: Self-organization of stable category recognition codes for analog input patterns, *Applied Optics, Vol. 26,* pp. 4919-4930.

Castellano, A. and Gray, B. (1990). Autonomous interpretation of side scan sonar returns, *IEEE Symposium on Autonomous Underwater Vehicle Technology,* Washington, DC, June 5 and 6, 1990.

Combes, J., Grossman, A. and Tchamitchian, P., Eds. (1989). *Wavelets: Time-Frequency Methods and Phase Space,* Springer-Verlag: Berlin.

Elman, J. (1988). Finding the structure of time, *University of California at San Diego, Center for Research in Language, Technical Report 8801.*

Friedlander, B. and Porat, B. (1989). Detection of transient signals by the Gabor representation, *IEEE Transactions on Acoustics, Speech, and Signal Processing, Vol. ASSP-37,* pp. 169-180.

Fu, L. (1990). Adaptive signal detection in noisy environments, *The Journal of Neural Network Computing, Spring Issue,* pp. 42-50.

Gorman, R. and Sejnowski, T. (1988a). Analysis of hidden units in a layered network trained to classify sonar targets, *Neural Networks, Vol. 1,* pp. 75-89.

Gorman, R. and Sejnowski, T. (1988b). Learned classification of sonar targets using a massively parallel network, *IEEE Transactions on Acoustics, Speech, and Signal Processing, Vol. ASSP-36,* pp. 1135-1140.

Grossberg, S. (1969). Some networks that can learn, remember, and reproduce any number of space-time patterns, I., *Journal of Mathematics and Mechanics, Vol. 19,* pp. 53-91.

Grossberg, S. (1970). Some networks that can learn, remember, and reproduce any number of space-time patterns, II., *Studies in Applied Mathematics, Vol. 49,* pp. 135-166.

Hecht-Nielsen, R. (1987a). Neural network nearest matched filter classification of spatio-temporal patterns, *Applied Optics, Vol. 26,* pp. 1892-1899.

Hecht-Nielsen, R. (1987b). Counterpropagation networks, *Applied Optics, Vol. 26,* pp. 4979-4985.

Hecht-Nielsen, R. (1990). *Neurocomputing,* Reading, MA: Addison-Wesley.

Homma, T., Atlas, L. and Marks, R. (1988). An artificial neural network for spatio-temporal bipolar patterns: application to phoneme classification, In D. Anderson (Ed.), *Neural Information Processing Systems,* (pp. 31-40). New York: American Institute of Physics.

Ito, T. and Fukushima, K. (1990). Recognition of spatio-temporal patterns with a hierarchical neural network, *Proceedings of the 1990 IEEE/INNS International Joint Conference on Neural Networks — Washington: Vol. I* (pp. 273-276). Hillsdale, NJ: Lawrence Erlbaum Associates.

Jha, S., Chapman, C. and Durrani, T. (1988). Investigation into neural networks for bearing estimation, In Lacoume, J., Chehikean, A., Martyin, N. and Malbos, J. (Eds.), *Signal Processing IV: Theories and Applications,* Elsevier Science Publishers, London.

Kanerva, P. (1988). *Sparse Distributed Memory,* Cambridge: MIT Press.

Khotanzad, A., Lu, J. and Srinath, M. (1989). Target detection using a neural network based passive sonar system, *Proceedings of the IEEE/INNS International Joint Conference on Neural Networks: Vol. I* (pp. 335-340). San Diego: IEEE Press.

Klimisauskas, C. (1989). Neural nets and noise filtering, *Dr. Dobb's Journal,* January 1989, pp. 32-48.

Kohonen, T. (1984). *Self-Organization and Associative Memory,* Springer-Verlag, Berlin.

Lippman, R. and Gold, B. (1987). Neural classifiers useful for speech recognition, *Proceedings of the IEEE First International Conference on Neural Networks: Vol. IV,* (pp. 417-426). San Diego: IEEE Press.

Maccato, A. and de Figueiredo, R. (1990). A neural network based framework for classification of oceanic acoustic signals, *Proceedings of Oceans '89,* Seattle, WA, September 18-21, pp. 1118-1123.

Mattison, T., Work, P. and Korbin, A. (1989). A theoretical model for undersea localization within a field of multi-element arrays using neural networks, *Proceedings of Under Seas Defense '89,* (pp. 128-137). October 23-26, San Diego, CA.

NAVSHIPS (1963). "Introduction to Sonar Technology," *Bureau of Ships - Navy Department, Technical Report NAVSHIPS 0967-129-3010,* Washington, DC, December 1965.

Nestor (1989). "Target recognition of sonar pings," *Nestor Marketing Brochure.* These results were presented at the Neural Networks for Defense Conference held in Washington, DC in June 1990.

Nigrin, A. (1990). "The real-time classification of temporal sequences with an adaptive resonance circuit," *Proceedings of the 1990 IEEE/INNS International Joint Conference on Neural Networks — Washington: Vol. I* (pp. 525-528). Hillsdale, NJ: Lawrence Erlbaum Associates.

Nuttall, A. (1988). "Wigner distribution function: Relation to short-term spectral estimation, smoothing, and performance in noise," *Naval Underwater Systems Center, Technical Report 8225,* 16 February 1988.

ONR (1989). "Cognitive and Neural Sciences Division 1989 Programs," *Office of the Chief of Naval Research, Technical Report OCNR 114289-22.*

Oppenheim, A., Ed. (1978). *Applications of Digital Signal Processing,* Englewood Cliffs, NJ: Prentice-Hall.

Planning Systems (1989). "Sonar classification with neural networks," *NeuralWare Connections, Vol. 1, No. 1.*

Ramani, N., et al. (1989). "Fish-detection and classification using a neural-network-based active sonar system — preliminary results," *Proceedings of the IEEE/INNS International Joint Conference on Neural Networks: Vol. II,* IEEE Press, San Diego, CA, 527-530.

Rastogi, R., Gupta, P. and Kumeresan, R. (1987). "Array signal processing with inter-connected neuron-like elements," *Proceedings of the International Conference on Acoustics, Speech, and Signal Processing,* pp. 54.8.1-4.

Reilly, D., Cooper, L. and Elbaum, C. (1982). "A neural model for category learning," *Biological Cybernetics,* Vol. 45, pp. 35-41.

Roitblat, H., et al. (1989). "Dolphin echolocation: Identification of returning echoes using a counterpropagation network," *Proceedings of the IEEE/INNS International Joint Conference on Neural Networks: Vol. I* (pp. 295-300). San Diego: IEEE Press.

Rumelhart, D. and McClelland, J. (1986). *Parallel Distributed Processing: Explorations in the Microstructure of Cognition: Vol. I,* Cambridge: Bradford Books/MIT Press.

Shaw, S. and de Figuerido, J. (1990). "Structural processing of waveforms as trees," *IEEE Transactions on Acoustics, Speech, and Signal Processing,* in press.

Simpson, P. (1990). *Artificial Neural Systems: Foundations, Paradigms, Applications and Implementations,* Elmsford Press: Pergamon Press.

Simpson, P. and Deich, R. (1989). "Battlefield surveillance using the GDE1 neural network," *Intelligent Systems Review, Vol. 1, No. 3,* pp. 3-13.

Specht, D. (1989). "Probabilistic neural networks (a one-pass learning method) and potential applications," *WESCON Conference Record.* San Francisco: IEEE Press.

Specht, D. (1990). "Probabilistic neural networks and the polynomial adaline as complementary techniques for classification," *IEEE Transactions on Neural Networks, Vol. 1,* pp. 111-121.

Taber, W., Seigel, M. and Deich, R. (1988). "Fuzzy sets and neural networks," *Proceedings of the First Joint Technology Workshop on Neural Networks and Fuzzy Logic,* NASA/University of Houston, Houston, TX.

Taber, W., et al. (1988). "The recognition of orca calls with a neural network," *Proceedings of the International Workshop on Fuzzy Systems Applications,* (pp. 195-201), Iizuka City, Japan, August 21-24.

Tank, D. and Hopfield, J. (1986). "Simple 'neural' optimization networks: A/D converter, signal decision circuits, and a linear programming circuit," *IEEE Transactions on Circuits and Systems, Vol. CAS-33,* pp. 533-541.

Waibel, A., et al. (1987). "Phoneme recognition using time-delay neural networks," *ATR Interpreting Telephony Research Laboratories, Technical Report TR-1-0006.*

Werbos, P. (1974). "Beyond regression: New tools for prediction and analysis in behavioral sciences," Ph.D. Dissertation, Harvard University.

Widrow, B., et al. (1975). "Adaptive noise cancelling: Principles and applications," *Proceedings of the IEEE, Vol. 63,* pp. 1692-1716.

Widrow, B. and Stearns, S. (1985). *Adaptive Signal Processing,* Englewood Cliffs: Prentice-Hall.

Williams, R. and Zipser, D. (1988). "A learning algorithm for continually running fully recurrent neural networks," *University of California at San Diego, Institute for Cognitive Science, Technical Report 8805.*

Willshaw, D., Buneman, O. and Longuet-Higgins, H. (1969). "Non-holographic associative memory," *Nature, Vol. 222,* pp. 960-962.

Witkin, A. (1984). "Scale space filtering: A new approach to multiscale description," In Ullman, S. and Richards, W. (Eds.), *Image Understanding,* Norwood, NJ: Ablex Publishing.

21

FAULT DIAGNOSIS

Robert L. Gezelter and Robert M. Pap
Introduction by Paul Werbos, Ph.D

21.0 INTRODUCTION: MAKING DIAGNOSTICS WORK IN THE REAL WORLD — A FEW TRICKS

Neural networks have been widely used for quality control and diagnostic applications in the past few years. Indeed, I would guess that these applications are *the* major useful engineering application of neural networks at the present time. On the factory floor (in proprietary applications), neural networks are already listening to truck engines to detect possible faults. They are being applied to the space shuttle, in a successful pilot project at NASA. They are working successfully on chip fabrication at the NIST experimental facility. And so on.

From an economic viewpoint, quality control and diagnostics are vital to the competitiveness of American industry. Back when I was lead analyst for transportation at the Energy Information Administration (EIA/DOE), we reviewed hundreds of papers on automobile sales, because we wanted to know whether fuel-efficient cars would sell better in the future, under what conditions. All the best, most realistic studies agreed: the key factor in selling cars was quality control or reliability, far ahead of any other factor. To sell more cars (or anything else), we need to give priority to these factors.

Most applications of neural nets to quality control take a very simple approach. The user begins by setting up a training set, consisting of examples where the product is working satisfactorily or falling short. In the training set, each input vector is the vector of sensor readings for one case of the product; the target vector is usually just a single variable, set to 1 for failure or low quality; or 0 for acceptable performance. The neural net learns the static mapping for the inputs to the output.

A slight variant of this — which sometimes works a little better — is to enumerate a list of possible failure modes, and set up a distinct target variable fore each failure mode, It is often easier to learn a *specific* or *concrete* concept like bad brakes than one like broken car, at least if one has enough data on the various failure modes. Likewise, in evaluating a dynamic system, like a running motor, one can build a time-lagged recurrent network, which accounts for earlier sensor readings to detect a patter. (Supervised learning with time-lagged recurrence is described in "Back-Propagation Through Time." cited in my chapter.)

These approaches are worthwhile, but they are often not quite enough in some applications. For example, in monitoring a complex system like a nuclear reactor or space shuttle, one always has to be alert to unforeseen failure modes. Test data are often very rich in normal-mode operation but very sparse in failure-mode operation.

In these applications, there is a very important supplementary technique available. One can simply build up a dynamic model of normal system operation, using the system identification techniques discussed in my chapter. Then, when the system deviates in any way from the expected dynamics — say, by 15 standard deviations — one can set off an alarm. Furthermore, one can give the operator of the plant more information than just a general alarm. Once could build up a three-layer warning board. Each column of the board would consist of a label (on top) and two variable intensity lights below it. the label would identify a particular sensor. The top light would shine in proportion to the deviation of that sensor from its predicted value. The bottom light (optional) would shine in proportion to the derivative of error with respect to the previous (time t-1) value of that sensor, calculated by back-propagation. (This is essentially a crisis warning approach that I suggested to DARPA in the late 1970s.)

This system identification network could be used in control, as well, The hidden and output nodes of the network could be used as inputs to a more conventional status fault detection network, trying to detect fault versus no fault. One could use this network to generate forecasts of system trends, to make sure that the normal expected behavior of the system is not heading towards undesirable regions in state space.

21.1 OVERVIEW

Fault diagnosis requires the collection and processing of large amounts of often incomplete data to determine the nature and severity of equipment and system malfunctions. Neural networks, by their very nature, are well-suited to such fault diagnosis tasks.

Neural networks are well-suited to processing performance data and searching for problems. First, neural networks are capable of recognizing an input pattern from partial or even fragmentary input data. Additionally, the learning aspects of neural networks permit them to learn what constitutes correct and normal device operation, without the need for extensive preprogramming.

Comparing neural networks with expert systems will show that an expert systems approach to fault diagnosis requires a detailed, extremely careful debriefing of a human expert to the knowledge base for the expert system.

Necessarily, fault diagnosis is composed of two distinct stages: fault detection and fault isolation. The fault detection stage is characterized by a single, distinct operation, the observation that the system, or a component subsystem, is not operating in a completely normal fashion. While in some sense this means that a component has malfunctioned, a more correct interpretation would allow for a looser definition — namely, a projected failure of a component, rather than the actual failure. Thus, a situation where an over-pressure is not detected in a coolant or hydraulic line should trigger a fault detection event even though the over-pressure has not yet caused an actual physical failure.

Fault isolation requires the identification of the underlying cause of anomalous behavior. Ideally, fault isolation results in an accurate, unambiguous identification of the component or components responsible for the root cause of the anomalous behavior.

In terms of utility, fault diagnosis represents one of the most critical applications of neural networks. However, as our need for critical systems increases, and the attendant safety margins and response time decrease, the importance of reliable, highly efficient monitoring, control, and diagnosis systems increases. For example, in aircraft such as the Grumman X-29 or the National Aerospace Plane (NASP or X-30) are examples of the types of systems we are capable of constructing.

In the case of both the X-29 and the NASP, the response times required to preserve the integrity and operation of the aircraft are shorter than most human capabilities. In the case of the X-29, the forward swept wings represent an aerodynamic configuration which is statically unstable and requires continuous computer control to fly. The hypersonic speed of the NASP requires continuous computer monitoring of numerous sensors and instruments, together with automatic diagnosis of problems in real-time before an equipment malfunction can cause aircraft failure.

In other spheres, our need for accurate, unambiguous fault detection and isolation is as critical as in aerospace applications. Consider the impact of failures in large complex systems, such as the nuclear industry, chemical complexes, and petrochemical facilities. In many of these cases, a single, incorrectly identified and diagnosed problem can lead to catastrophic problems involving the lives of thousands of people living many miles from the site of the incident.

21.2 TECHNIQUES

There are several different ways of approaching the fault diagnosis problem. Generally speaking, all fault detection and isolation schemes fall into two broad categories, depending on whether they are based upon fault models. The model-based methods are further subdivided into state-space methods, input-output methods, and component-connection methods. (See Figures 21-1, 21-2, 21-3.)

Generally, model-based fault detection and isolation schemes are based upon the concept of *analytical redundancy*. These models are used by comparing actual parameters measured by a sensor suite with the values predicted by an analytical model of the system.

In a perfect environment, the differences between predicted and measured values would be zero. In reality, such perfection is impossible. The discrepancies are caused by inaccuracies in the analytical model of the system.

The variations induced by errors in the analytical model depend upon the comprehensiveness of the analytical model used. For extremely well-understood phenomena, the discrepancy between predicted and actual values should be small.

$$Y(s) = T(s)U(s)$$

Figure 21.1 State-space models

21.2.1 Sensor Fusion

Sensor fusion is a critical issue in fault detection and isolation. In complex technical systems, multisensor fusion lies at the heart of many aspects of fault diagnosis. Complex mechanical systems typically contain a myriad of monitoring points measuring the performance of the system. In any such large collection of sensors, there are many variables which directly affect the reliability and accuracy of the information collected by these points.

It is important to note that any sensor fusion scheme must make adequate provisions for the detection and suppression of incorrectly functioning sensors. In this sense, it is critical for all sensor outputs to be cross-checked with other transducers measuring related parameters to guard against precipitous actions grounded in incorrect data.

$$y = Su \quad b = Za \quad \begin{bmatrix} a \\ y \end{bmatrix} * \begin{bmatrix} L_{11} & L_{12} \\ L_{21} & L_{22} \end{bmatrix} \begin{bmatrix} b \\ u \end{bmatrix}$$

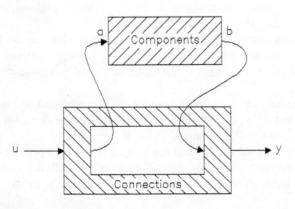

Figure 21.2 Input-output methods

$$x(t+1) = Ax(t) + Bu(t) \text{ and } y(t) = Cx(t)$$

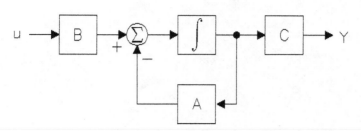

Figure 21.3   Component-connection methods.

Sensor errors occur when a transducer or related wiring is not operating as specified. A connection may be loose, physical damage may have occurred, or an electronic fault may have occurred in the sensor itself. Whatever the underlying cause, a fault diagnosis system should report a fault in the sensor, not in the actual subsystem.

Random noise is present in all systems. So long as it is truly noise, the fault diagnosis system should ignore it. However, it is important that sources of non-random noise be identified as they may very well be faults, not noise.

Actual faults represent real failures in the subsystem being monitored, as opposed to artifacts of the monitoring process. They are the actual targets of the automated fault detection and isolation process.

21.3   APPLICATIONS

Neural networks are also applicable to nonreal-time diagnosis of complex systems. Recent work by General Dynamics [McDuff et al., 1990] on the development of a neural network based diagnostic system for the F-16 Fighting Falcon illustrates this point well.

Servicing modern, high-performance aircraft requires many technicians, each of which must have an in-depth understanding of a particular aircraft subsystem and the way that subsystem interacts with the rest of the aircraft. Since many deployed systems are constantly undergoing revision, maintaining the expertise required to isolate, diagnose, and correct system faults is a constant problem.

Patrick Simpson and his colleagues at General Dynamics have developed a neural network using the binary adaptive resonance theory [Carpenter and Grossberg, 1987] network (see Figure 21-4] to diagnose failures in F-16 aircraft based upon reported symptoms.

This network was trained using 21 weeks of live date collected from real events in the F-16 fleet. After training, subsequent weeks of data collection were presented to the networks for analysis. Out of the 191 faults presented for diagnosis, the network produced 129 high-confidence problem evaluations. Of these 129 diagnoses, 115 were

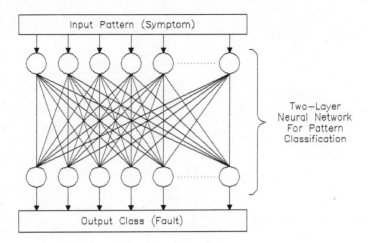

Figure 21.4   Binary adaptive resonance theory network

completely correct, 13 were partially correct, and 1 was incorrect. Thus, the network's overall score was 89 percent.

Considering the small data set used for training, this application shows excellent results on a problem of significance far beyond the F-16 program.

The NASP (X-30) is a revolutionary aerospace vehicle being designed to take off from a conventional runway, climbing to altitude and either cruising at hypersonic speed to its destination or, entering a low earth orbit on its own power.

Hypersonic flight is extremely demanding on both equipment and air crew. At such speeds, a small miscalculation can result in loss of control of the vehicle. The NASP can not afford to experience the malfunction of a flight control surface at speeds that have almost no safety margin.

A team at Accurate Automation Corporation, led by Robert Pap and Clifford Parten, is presently studying the use of neural networks for real-time monitoring and diagnosis of the airframe and engines on the NASP. This project is studying the application of neural networks to the State Space and the Component Connection models for the NASP, in addition to traditional pattern matching techniques. (See Figures 21-1 and 21-3.) A multilayer Perceptron with back-propagation is being used for part of this project.

To successfully operate in this environment, the engineering sensor suite for the NASP is anticipated to be thousands of sensors. Observing and integrating this torrent of information is a challenge to human capability. For the most part, the analytical task required to analyze this information in real-time and take action based upon this information is well in excess of the capabilities of conventional computer systems.

Rocket and jet engine diagnosis is being studied by Williams Dietz and Earl Kiech of Calapan, jointly with M. Ali of the University of Tennessee Space Institute [1989]. These applications use a three-layer feedforward neural network with nonlinear hidden-layer and output units. They have observed that the network produces highly reliable results when the input pattern falls into classes for which it has been trained. Also, the system renders accurate and reliable diagnoses when presented with incomplete data. This system performed engine diagnosis in near real-time.

21.4 POWER GENERATION FACILITIES

Recently, a team led by Robert Uhrig of the University of Tennessee-Knoxville [Uhrig and Guo, 1989] has investigated the application of neural networks to the analysis of engineering data reflecting the operation of a nuclear power facility.

Uhrig used a three-layer back-propagation neural network with 40 neuron in the input layer and 3 neurons in the output layer to analyze the operation of a steam generator. The network was trained using simulator data representing normal conditions and 10 percent variances on normal conditions.

Modern nuclear power facilities are richly instrumented environments. The large amount of information generated by these extensive instrumentation suites is both a blessing and a curse. The hundreds of thousands of data points reported by the sensor suite of a nuclear power plant presents a large, but potentially incomplete picture of plant operation. The data presented by this sensor suite can be misleading in two general ways.

First, instrumenting each inch of the plant is technically infeasible. Thus, by its very nature, there must be points within the plant which are not actually monitored. The plant operators are expected to extrapolate actual conditions from the status of components which are actually monitored. Secondly, sensors and gauges do fail. A malfunctioning gauge or display must presently be identified by comparing its reading with the readings of similar units at related points in the facility. If the suspect gauge is determined to be in error, then it is ignored. However, what happens when that gauge is the one displaying the true point of failure?

Both of the above situations must also be viewed in context. Human operators are expected to make these decisions 24 hours a day, 7 days a week. Thus, while these thousands of data points do represent a fairly complete picture of plant conditions, the sheer volume of data also represents an overload fro the operators responsible for safe operation.

Neural networks monitoring of much facilities take the form of pattern matching. A correctly power plant produces a set of instrument readings reflecting normal operating conditions. A malfunctioning plant produces a different set of readings, with the differences indicating the location and severity of the malfunction.

Consider diagnosis of steam generator problems. Changes in inputs and control setting alter the output and condition of the steam generator. For example, increasing the temperature of the coolant entering the steam generator increases the steam pressure.

After training, the network was fed data 5 percent, 15 percent, and 20 percent perturbations. It correctly identified the root cause of each of the perturbations. Additionally, the network was able to still correctly identify the cause of the problem when random noise of up to 30 percent was introduced into the readings.

This ability to correctly identify a problem, in the face of significant noise, is a major strength of neural networks in industrial environment.

21.5 SUMMARY

In summary, neural networks have been shown to be successful tools for the diagnosis of real-time and nonreal-time data.

In these roles, neural networks have correctly diagnosed problems in environments ranging from nuclear power plants to high-performance aircraft. Furthermore, they are able to perform these diagnoses without the need for the large scale programming normally associated with more conventional technologies.

Additionally, neural networks have been demonstrated to improve as they are given greater experience in their monitoring function. This self-organizing and self-improving capability holds great import for the future of operation and diagnosis of critical systems.

REFERENCES

Carpenter, G. and Grossberg, S. (1987). "A massively parallel architecture for self-organizing neural network pattern recognition machine," *Computer Vision, Graphics, and Image Understanding, 37*, pp. 54-115.

Dietz, W. E., Kiech, E. L., and Ali, M. (1989). "Jet and Rocket Engine Fault Diagnosis in Real Time," *Journal of Neural Network Computing*, Summer, pp. 5-18.

Isermann, R. (1984). "Process Fault Detection Based on Modelling and Estimation Methods — A Survey," *Automatic, 20:4-A*, pp. 387-404.

McDuff, R. and Simpson, P. (1990). "An Investigation of Neural Networks for F-16 Fault Diagnosis I: System Description," *Conference Record of IEEE Anktestcon 1989*, Philadelphia, PA., pp. 351-357.

McDuff R. and Simpson, P. (1990). "An Investigation and Neural Networks for F-16 Fault Diagnosis II: System Performance," *Proceeding of the SPIE Technical Symposium on Aerospace Sensing, Applications of Neural Networks*.

McDuff R. and Simpson, P. (1990). "An Adaptive Resonance Diagnostic System," *The Journal of Neural Networks*, Summer 1990. (preprint)

Uhrig, R. and Guo, Z. (1989). "Use of Neural Networks in Nuclear Power Plant Diagnosis," *Proceedings of the SPIE Technical Symposium on Aerospace Sensing, Applications of Artificial Intelligence VII*.

22

NEUROCONTROL AND RELATED TECHNIQUES

Paul J. Werbos[1]

22.0 OVERVIEW

The use of neural networks in control applications — including process control, robotics, and aerospace applications, among others — has recently begun a pattern of very rapid growth. In some cases, neural networks have been used to perform very narrow, specific tasks within a control system, such as pattern recognition or feature extraction. Neural networks have been used to detect attributes of objects, which are then input to the knowledge base of an expert system. In other cases, neural networks have been used *directly* to output the signals which control motors, or actuators, or whatever else needs to be controlled. The term *neurocontrol* refers to the last case. This chapter will discuss the current state of the art in neurocontrol, and will also discuss a few related issues, such as the problem of modeling the environment to be controlled and the relation to conventional control theory and expert systems.

Neurocontrol is an important field of research for two reasons:

- Because of the contribution it may ultimately make to our understanding of the human mind
- Because it can *enable* the automated control of systems which could not be controlled in the past, for one of two reasons — the physical *cost* of implementing a known control algorithm, or the difficulty of *finding* such an algorithm for complex, noisy, nonlinear problems.

This section will elaborate briefly on these two points, and then summarize the capabilities and characteristics of the five major methods used in neurocontrol today.

1 The views expressed here are purely the current impressions of the author, and in no way reflect the official views of his employers; they are very much subject to possible change.

The following sections will discuss the potential applications in more detail, and review the concept of supervised learning and the interface with expert systems. The final sections of this chapter will discuss the five major methods in more detail, link them to conventional control theory, and discuss the issue of how to model the system to be controlled. The last section will contain the actual equations for robust estimation methods which turn out to be crucial in practice to many real-world control applications, and will conclude with a quick discussion of biological parallels and future experimentation.

22.1 INTRODUCTION

The mammalian brain is living proof that it is possible to design a control system, based on parallel analog hardware, capable of handling millions of actuators, a heavy degree of nonlinearity and noise, and still achieve a high degree of long-term optimality, robustness and stability. Most of what it knows about its environment it learns in real-time (though it does require a sleep cycle). These capabilities do not require the use of language or expert systems, since few mammals are capable of formal symbolic reasoning.

From an engineering point of view, there are two critical points here — the parallel analog hardware, and the ability to control complex systems.

From a hardware point of view, most engineers realize that a typical CPU chip contains hundreds of thousands of transistors, at least, in order to enable a *single stream* of digital calculation. Even if hundreds of these chips are connected in parallel, there is still a certain limit to the computational throughput per chip. Huge numbers of transistors are needed, because of the need to support a large, flexible instruction set, and because of the need to use 32 bits or so (i.e., 32 different transistors) to represent a number in digital form. What if we could limit ourselves to a single transistor, using an analog representation of a number, and use fixed hard-wired mathematical operations? Throughput could increase by several orders of magnitude. This is one reason why Carver Mead (1989), who is often described as the father of VLSI technology, has become active in the neural network field. Neural network research — the attempt to develop mathematically valid algorithms consistent with the *constraints* of fixed, analog distributed hardware — is crucial to taking advantage of this improvement in throughput. It is also crucial to exploiting the full potential of optical computing.

In practical terms, this means that it makes sense to translate general mathematical algorithms into neural net (i.e., constrained) algorithms whenever possible, in order to exploit the high throughput advantage *already* available with neural network chips, described elsewhere in this book. Even in cases where a conventional algorithm works fine in the laboratory, it may be underused in real life, because aircraft designers cannot afford to carry two Crays aboard every airplane, and manufacturers cannot afford to put five Crays on every workstation; if the same algorithm (or a close approximation) can be put onto general-purpose neural network chips, with the *same* degree of computational complexity, a wider market may open up instantly.

The ability to control complex systems is also very critical from an engineering point of view. Increasingly, as engineers try to build highly complex, automated systems, the cost and difficulty of control becomes crucial to the feasibility of such systems. *If* one could duplicate those capabilities of the brain described above, systems which were risky or impossible to control in the past may now become controllable. (The next subsection will describe how close we are to achieving this.) If one builds systems which can truly learn in real-time, one can also achieve much greater flexibility in manufacturing, which is crucial to competitiveness in many economic sectors. (Actually, there are two ways of seeking flexibility when using neural nets — true real-time learning, and the use of "memory" nodes which detect changes in the parameters in the environment. Both are possible, but the latter may be more practical in many applications.) Neural nets can also be used in optimal control, so as to squeeze more efficiency out of nonlinear dynamic systems which are already being controlled to an adequate degree.

Finally, our understanding of neurocontrol in engineering applications may ultimately be crucial to our ability to understand the human mind. Control is not just *part* of what the human brain does; the biological function of the *whole system* is to input sensor data, output motor control signals, and achieve long-term results in a complex environment. Until we know at least *one way* to build such a system, we will not be able to construct a *model* of the brain which reproduces the important, aggregate capabilities cited above. Once we know *one* way of building such a system, our ability to construct alternative models should grow very quickly. The mathematics under discussion here are truly general; they can be applied (with some refinement) to brain cells, to silicon, or to any other substrate we are capable of imagining to sustain intelligence.

22.2 THE FIVE BASIC DESIGNS

Underneath the apparent complexity (required by specific applications), there are really only five generic designs now used in neurocontrol. Each of the five — and the adaptive critic approach, especially — does allow for some degree of variation. Each of the five — with the possible exception of neural adaptive control — is essentially a *general* method in control theory, which can be applied to *any* network made up of differentiable functions; there is no mathematical reason to limit one's attention to functions which look like neurons. For practical reasons, however, most of the applications to date do involve artificial neural networks.

Figure 22.1 compares the five basic methods of neurocontrol against four of the capabilities of the brain, as discussed above.

In *supervised control*, a neural net learns the mapping from sensor inputs to desired actions, by adapting to a training set of examples of what it should do. If these examples come from a human expert who can perform the desired task, one can "clone" that expert. One could train a human to fly an airplane simulator in slow motion, and later run the neural net clone at realistic speeds. Roboticists have long used "pendants" to teach robots to copy human motions on the factory floor; supervised control is more general, because the robot also learns how to respond to sensory cues. Widrow (1963) made this work in the 1960s, and Fuji has applied it to working robots. I know of no connection to "supervisory control," a term now used in the control theory literature.

In *direct inverse control*, a neural net learns the inverse dynamics of a system, so that it can make the system follow a desired trajectory. Miller of the University of New

	Many Motors	Noise	Long-Term Optimum	Real-Time Learning
Supervised Control	+	+		+
Inverse Control	(+)	+		+
Neural Adaptive Control	+	+		+
Backpropagating Utility	+		+	
Adaptive Critics 2-Net		+	+	+
Advanced	+	+	+	+

Figure 22.1 Neurocontrol designs versus brain capabilities

Hampshire has built a neural net board which makes a Puma robot follow a desired path in three-dimensional space, with errors well under 0.1 percent (Kraft and Campagna, 1990). This approach may be useful whenever there exists a desired "trajectory" — a desired pathway in n-dimensional space — and n actuators available to move the system in that space. As with supervised control, there is no real planning or optimization involved; the planning is done by the person or the computer program which supplies the desired trajectory. (In some systems, the desired trajectory is simply a desired point.)

In *neural adaptive control*, linear mappings used in standard designs like Model-Reference Adaptive Control are replaced by neural networks, resulting in greater robustness and ability to handle nonlinearity. Narendra of Yale — famous for his textbooks on adaptive control — has recently demonstrated the viability of this approach. In Model-Reference Adaptive Control (MRAC), the goal is to control a subsystem so that it continues to match a proposed "Model," which really acts like a design specification for the subsystem; the controller makes internal compensations, so as to keep the actual behavior of the subsystem consistent with the specifications. Adaptive control may also be used in tasks like maintaining a desired equilibrium or following a trajectory. Proofs of stability are an important part of adaptive control theory.

The *back-propagation of utility* (not error!) allows you to derive a *schedule* of actions, or adapt an optimal Action network, so as to maximize any performance index or utility function which you choose to specify, over multiple time-periods. By specifying different utility functions or cost functions, you can apply this method to a wide range of applications. For example, Jordan (1989) and Kawato (1990) have shown that you can train a robot arm to follow a desired trajectory, at the minimum possible energy cost or jerkiness, even when the arm has more than 3 degrees of freedom. Nguyen and

Widrow (1990) developed a simulated truck-backer-upper using this principle, and I developed an official DOE model of the natural gas industry (which maximizes profit) (Werbos, 1989). Other chapters of this book describe more recent applications in Tennessee.

Because it is so straightforward, proven and flexible, the back-propagation of utility is suitable for a wide range of applications, but there are three crucial limitations to keep in mind:

- You can only use this method *after* you have a differentiable model to describe the process you are trying to control. (The model could itself be a neural network, updated in parallel with the Action network.)
- If your model involves any significant dynamics (i.e., inputs at time t from earlier time periods), then you have to adapt your Action network in *off-line* mode, instead of in *real-time*. (This could be done in a "sleep cycle," in theory, but only if data from the previous activity cycle were stored. Real-time versions exist for tiny problems, but do not really qualify, in my view, as neural networks, and they may lead to stability problems.) Still, it is possible to learn *offline* how to be adaptive in *real-time*, if you set up your network and training data correctly.
- The method assumes in theory that your model is *exact*, that there is no noise. If you are adapting an Action network — rather than a schedule of actions — then the results may be relatively robust against small amounts of noise, but large amounts might be a problem.

The back-propagation of utility may or may not involve the use of steepest descent methods to adapt the weights in an Action network; in the extreme — as in the work of Cotter and Conwell (1990) — one may use a direct search on the space of possible weight vectors, which forces one to simulate the system many times, but avoids the need for calculating derivatives.

Adaptive critic methods — like the back-propagation of utility — allow you to maximize or minimize any utility function or cost function over time. These methods are actually a complex family of methods; only the simplest (2-Net) versions have been widely tested to date. The simplest versions do *not* require a model of the system to be controlled, and — as shown in Figure 22.1 — they are provably consistent with noise and real-time learning. They work very well in tests involving a small number of variables, but become unacceptably slow when the number of variables grows larger. The more complex versions should overcome this problem, but require more testing. The complex versions, like the back-propagation of utility, require a model of the process to be controlled; however, this model need not be deterministic. (For example, one may use a stochastic neural network, so long as its transfer functions are deterministic.)

22.3   AREAS OF APPLICATION

Broadly speaking, neurocontrol has been used or proposed in four major areas:

- Vehicles and structures
- Robots and manufacturing
- Teleoperation and aid to the disabled
- Communications, computation and general-purpose modeling (e.g., economics)

This chapter will not review everything done in these areas, for several reasons. First of all, the literature is quite large; Allon Guez, in late 1987, was able to classify the fifty-odd published papers in neurocontrol into design categories (similar to the five categories mentioned earlier); by 1990, at least twice as many papers had appeared, and many of the most important applications have not been published. Indeed, the most exciting applications I am aware of are highly proprietary. (This pattern might be expected in a field where many designs are patentable, and most of the funding comes from market-oriented corporations and applications-oriented work in the military.) In some cases, neural networks have been used inside of software which is widely used, but are totally invisible to the user.

This section will emphasize *potential* applications, rather than past applications, because of their great importance. Many of my conclusions here were based on a workshop described in Miller, Sutton and Werbos (1990).

22.3.1 Vehicles and Structures

The most important applications of neurocontrol are those which may make it possible to do something which could not be done at all without them. Applications of this sort may arise at times in high-technology areas where control is extremely difficult.

Perhaps the most exciting example of this possibility concerns the National Aerospace Plane (NASP), and other forms of hypersonic vehicles.

Years ago, President Reagan announced the NASP program as a major new national priority, in his State of the Union speech. Despite various fluctuations and attacks from various sources, the present Administration has continued to support the program. The main goal of NASP is to build two or three prototype vehicles, in the 1990s, which could reach speeds as high as Mach 25 in an airbreathing, reusable vehicle, which takes off and lands like an airplane. Success in this program could lead directly to high-speed intercontinental passenger jets, to military vehicles capable of backing up U.S. assets in space, and to civilian vehicles which dramatically reduce the cost of transport to earth orbit.

The critics have argued that NASP faces such difficult obstacles that there is at least a 50 percent chance that the prototypes cannot be built successfully, even if we use the best technology known today. The advocates of NASP have mostly not disputed that point; instead, they have pointed to the large benefits which we would receive if NASP should succeed. The National Space Society (1989) has argued that the entire solar system could become ripe for large-scale, economically profitable human settlement, if only we could overcome three fundamental hurdles; the high cost of transportation to earth orbit is the first of these hurdles. (The other two involve activities in earth orbit and nonterrestrial materials, which is also addressed below.)

With the space shuttle, costs now run about $4,000 per pound to earth orbit. With commercial rockets, it is more like $2,000 per pound. At these high costs, space may still be profitable in some areas, but there is not the multiplier dynamic (or "input-output capture") which leads to an economic takeoff effect in the local economy. On the other hand, if costs were reduced to $50 per pound, the Russians have recently estimated that it would make economic sense to orbit gigantic arrays of solar cells and beam the electricity down for sale to earth. This would be especially useful for poor nations where the local cost of building power plants is very high, and the U.S. has a strong national interest in discouraging the nuclear alternative [Werbos, 1990b]; furthermore, this is only one example of what could become possible at lower costs. Many people believe that success with NASP could ultimately lead to costs in this range or lower. (Cost studies by George Mueller of NASA in the late 1960's support this conclusion, and the technological options now are far better than they were then.) If this is true, then success with NASP — however risky — may be crucial to an entire economic frontier, larger than the earth in its potential. I am reminded of the well-known economist Schumpeter, whose studies showed that the (government-subsidized) transcontinental railroad was the primary factor explaining the very rapid rise in productivity in the U.S. in the late nineteenth century.

What are the crucial obstacles which make NASP so risky? The list of obstacles varies somewhat from study to study, but may be summarized as follows: propulsion; materials; miscellaneous control problems. Because NASP uses a very advanced and delicate propulsion system, the challenge of adequate real-time control and adjustment of that system is central to the entire problem. Other control problems involve avionics (stabilizing the vehicle at speeds ten times what humans have handled in the past in atmosphere), thermal control, real-time diagnostics, and the like. In brief: more than half the possibility of failure can be traced to the difficulty of controlling highly nonlinear, noisy complex systems, at speeds so great that automatic control is required. Neural networks should be capable of coping with such challenges, if Figure 22.1 and the known properties of the human brain are a valid guide.

From a management viewpoint, it would be risky in a different way to simply write neural networks into the baseline schedule for all aspects of NASP. However, the NASP management plan — recognizing the key role of risk and new technology to that program — does include a major component of "technology enhancement" and "generic hypersonic flight," so as to encourage team B technologies which provide a backstop or insurance against the failure of baseline technologies. Considering the high cost of failure, the extra insurance is well worth the limited costs; neural networks are a perfect candidate for that kind of backup effort. Neural networks have already been applied successfully to one of the critical, risky control systems in NASP, but the details are proprietary to the large corporation which has achieved this.

Beyond NASP, there are many other potential applications for neural networks involving vehicles and structures. There are other high-performance military vehicles which have presented severe control challenges. There are applications to conserving energy both in commercial aviation and in automobiles (e.g., transmission systems), using nonlinear optimal control. The Department of Energy (DOE) has shown great interest in intelligent buildings, and has discussed the possibility of using neural networks to conserve energy in that sector. (The DOE Conservation Office has held major conferences on intelligent buildings.) In the space station program, NASA ultimately hopes to build very complex structures, which could take advantage of new control techniques in order to enhance stability. In the ocean science area, there are problems in steering autonomous vehicles (discussed by Kuh of the University of Hawaii) through upwells of warm water, which generate noise and nonlinearity. All of these control problems involve the continuous, nonlinear and noisy type of problem where neural networks should have their greatest competitive advantage.

22.3.2 Robotics and Manufacturing

In the near term, the largest application of neurocontrol to manufacturing will probably be in chemical process control. This may involve everything from oil refineries to advanced bioreactors.

In oil refineries and chemical process plants, a key goal is to manage transitions from one flow state to another in an efficient manner, so as to maximize profit over time; back-propagating utility and adaptive critic networks are being investigated. Interdisciplinary teams with very close links to the industry are well under way in practical work here; for example, McAvoy at the University of Maryland (in collaboration with the Systems Research Center there) and Ydstie of Amherst (in collaboration with Barto) are prominent in this area. (Bhat and McAvoy, 1989). DuPont is also a player here. Ungar, in the chemical engineering department at the University of Pennsylvania, has argued that millions or even billions of dollars could be saved here, because the flow of product through the chemical and petroleum industries is worth hundreds of billions of dollars per year; even a half of a percent improvement in efficiency would be worth a great deal (Ungar, 1990).

For similar reasons, applications are also being explored in the electric utility sector. In the electric utility sector, it is even possible that optimal neurocontrol schemes might have a clean interface with new, more competitive regulatory regimes, because the central concept of Lagrange multipliers plays a crucial role both in neurocontrol methods and in classical market economics. A recent report from the Office of Technology Assessment (OTA) has described the difficulty of such an interface (between regulation and control) as one of the key difficulties in deregulating that industry.

The application to bioreactors may ultimately prove more important. Bioreactors are like chemical reactors, for biotechnology products. Bioreactors currently produce about $4 billion per year of product, but forecasters have claimed that the potential value of product may be one or two orders of magnitude larger if the technical problems can be solved. Products might include drugs, chemicals, stimulants for food production, and eventually even whole tissues for grafting into the sick or elderly, so as to extend the healthy and productive period of their lives. More complex products are trickier to grow, however, because it is necessary to control the environment in a very sophisticated way. Once again, there is a complex nonlinear control problem which is vital to the very existence of new productive processes. More advanced sensors and reactor concepts will be necessary here, but they will probably not be sufficient. Ungar, among others, is exploring the possibility of neurocontrol in this application.

The biggest target of current efforts — for obvious economic reasons — is the area of robotics and automation. Most economists agree that the major determinant of our economic well-being is the rate of growth of productivity in our economy. Productivity is the bottom line; capital investment, the national debt, and other such issues (except for the balance of payments) are important in the long run only because of how they affect productivity. Productivity also has a direct effect on the competitiveness of industry, which tends to determine the trade deficit. The primary economic argument for government support of basic R&D is the hope that such support will boost productivity, directly or indirectly. Productivity can be increased by the development of whole new industries — as in the previous examples — or by improving productivity in the core existing industries. Productivity in the existing manufacturing base is clearly a central issue here, an issue which neural net research can address fairly directly.

Realistic efforts to boost productivity need to confront the concrete issue of why productivity has grown so much faster in Japan in recent years than in the United States. Concretely, the rapid growth in robotics and automation in Japan is the obvious cause, but there is more to the story than this. A few years ago, major U.S. manufacturers began a similar buildup in robotics, which flattened out after certain difficulties became apparent. According to leading experts in robotics and automation (at the neurocontrol workshop cited earlier), the key difficulty was in *adapting* robots to the reality of the factory floor. Oftentimes, 80 percent of the cost of using robots came *after* the plant was built and the robots were installed. These costs were prohibitive for many manufacturers.

Whatever the reason for these prohibitive costs, it is clear that *adaptive* robots would be less expensive to retrain than robots based on fixed programs. If neurocontrol could cut these costs significantly, it could have a crucial effect on the economic viability of robotics in the United States.

Managers of some of these companies have sometimes pointed toward different explanations of their difficulties in keeping up with Japan. For example, they have pointed towards the quality of labor in downtown Detroit versus the quality of labor in the Far East. If this were the crucial point, it would be all the more useful to be able to "clone" the most competent workers with neural networks, or to build adaptive robots which do not require such sophisticated reprogramming to perform new tasks. One might expect high initial costs in the first adaptive factory, followed by very low costs in subsequent plants, simply because it is easier to transfer computer programs and memories than it is to transfer skills to a new human labor force.

Many current discussions of the future of the automobile industry and the aerospace industry seem to revolve about the issue of who can get the most innovative designs (both advanced and reliable) to the market fastest. As flexibility becomes more and more crucial to competitiveness, *adaptive* robots and manufacturing systems will become crucial in turn. Precision engineering and smart sensors will also be necessary — in addition to intelligent controllers — to achieve all this; however, neural networks may also be used as components of smart sensors.

Some critics have argued that intelligent robots and factories might become so capable and so independent that they might actually pose a threat to human beings. These criticisms should not be taken lightly. For example, an automated factory designed to maximize profits over time, equipped with a high effective foresight horizon and many degrees of freedom, might well develop strategies of production which yield huge outputs but pollute the local environment to an extreme degree. Human managers, left to their own devices, have often had similar failings. This underlines the need for competent and ethical higher management, and for care in instructing the automated system about what it is supposed to maximize. For the time being, even the most advanced adaptive critic systems do not have such a long foresight horizon and imagination.

In the end, our only real security against the abuse of technology is the ability of human beings to understand themselves and their technologies at a higher level. Hopefully, the development of neurocontrol — with a firm link to cognitive science, to humanistic psychology and to neuroscience — can contribute to such understanding. There are tremendous short-term pressures to downplay these links; this makes it all the more important for us to counter these pressures by steadfastly maintaining a broader view.

22.3.3 Teleoperation and Aid to the Disabled

Teleoperation is one of the fundamental technologies now in use in remote or hazardous environments, such as nuclear power plants, outer space, battle zones, or the depths of the ocean. In teleoperation, a robot does the work, but a human controls the robot as closely as possible, as if the robot were an extension of his or her own body. Researchers in teleoperation have sometimes tried to create a kind of "artificial reality," in which the human sees, feels, hears and smells exactly what the robot senses in the remote location. Devices like the "Dataglove" have already been developed, which go a long way toward creating such a total interface between human and machine. Very similar techniques are being developed for the disabled, to let them control artificial limbs or artificial voiceboxes.

There is a crucial gap in today's teleoperation technology, which may affect our ability to exploit the enormous economic potential of hazardous environments. There is a need for more of an intelligent interface between the human controller and the robot. For example, researchers at the Space Science Institute in Princeton have argued that the use of teleoperation could make it possible to mine the moon in an economic fashion, at one-tenth the cost of a manned effort to achieve the same objective; however, many other researchers argue that humans perform very badly in controlling robots when there is a 3-second time delay (due to the speed of light and a round trip of half a million miles). In the production of speech for the disabled, the slow interface between the human and the sound producer has also led to severe quality (and intelligibility) problems.

The problem of providing an intelligent buffer between a slow, higher-level controller and a fast, high-frequency actuator has already been faced — and solved — in the mammalian brain. The higher-level controller, including the cerebral cortex, runs at a frequency of about ten frames per second (like a movie). The muscles operate at more like 100-200 frames per second. The buffer between them is a *subordinate* neurocontroller, the cerebellum, without which we lose much of our normal coordination. Some sort of artificial cerebellum would be tremendously useful in applications like teleoperation and aid to the disabled.

How could one build an artificial cerebellum? Presumably, one could try one of the five basic designs for neurocontrol now in use. A critical area — which requires more research — is the area of how to interface these designs with a human supervisor. One possibility is to have the human control the utility function — or one term of the utility function — which the neurocontroller tries to maximize, in real time; additional terms could be added to represent the jerkiness or energy efficiency of the motion. Biological research into the cerebellum, if closely linked to the neurocontrol literature, might give useful clues. (In a later section, we will mention a few possibilities, building on the designs discussed in the rest of this chapter.) On a more mundane level, there are other ways of using neural networks which might be useful in teleoperation as well; for example, the stream of sensory feedback going to the human might be modified by artificial neural networks.

There are some applications where an artificial cerebellum may not be the best way to go. After all, a human operator has a natural cerebellum which — with sufficient training — can take over many of these tasks. Still, the task of training the human operator may be nontrivial in itself, especially if the task to be performed is not "natural." Because it must operate at high speed using slow components, the natural cerebellum has certain limitations. For many of the disabled, the natural nervous system may be impaired as well.

It is intriguing to take all this a step further, and to speculate on the possibility of a great chain of neurocontrol, with primary controllers driving a set of subordinates, driving another set of subordinates, and so on. It is better, in the mathematics, to have *one* integrated, distributed controller, but radical differences in hardware constraints and cycle times sometimes prohibit this.

It is difficult to assess the potential of this area more concretely and more precisely, because the current level of funding and effort is quite low, at least in the United States. (In Japan, the linkages between biological research into the cerebellum and engineering approaches has been somewhat stronger.)

22.3.4   Communications, Computation and General-Purpose Modeling

There are many other potential applications for neurocontrol besides those discussed above. This section will briefly mention a few others, which have only recently begun to excite serious discussion.

Large-scale communications networks and communications systems present control challenges which are the ultimate in nonlinearity and noise with a large number of variables. Other authors in this book have studied the tremendous challenges which will arise as we move to truly high-density, multiformatted networks based on fiber optics and global interconnections. Powerful, distributed computing networks and super-computers involve similar challenges. Neurocontrol, by definition, tries to achieve effective control *within* the constraint of a highly distributed architecture — exactly the kind of architecture which would seem most appropriate in controlling a highly distributed global network in a fault-tolerant manner. A few examples of dynamic control problems in communications are given by Ephremides and Verdu (1989), but they do not purport to cover the whole range of issues involved. Because another author has discussed these applications in detail, I will say no more.

Neurocontrol and related techniques can also be used in general-purpose modeling, which can include everything from socioeconomic modeling and global climate modeling through to device characterization. In fact, the very first application of back-propagation was to estimating a political forecasting model developed by Karl Deutsch (Werbos, 1974). The first application of back-propagating utility through time was to building a model of the natural gas industry, based on profit maximization (Werbos, 1989). Given a dynamic model, or a fuzzy logic model, techniques borrowed from neural network theory can be used to tune parameters, to calculate the sensitivity of desired values to present actions, to suggest changes in actions, to pinpoint the most critical policy variables or system variables out of hundreds or thousands, and so on. The use of these capabilities could be made transparent to the user; for example, Werbos (1986) describes just one way (if not the best way) to make adaptive critic capabilities available within a standard econometric package.

22.4 SUPERVISED LEARNING AND EXPERT SYSTEMS

Supervised learning is usually defined as the problem of learning the map from a vector $X(t)$ to another vector, $Y(t)$, when we have a set of training data which consists of T observations in which both vectors are known. The vector $X(t)$ is made up of m numbers, labelled $X_1(t)$ through $X_m(t)$, and is called the input vector. The vector $Y(t)$ is made up of n numbers, labelled $Y_1(t)$ through $Y_n(t)$, and is called the target vector (or the vector of targets).

This book has already made some reference to the myriad of neural network techniques (like basic back-propagation) available to solve this problem. This section will make some additional comments, relevant to neurocontrol and the link to expert systems.

22.4.1. Supervised Learning and Neurocontrol

First of all, you should note that the very brief definitions of supervised control and direct inverse control given in an earlier section are *already enough to specify how to implement those methods*. Once you know how to build a supervised learning system, all you need to know are the *inputs* and the *targets*.

Admittedly, there are some fine points in deciding how to *encode* the inputs. For example, Kuperstein's direct inverse control scheme involves a very complex biologically based preprocessor (Kuperstein 1987). The past work of Bullock is somewhat similar. Miller — whose results may be the best to date — uses a CMAC system to perform supervised learning, and appends the previous-time spatial coordinates to the input vector (Kraft and Campagna, 1990). The first "C" in "CMAC" stands for "cerebellum," but — despite the original biological motivation (1971) — "CMAC" now refers to a supervised learning scheme, not a control scheme.

Strictly speaking, these applications may be viewed as examples of system identification, where our goal is to adapt a model of the human controller or of the actuator angles; therefore, the equations given in a later section may be applied directly. Miller, however, provides a way of allowing a simpler kind of memory, without resorting to any of the complexities of back-propagation. (Miller's new "tapered" variant of CMAC is differentiable, as well, which makes it usable in some of the other neurocontrol designs.)

Researchers such as Bavarian (1989) have gone further, by using simultaneous-time recurrence in such networks. It is possible to perform supervised learning using back-propagation through time, in networks which *combine* time-lagged memory and simultaneous-time recurrence, and are built up from *any* set of differentiable functions, but the mathematics are a bit complicated (Werbos, 1988).

The decision to add memory or not in supervised learning is equivalent to the decision to add lagged terms in a classical time-series model. Box and Jenkins (1970) have provided many heuristic guidelines there; often the best practice is to use one's intuition, based on what one knows about the particular application, and to test out additional connections if they seem potentially useful. If the state of the system is determined almost entirely by current control actions, memory may not be useful. When there is considerable inertia in the system, in variables not directly observed by one's sensors, memory tends to be crucial.

22.4.2.   The Link to Expert Systems and Fuzzy Logic

This section will also discuss the links between neural networks and expert systems, links which are easiest to understand in the context of supervised learning.

There are three obvious ways in which expert systems could be linked together with neural networks in solving a supervised learning problem:

- Either one could serve as a preprocessor for the other. For example, if an expert system needs to know about an intermediate variable R_1 in order to reason effectively about the problem, one might train a neural network to predict R_1 as a function of X; the expert system might take the problem from there. Alternatively, the expert system could use its logic to evaluate some other system variable, R_2, which is then treated as an additional input variable by the neural network.
- The expert system could be used to set up *initial values* for the weights and structure of a neural network. To do this, one would first use the expert system to generate a chain of propositions stretching from the vector X through to a prediction for y; each node in this chain — each proposition — could be associated with a neuron, and the associated inference rule translated into the weights used to determine the output of this neuron. *If* the same chain of inference (with different truth values in different observations) could apply to *all* examples in the training set, one could then go ahead and *adapt* the weights, starting from these initial values, so as to match the training set better. One could even use a modified error function, with an additional term representing the square error between the actual weights and the initial guesses (as described by Ikuo Matsuba of Hitachi).
- *If* the expert system were based on some kind of continuous-valued logic or fuzzy logic, such that each rule corresponds to a differentiable function (differentiable almost everywhere, at least) with continuous adjustable parameters, one could use back-propagation to *adjust these* parameters — without ever building a neural network as such — *even* if the inference structures were different for different observations, and even if the inference structure were a recurrent net. (At each observation, one merely back-propagates through the current inference structure, using the methods of Werbos (1988), which require *no* additional storage here.)

A variety of authors have applied the first of these options to practical problems, such as automotive engine diagnostics (by Hornig of the Carnegie Group in Pittsburgh in association with Ford), and Fu and others have proposed ideas related to the second; however, I believe that this is the first mention of the third. The third approach is similar in spirit to Holland's "bucket brigade," which learns to classify patterns by adapting the "strength" of rules in response to local feedback, calculated by something very similar to back-propagation.

All three of these options assume that expert systems give you useful prior knowledge, that the training set gives you useful additional information, and that the optimal predictor of y is based on combining both pieces of knowledge together. The first option is very clumsy, in that it requires us to choose *either* prior knowledge *or* the knowledge embedded in the training set in predicting *each* intermediate variable or final variable. For each such variable, we have to choose one or the other, based on what we know about the system to be predicted. (For example, we might find, *empirically*, that expert systems are good at some things and bad at others, and proceed accordingly.) The second option lets us *combine* the two kinds of knowledge in forecasting any individual variable, but it only works if the inference structure has a fixed form across the training set. (This is not so severe a constraint as it first appears; for example, one can always *expand* inference structures to include the evaluation of propositions which are important in one observation but unimportant in another.)

Still, the real strength of expert systems lies in their ability to use *different* inference structures when confronted with different problems. Only the third of these options fully preserves this capability. The third of these options permits the development of a completely integrated *hybrid* supervised learning system, which can be used *in place of* neural networks, within *any* of the neurocontrol architectures discussed in this chapter. This hybrid is very complex, and should only be used when the extra cost is justified by the application (i.e., where variable inference structures are truly essential).

It may well be that further research in cognitive science is needed to figure out what *kinds* of differentiable rule structure correspond best to what can be elicited easily from human experts; thus instead of choosing between fuzzy logic and Dempster-Shafer logic, it may be better to choose a more neural-like hybrid, or a more complex interview strategy.

Given the possibility of inserting a hybrid system in place of a neural net component, everything in the remainder of this paper could be used immediately as a trainable fuzzy control system. Alternatively, one could replace *some* of the neural networks in what follows with hybrid or purely expert systems, so long as they are continuous and are differentiable almost everywhere. When people publish papers based on this approach, they may sometimes give new names to back-propagation as applied to various kinds of special networks; however, it will be easier for engineers to exploit such work if the authors make the generality of the mathematics and the logical connections more explicit.

In the long-term, one may even treat the problem of inference itself as a control problem, and use neural networks to guide inference. After all, that is how symbolic reasoning is done in the human brain itself, and there is a close parallel between the "evaluation functions" used in inference in expert systems and adaptive critic networks.

22.5   FURTHER DETAILS ON THE FIVE BASIC DESIGNS

Out of the five basic designs discussed, two of them — supervised control and direct inverse control — are a straightforward application of supervised learning, discussed in the previous section. A third — neural adaptive control — involves very recent work by Narendra and Parthasarathy (1990), which mainly speaks for itself. This section will therefore add only a few brief comments on these designs, and focus more attention on the two approaches — back-propagating utility and adaptive critics — which are suitable for planning or optimization over time.

Throughout this section, I will use the expression $X(t)$ to denote the vector of sensor inputs at time t, and $u(t)$ to denote the vector of actions taken by the system at time t (*after* $X(t)$ has been observed).

22.5.1. Supervised Control

In supervised control, we begin with a training set of $X(t)$ and $u^*(t)$, where u^* is a vector of targets for the action vector. The target actions usually come from recording the actions of human beings solving the desired control problem, but they may also come from recording other control systems.

The motivation behind supervised control is approximately the same as the motivation behind expert systems: to "clone" the skills of a human being. However, instead of copying what a person *says*, we copy what he or she *does*. If we were building a computer to ride a bicycle, for example, it would be absurd to depend only on what a person *says*, since much of the skill is quantitative and nonverbal. It is possible to *start* from what a person says, and then *adjust* our network to reflect what the person actually does, when we think it is important to exploit both kinds of information.

When it is possible to find a human or a computer program which can perform a task adequately, supervised control has the advantage of being relatively straightforward and relatively certain of success. If we already have a human or a computer program capable of performing a task, neurocontrol may still be worthwhile if our supply of such humans is limited, if the humans or computer programs are expensive to use (compared with a neural net chip implementation), or if the humans or computer programs can only control the system safely when it is simulated and slowed down.

In some applications, it may be best to *initialize* a neural net with supervised learning, so as to copy a human flying a simulator, and then *later* adapt a more efficient controller using an optimizing method but *starting* from those earlier weights. This is especially useful in applications where the system undergoing real-time learning needs to *start out* at a certain minimal level of performance (e.g., not crashing an airplane). When this is a serious issue, it may even pay to go through a three-stage approach, starting with supervised control, moving up to a more difficult simulation incorporating noise, and then only in the last stage moving on to real-time learning. In extreme cases — as with nuclear power plant control — it may be desirable to put in "dumb" override circuits for neurocontrol systems, just as one does with human operators.

The earliest example of a supervised control system was Widrow's broom balancer in the 1960s. Guez and Selinsky (1988) provide a more sophisticated, updated version of the same idea. Jorgenson and Schley (1990), of Thompson-CSF, describe an application to aircraft control in Miller, Sutton and Werbos (1990). Dozens of new applications of supervised control and direct inverse control have appeared in the various *Proceedings* of the International Joint Conference on Neural Networks.

22.5.2 Direct Inverse Control

Miller's robot controller is described in more detail by Kraft and Campagna (1990). Those authors compare direct inverse control against traditional forms of adaptive form in the same application, and find a mixed picture of advantages and disadvantages to both. In more recent, unpublished work, they have begun to arrive at stability results as well for this class of control. Miller's neurocontrol board may be at or close to the point of commercial viability. Direct inverse control can have problems when there is not a unique mapping back from trajectory coordinates to actuator coordinates. Therefore, Jordan (1989) and Kawato (1990) have used optimal control, instead, as discussed. Their approach is sometimes called "indirect inverse control," because it solves the *same* problem of following a trajectory, but in a totally different way.

Kawato, in earlier papers, has also proposed a third solution to such problems, called "feedback error learning," which *presupposes* the availability of a fixed feedback control system adequate to stabilize the system (Kawato 1990). In a formal sense, this method is equivalent to a kind of adaptive critic design, in which the critic has been *fixed in advance*. Its performance is highly sensitive to the quality of the fixed feedback system. Wilhelmson and Cotter (1990), among others, have successfully replicated this design.

22.5.3 Neural Adaptive Control

Neural adaptive control has been defined as the use of artificial neural networks in place of more classical mappings, within the classical designs of adaptive control theory.

There is a wide variety of such designs, as described in Narendra and Annaswamy (1989) and Narendra and Boskovic (1990).

Narendra and Parthasarathy (1990) focus on the example of Model-Reference Adaptive Control. A number of other authors have focused on Self-Tuning Regulators, where the controller is *fixed* as a function of certain parameters describing the system to be controlled; neural networks are used to estimate these *parameters* in real time. Strictly speaking, this latter approach does not meet the definition of neurocontrol given earlier; it is more of an exercise in *neuroidentification*, in the use of neural networks to identify system dynamics. Lapides and Farber of Los Alamos are often cited as noteworthy examples of this approach, but the details are beyond the scope of this chapter,

Stability theorems are a major part of this research, because even a rare breakdown in control can have serious costs in many applications.

In recent, unpublished work, Narendra has argued that the most flexible way now known to implement an MRAC system is by back-propagating utility, using a design which is very similar in spirit to Jordan's and Kawato's. The idea is to define a cost function as the difference between the desired trajectory, as output by the reference model, and the actual trajectory; the back-propagation of utility (cost) can be used in a straightforward way to minimize that cost function. Narendra has stated that there may be dangers of instability if the back-propagation of utility is done using real-time, time-forward methods; however, the offline version — to be discussed here — does not have this problem.

There are many engineering problems where people use adaptive control as a way of coping with systems governed by parameters which change slowly but invisibly over time. Thus they are seeking a *real-time* kind of adaptation. Adaptive critics do offer a real-time adaptation capability, but for now it is probably easier to train a net *offline* how to be adaptive *online*, using the back-propagation of utility. In an ideal system, one would combine both approaches, by using adaptive critic methods with Model networks like those discussed in this chapter; however, such complexity is probably not necessary for near-term applications.

22.5.4 Back-propagating Utility and Adaptive Critics: Some Points in Common

Back-propagating utility and adaptive critics have three points in common which should be discussed before I get into the details of the two methods: (1) both require a utility function or reinforcement variable U; (2) both lead to modular designs, made up of different neural network components performing different functions; (3) both may require a model of the system to be controlled, but the details vary.

The idea of utility maximization has been very controversial in the social sciences. It seems to indicate a degree of rationality and intelligence on the part of human beings which seem to conflict with the notion of studying human beings the way one studies electrons or pieces of rock. But in actuality, scientists studying electrons have found it very useful to assume that the universe itself maximizes or minimizes a "Lagrangian function," over all space and time, which functions very much like a utility function. (The analogy is so strong that some control theorists use the letter "L" for "Lagrangian," instead of "U," to denote a utility function.) If organic brains were designed like adaptive critic systems, this would not imply that they always behave in a rational way; rather, it would imply that they *approach* rationality, as they *learn*, gradually, to maximize utility in a variety of problem domains.

Because of such historical disputes, classical psychologists like B.F. Skinner have traditionally written about *reinforcement* (or "primary reinforcement") as a function of time (U(t)) rather than *utility* as a function of current sensor inputs (U(X)). From a mathematical point of view, however, the former is just a *special case* of the latter. Even if reinforcement were a totally mysterious quantity, we could treat U(t) as an *additional component* of the vector X(t). Using that expanded vector, U(X) is simply that function which picks out that component of the vector. From a pragmatic point of view, some of the methods of neurocontrol require that we have a *model* of the system to be controlled, *which would include* the utility variable; methods of that sort tend to work best if utility is expressed as explicitly as possible in terms of known, predictable sensor variables.

For the same historical reasons, adaptive critic methods and the back-propagation of utility have both been described as "reinforcement learning" methods, as have a few other methods which are not relevant to optimization over time or complex control applications.

In adaptive critic networks and in the back-propagation of utility, there is almost always one neural network — the Action network — which inputs information about the current state of the world, and outputs the actual control signals, $u(t)$. Because there are no targets, u^*, for the output of this network, there must be *other* networks in the system, adapted by *different* learning rules, which somehow govern the adaptation of the Action net. This kind of multiple-network modular structure is talked about at length in Chapter 10 of this book.

There are three other kinds of network which may appear in these designs — Utility networks, Model networks, and Critic networks.

Utility networks are simply utility functions expressed (or treated) as neural networks. These networks are not adapted, but one sometimes needs to know the *derivatives* of utility with respect to other variables; it is convenient to think of those derivatives as a kind of back-propagation output from a Utility network.

Model networks (or "Emulators") are networks which input current sensor readings and actions, and output simulated values for the sensor readings at the next time period. The problem of how to *adapt* such networks will be discussed in this Chapter. In neurocontrol, the Model network is usually treated as if it is fixed, as if it has *already* been adapted. It can be any network of differentiable functions, such as a fuzzy logic net, a fluid dynamics code, an econometric model, or a neural network. The back-propagation of utility always requires a Model network, but the simplest adaptive critic designs do not. In theory, the Model network is the only network which needs to have any "memory"; however, this assumes that the other networks are allowed to input the outputs of intermediate neurons in the Model network. The Model network may itself be a modular structure, made up of subnetworks, as discussed in Werbos (submitted).

Critic networks — as one might expect — are used only in adaptive critic designs.

22.5.5 The Back-propagation of Utility

The back-propagation of utility involves a *direct* and *exact* maximization of utility, summed over future times, based on a known model. Strictly speaking, this might be called "direct model-based utility maximization," so as to include equivalent methods like that of Cotter and Conwell (1990), which do not involve back-propagation per se; however, the bulk of the work so far in this category does include the use of back-propagation through time.

There are two major variants of this method. In the most important variant, the user specifies a utility function, $U(X)$, a Model network, and an Action network. The goal is to adapt the weights or parameters in the Action network. As in basic back-propagation, the user supplies initial guesses (which may be random) for the weights in the Action network, and iterates through a three-step sequence:

1. Evaluate the current guess for the weights as follows: using the current weights, and the initial values of $X(1)$, use the Action network and the Model network to generate a chain of forecasts of $X(t)$ through to some final time t=T, and calculate the resulting sum of utility over time.

2. Use back-propagation through time to calculate the derivatives of the sum of utility across time with respect to all of the weights in the Action network.

3. Use steepest descent or some other gradient-based method to adjust the weights, and go back to step 1, unless you have converged and can quit. (As in the example of Cotter and Conwell, this step could also exploit information about total error or — in the extreme — *only* use such information.)

If the goal is to achieve a special end state, as in Nguyen and Widrow's truck backer-upper (1990), then the major component of utility comes at the end of the process (though wasting time may be included as a cost), and it pays to *begin* training on initial values near the end state. It is also common to train on multiple "strings" or "episodes," starting from different initial values, instead of just one time-series; however, the generalization to that case is trivial.

This kind of back-propagation of utility was first proposed in Werbos (1974). In 1988, the two leading examples were the truck backer-upper and Jordan's indirect robot controller, described in Jordan (1989).

The one difficult part of this method lies in step 2 above — calculating all the relevant derivatives. Jordan explains what he does by showing how the Action net, the Model net, and the Utility net, from time 0 to time T, can be thought of as one enormous feedforward network, straddling all the time periods; one can then apply conventional back-propagation to this global network, to calculate the required derivatives. When actually programming this kind of procedure, however, it is easier to use a modular design, following Figure 22.2, in which the Action network, the Model network, and the Utility network are located in separate subroutines. The back-propagation can be done by using *dual* subroutines; for example, the dual subroutine for the Utility net would input the contents of the Utility network at every time t, and output the gradient of U(X) with respect to the vector X(t). The dual to the Model network would input the derivatives of utility with respect to X(t+1), and output the derivatives of utility with respect to X(t) and u(t); the former derivatives are simply worked back to calculate the latter, using back-propagation. (See Werbos (1990) for a more detailed example, somewhat generalized, or see Section 22.6 for a similar set of calculations.)

Nguyen and Widrow use a modular approach similar to Figure 22.2, but they do not publish the dashed lines, which indicate which derivatives to send back to the various dual subroutines, and are crucial to success with this method. There is no real *mathematical* difference between the formulations by myself, by Jordan, and by Nguyen and Widrow, on this point; the difference is entirely one of *labelling* the same kinds of calculations. Jordan's paper tends to presuppose a knowledge of this basic methodology, and devotes most of its attention to the problem of how one formulates a utility function appropriate to the trajectory-tracking problem.

Kawato's cascade method and my own natural gas model (in Werbos 1989) use a slightly different arrangement. Instead of an Action network, they directly adapt the *action vectors*, u(1) through u(T), starting from the initial state X(1). The procedure is much the same, in practice. Werbos (1989) gives a very basic tutorial on how to do this *when* the Model is in fact an arbitrary set of differentiable functions. (However, it does not allow for instantaneous recurrence, as in differential equation models, treated in a

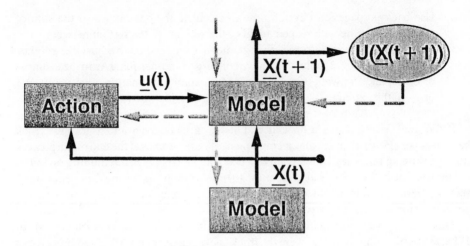

Figure 22.2   The Back-propagation of Utility

harder paper, Werbos (1988).) Kawato does all this in the *second* phase of his "cascade method"; the first phase is one of neuroidentification, which adapts the Model network.

Werbos (1990) briefly mentions variants of back-propagation through time, which calculate the required dynamics in a time-forward manner, but which multiply the run time by a factor proportional to the complexity of the problem; since these variants are expensive for large problems, and therefore nonbiological, I will discuss them no further here.

The back-propagation of utility is very closely related to the calculus of variations, a well-known method in control theory, designed to calculate the optimal schedule of actions in precisely the same situation as above. The only real difference, in fact, lies in issues such as computational cost and the notion of an Action network.

Jacobson and Mayne (1970) proposed a method called "differential dynamic programming," which is probably the closest method in classical control theory to the back-propagation of utility. Jacobson and Mayne did not discuss ordered derivatives or the full exploitation of sparse structure (the foundations of back-propagation); however, they *did* propagate derivatives, in effect, from a later time period back to an earlier time period, and they discussed "policy models" which are more or less equivalent to Action networks. Jacobson and Mayne generalized their procedure to the stochastic case, using impressive formal mathematics which could be used in many other niches in neural network research; however, their approach — like the obvious stochastic generalization one might use with Cotter and Conwell — may work best in situations where T is not very large. It appears to have features in common with an old black-jack playing program by Widrow (1973), which has since been superseded by more modern adaptive critic designs; however, more research is needed to establish this.

It is straightforward and obvious how to modify the back-propagation of utility to exploit the stochastic methods of Jacobson and Mayne, but the underlying insights have other uses as well. Differential dynamic programming requires that you have a good *stochastic* model of the system to be controlled. This is not a trivial requirement.

When the back-propagation of utility is used to train a system offline how to be adaptive online, it is crucial that the Model network contain neurons which somehow represent the slowly-varying parameters of the system. It is also crucial that the output of those hidden units be available as an extra input to the Action network, so that the Action network can in fact change its behavior in response to these parameters.

22.5.6   Adaptive Critic Designs: Underlying Concepts

The term "adaptive critic" was first published in Barto, Sutton and Anderson (1983); however, the adaptive critic family of methods includes many other designs, with deep roots in the past and considerable room for new development in the future. (Widrow (1973) was probably the first actual implementation of a neural-network adaptive critic.) At present, adaptive critic designs are the *only* known designs which have a serious hope of explaining the basic capabilities of the mammalian brain indicated in Figure 22.1.

Adaptive critic methods — like the back-propagation of utility — try to maximize some measure of utility over future time periods. Instead of *directly calculating* the impact of present actions on future utility, by *directly* using a model of the external environment, they adapt a new network — the Critic network — which more or less predicts what future utility will be — summed over future time — as a function of present actions. Grossberg and Levine (1987) and myself have sometimes called this network a "secondary reinforcement system" or a "strategic assessment network."

Adaptive critic methods may be defined more precisely as methods which attempt to approximate dynamic programming, as described in Werbos (1977). Dynamic programming is the *only* exact and efficient method available to control actions or movements over time, so as to maximize a utility function in a noisy, nonlinear environment, without making highly specialized assumptions about the nature of that environment. Figure 22.3 illustrates the trick used by dynamic programming to solve this very difficult problem.

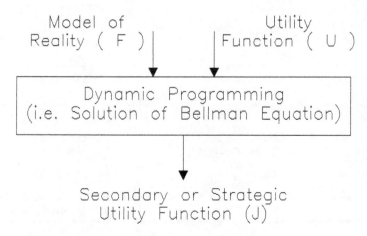

Figure 22.3   What dynamic programming inputs and outputs

Dynamic programming requires as its input a utility function U and a model of the external environment, F. Dynamic programming *produces*, as its major output, *another* function, J, which I like to call a secondary or *strategic* utility function. The key insight in dynamic programming is that you can maximize the function U, in the *long-term*, *over time*, simply by maximizing this function J in the immediate future. When you know the function J and the model F, it is then a simple problem in function maximization to pick the actions which maximize J.

The notation here is taken from Raiffa (1968), whose books on decision analysis may be viewed as a highly practical and intuitive introduction to the ideas underlying dynamic programming.

Unfortunately, we cannot use dynamic programming *exactly* on complicated problems, because the calculations become hopelessly complex. However, it *is* possible to *approximate* these calculations by using a *model* or *network* to estimate the J function or its derivatives (or something quite close to the J function, like the J' function of Lukes et al (1990)). Adaptive critic methods may be defined more precisely as methods which take this approach.

If this kind of design were truly fundamental to human intelligence, as I have suggested, one might expect to find it reflected in a wide variety of fields. In fact, notions like U and J do reappear in a wide variety of fields, as illustrated in Figure 22.4 (taken from Werbos (1986)).

Please note that the last entry in Figure 22.4, the entry for Lagrange multipliers, corresponds to the *derivatives* of J, rather than the value of J itself. In economic theory, the prices of goods are supposed to reflect the *change* in overall utility which would

Domain	Basic Utility (U)	Strategic Utility (J)
Chess	Win/Lose	Queen = 9 points, etc.
Business Theory	Current Profit Cash Flow	Present Value of Strategic Assets (Performance Measures)
Human Thought	Pleasure/Pain Hunger	Hope/Fear Reaction to Job Loss
Behavioral Psychology	Primary Reinforcement	Secondary Reinforcement
Artificial Intelligence	Utility Function	Static Position Evaluator (Simon) Evaluation Function (Hayes–Roth)
Government Finance	National Values, Long–Term Goals	Cost/Benefit Measures
Physics	Lagrangian	Action Function
Economics	Current Value of Product to You	Market Price or Shadow Price ("Lagrange multipliers")

Figure 22.4 Examples of Intrinsic Utility (U) and Strategic Utility (J)

result from changing your level of consumption of a *particular good*. Likewise, in Freudian psychology, the notion of emotional charge associated with a *particular object* corresponds more to the *derivatives* of J; in fact, the original inspiration for back-propagation came from Freud's theory that emotional charge is passed *backwards* from object to object, with a strength proportionate to the usual *forwards* association between the two objects (Freud, 1961). The word "pleasure" in Figure 22.4 should not be interpreted in a narrow way; for example, it could include such things as parental pleasure in experiencing happy children.

22.5.7   Adaptive Critic Designs: Alternative Architectures

To build an adaptive critic controller, we have to do two things: (1) specify how to adapt the Action network in response to the Critic; (2) specify how to adapt the Critic network.

There are three alternative approaches used to adapt the Action network, illustrated in Figures 22.5, 22.6 and 22.7. In the 2-net approach of Barto, Sutton and Anderson (1983) — shown in Figure 22.5 — there is no need for a model of the process to be controlled. The estimate of J is treated as a gross reward or punishment signal. This design has worked well on many small problems (as in Franklin 1988), but there is a serious information bottleneck (or slow learning) with larger problems, because the gross reward signal doesn't "tell" the Action network which components of the action vector need changing, in which direction. The Backpropagated Adaptive Critic (BAC), first published in Werbos (1977) and illustrated in Figure 22.5, overcomes this problem by using the back-propagation of *strategic* utility, J, through the Model network. In effect, it exploits the cause-and-effect knowledge embedded in the Model, in order to pinpoint *which* actions need changing, in *which* direction. The Action-Dependent Adaptive Critic (Figure 22.7), used by Lukes et al, makes it possible to use back-propagation here *without* a model of the environment. All of these approaches have their advantages and disadvantages, which require further study, and there is reason to try to blend the best of all three.

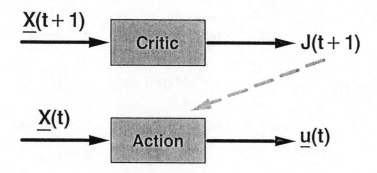

2-Net Adaptive Critic

Figure 22.5   A Simple 2-Net Adaptive Critic Controller

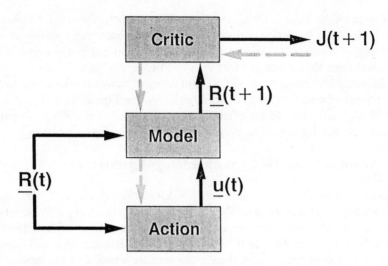

Backpropagated Adaptive Critic

Figure 22.6 The Back-Propagated Adaptive Critic (BAC)

If we choose to adapt a Critic which estimates the *derivatives* of J, instead of J itself, it is easy enough to modify Figures 22.6 and 22.7 accordingly. When we backpropagate the derivatives of J through the Model network, we start out by obtaining the derivatives of J(t+1) with respect to X(t+1); we may obtain these derivatives by backpropagating through a J-type critic (as in Figure 22.6) or we may obtain them as the *output* of a

ADAC

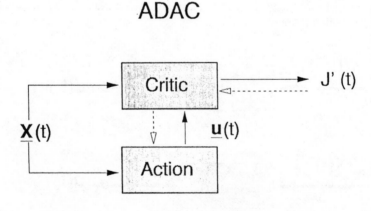

Action Dependent Adaptive Critic

Figure 22.7 The Action-Dependent Adaptive Critic (ADAC)

derivative-type critic at time t+1. In either case, we then work our way back through the Model network to calculate the derivatives of J(t+1) with respect to u(t). Finally, we backpropagate these derivatives through the Action network, to the weights in that network, which are changed in response to these derivatives.

To adapt the Critic network, we again face three choices, involving three generic approaches. The method of temporal differences (Barto, Sutton and Anderson 1983) is a special case of Heuristic Dynamic Programming (HDP) (Werbos, 1977), which involves the following sequence of operations in the simplest formulation:

- At each time t, define the inputs to the Critic as X(t), the current sensor readings. Use the Critic to estimate J(t), and *save* the intermediate calculations, *without* adapting the network.
- At time t+1, use the Critic to estimate J(t+1), and measure U(t), the reward which resulted at time t from the situation X(t) and the actions u(t). Calculate J(t + 1) + u(t).
- Go back to the state of the net at time t, and use J(t+1)+U(t) as the *target* value of J(t). Adapt the Critic network "at" time t, so as to make J(t) closer to J(t+1)+U(t) (treating J(t+1)+U(t) *as if it were fixed*).

These steps are illustrated in Figure 22.8, for the case of Action-Dependent HDP, which is very similar to HDP.

HDP works well when there are only a few variables to evaluate, but, like Figure 22.5, it *still* leads to an information bottleneck. If you combined HDP with BAC, the information bottleneck in adapting the Critic itself would still slow down the learning of the overall system. To solve this problem, Werbos (1977) proposed a *derivative-type* Critic, based on Dual Heuristic Programming (DHP). The steps for adapting the Critic in DHP are similar to those of HDP, except for the middle step: instead of using

ADAC LEARNING

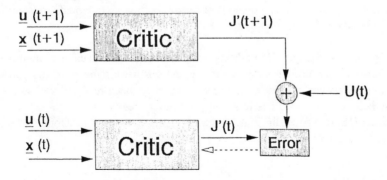

Figure 22.8   Action-Dependent HDP

J(t+1)+U(t) as target, we need a *vector* of targets which, again, cannot be calculated until we reach time t+1. In DHP, the target vector equals the gradient of $U(X(t))$ with respect to $X(t)$ — which is easy to calculate — *plus* the gradient of J(t+1) with respect to $X(t)$. To calculate the latter, we begin by using the Critic at time t+1 to output the gradient of J(t+1) with respect to $X(t+1)$; we then *backpropagate* these derivatives *through* the Model network to get the gradient with respect to $X(t)$.

Note that we need to use back-propagation in order to calculate the *targets* in DHP. With HDP or DHP, we can use *any* supervised learning method to adjust the weights in the Critic, so long as we aim at the right targets. For example, Lukes et al. — using an action-dependent version of HDP — used a simple associative memory for the Critic.

My paper in Miller, Sutton and Werbos talks at greater length about the vicissitudes of HDP and DHP, and a related method called GDHP, which tries to combine the two. There are five vicissitudes of particular importance:

- Even though both methods can be applied in strict real-time learning, they work far better if there are real or simulated explorations near important terminal states or goal states, especially in the early stages of adaptation. (Such "dreaming" may interfere with real-time performance, but one can build dual systems where one side dreams while the other acts, as in the dolphin.)
- In principle, the Critic and the Action network need to input a more complete description of reality, $R(t)$, instead of just $X(t)$. This can be done by allowing them to input intermediate information in the Model network, and modifying the discussion above accordingly.
- When the model F is truly stochastic, the simulation of F involves the calculation of random numbers; one can backpropagate through such a scheme simply by treating the random numbers as if they were exogenous constants. (Fixed-distribution random numbers would suffice.)
- The effective foresight horizon (future time periods truly accounted for) grows only slowly in HDP, unless one uses an accelerated learning method; I have discussed a few possibilities here, but none have been tested as yet.
- When the time horizon is infinite in HDP (but not DHP), there is a danger of divergence in the weights, which can be avoided by using a slight discount factor (as in Barto, Sutton and Anderson) or by subtracting an (adapted) constant from U(t) to make it average out to zero.

BehavHeuristics of College Park, Maryland, has stated that it is using a full three-net adaptive critic design, in true commercial applications netting the company considerable revenue. However, no academic applications have been reported yet, so far as I know. Jameson of Lockheed has reported preliminary results, to be published in the 1990 summer IJCNN Proceedings. As this goes to press, White of McDonnell-Douglas has reported important successes.

22.6 ROBUST NEURO-IDENTIFICATION

Both in back-propagating utility and in advanced forms of adaptive critic design, the quality of control depends heavily on the quality of our model of the system to be controlled. In those few cases I know of where neurocontrol has *not* succeeded, in realistic applications, the problem can be traced back to an inadequate model.

This section will provide the equations which are needed to solve this problem, to an adequate degree, in current engineering applications. It will *not* discuss the more difficult problem of adapting a *true* stochastic model (described in Werbos (submitted), which gives more of a complete, theoretical overview of this subject). It will not review the usual literature on how to cut out unused connections and speed up convergence; that has been discussed elsewhere in this book. Those users who are fully confident of their existing models, and feel confident in backpropagating through them, need not worry about the material in this section.

22.6.1 A General Network Structure

The problem here is how to build a network which inputs $X(t)$ and $u(t)$ at time t, and outputs a good prediction of $X(t+1)$, using some memory of the previous time, t-1. I will assume that we have data for t=1 to t=T, forming a single time-series; multiple time-series are a trivial generalization. $X(t)$ has m components, $X_1(t)$ through $X_m(t)$, and $u(t)$ has n components, $u_1(t)$ through $u_n(t)$. The vector $x(t)$ contains all the inputs available to cells in the network, including the components of $X(t)$, $u(t)$, and the outputs of other neurons.

A classical network to make such predictions can be written as the system:

$$x_i(t) = X_i(t) \qquad i = 1,...,m \qquad\qquad \text{Eq. 22.1}$$

$$x_{m+i}(t) = u_i(t) \qquad i = 1,...,n \qquad\qquad \text{Eq. 22.2}$$

$$net_i(t) = \sum_{j=1}^{i-1} W_{ij}s_j(t) + \sum_{j=1}^{N-m} W_{ij}s_j(t-1) \quad i = m+n+1,...,N) \qquad \text{Eq. 22.3}$$

$$x_i(t) = s(net_i(t)) \qquad\qquad i = m+n+1,...,N \qquad \text{Eq. 22.4}$$

$$\hat{X}_i(t+1) = x_{N-m+i}^{(t)} \qquad\qquad i = 1,...,m \qquad\qquad \text{Eq. 22.5}$$

where N is the number of neurons in the system (including input neurons), where x is the vector of outputs of all neurons, where equations 1 and 2 simply read in the inputs, and where equation 5 reads out the outputs. In equation 4, s is simply the usual "sigmoid" function:

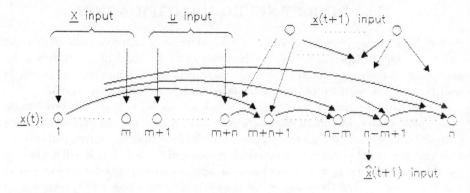

Figure 22.9 The network defined in Equations 22.1–22.5

$$s(z) = 1 / (1 + e^{-z}), \qquad \text{Eq. 22.6}$$

though Werbos (1990) mentions a few simple alternatives to speed up performance. In equation 3 (which determines the "excitation" or "voltage" at neuron number i), most of the weights can actually be set to zero; I have deliberately described a fully connected network (illustrated in Figure 22.9), for the sake of generality, knowing full well that it is usually good to use a sparser structure, depending on one's application. At the end of this section, I will describe a more conventional special case. It is usually best to treat $x(0)$ as zero (except when there is only one working version of the system, working in an ongoing way — like the U.S. GNP — in which case one may treat $x(0)$ as an additional set of weights).

22.6.2   Conventional Least-Squares Adaptation

The usual way to adapt such a network is by trying to minimize:

$$E = (1/2) \sum_{t=1}^{T} \sum_{i=1}^{m} (\hat{X}_i(t) - X_i(t))^2$$

$$\text{Eq. 22.7}$$

In adapting this system, one uses equations 1 through 7 to calculate the error for the current weights. To calculate the gradient of error with respect to the weights, one must first calculate a kind of intermediate feedback ("F_"), which, in formal terms, represents the ordered derivative of error with respect to the intermediate quantities. One does this by first using the following equations, to be calculated as a *group* for all i and t=T-1, then as a group for t=T-2, on down to t=1:

$$F\_x_{N-m+i}(t) = (x_{N-m+i}(t) - X_i(t+1)) + \sum_{j=i+1}^{m} F\_net_{N-m+j} * W_{N-m+j,N-m+i}$$

$$\text{Eq. 22. 8}$$

and

$$F\_x_i(t) = \sum_{j=i+1}^{N} F\_net_j(t) * W_{ji} + \sum_{j=m+n+1}^{N} F\_net_j(t+1) * W'_{ji}$$

Eq. 22.9

where

$$F\_net_i(t) = s'(net_i(t)) * F\_x_i(t) \qquad i = N,...,1$$

Eq. 22.10

Note that we start by using equations 8 and 10, in alternation, for the last neuron, and work our way back to neuron 1 (using equation 9 instead of 8 when we get down to neuron number N-m). Equation 8 can be used first, for the case i=m, because in that case the summation on the far right is null, and one does not need to know F_net. Of course, F_net(T+1)=0.

These equations follow directly from the chain rule for ordered derivatives (Werbos 1989) applied to equations 1 through 7.

Finally, one can calculate:

$$F\_W'_{ij} = \sum_t F\_net_i(t) * x_j(t)$$

Eq. 22.11

$$F\_W'_{ij} = \sum_t F\_net_i(t) * x_j(t-1)$$

Eq. 22.12

These are the gradients one uses in adapting the weights. To minimize error, of course, one adapts the weights in the direction opposite the gradient.

22.6.3   The Pure Robust Method

The approach above represents the current state of the art in neurocontrol at present. The memory of previous time periods (through the weights W'_{ij}) allows one — in theory — to estimate *time-varying* system parameters, or to account for quantities like fluid levels inside of an unobserved holding tank.

As a practical matter, the predictions from this approach tend to deteriorate very quickly with time. In some cases, ad hoc fixes can be good enough; for example, one can force a few neurons, x_i, to have *fixed* memory weights $W'_{ii}=1$, and disable all other memory weights for those cells, and then see what kind of long-term memory develops. (This can help in overcoming some local minimum situations related to memory, as well; in such cases, one may later relax the system and let the W'_{ii} be adapted.) Very often, however, it is important to move on to a form of estimation which I would call dynamic robust estimation.

The simplest version of dynamic robust estimation is one which was proven successful in forecasting applications in Werbos (1974,1977) and in robotics applications by Kawato. I called this the "pure robust" method and Kawato calls it the "cascade method first phase." The underlying idea is to calculate the entire *trajectory* of forecasts predicted by one's model, *starting from* $X(1)$, *without* using any subsequent data on X; one minimizes the error between that entire *trajectory* and the actual history of X.

The underlying theory here is well beyond the scope of this paper, but the implementation is simple. When you calculate the error, E, associated with the current set of weights, use the following equation instead of equation 1, for $t > 1$:

$$x_i(t) = \hat{X}_i(t) = x_{N - m + i}(t - 1) \qquad \text{Eq. 22.13}$$

$$F\_x_{N - m + i}(t) = (x_{N - m + i}(t) - X_i(t + 1)) + F\_x_i(t + 1)$$

$$+ \sum_{j = i + 1}^{m} F\_net_{N - m + j} * W_{N - m + j, N - m + i} \qquad \text{Eq. 22.14}$$

This change is simple enough in appearance, but the effects are profound. If T is large, especially, it is much harder to get the weights to converge than it was before; therefore, I have often had to *begin* by adapting the conventional version, and then using the resulting weights as initial guesses here. (Alternatively, one can adapt the weights over a smaller time interval, and gradually extend it back.) In extreme cases, I did the conventional version, moved on to compromise versions, and *then* moved back to the pure robust version.

22.6.4. The Compromise Method

The pure robust method works extremely well in any situation which is essentially orderly and predictable underneath, even if one cannot observe that order. However, when there is noise or slippage underneath, and one looks over long time intervals, it can break down. Werbos (1977) proposed a solution to this problem, which needs to be modified a little in order to work; Werbos (1978) was the first of several successful tests of a "compromise method," in which equations 1 and 13 are replaced by:

$$x_i(t) = w_i X_i(t) + (1 - w_i) \hat{X}_i(t)$$

$$= w_i X_i(t) + (1 - w_i) x_{N - m + i}(t - 1) \qquad \text{Eq. 22.15}$$

where w_i is a "relaxation constant." When I used equation 15, minimizing E as defined in equation 7, the results were *not* robust, for reasons closely related to the failure of the classical formulation. (The results were a little better at times, but not by much with real data). Werbos (1978) used:

$$E = (1/2) \sum_{t = 1}^{T} \sum_{i = 1}^{m} (\hat{X}_i(t) - X_i(t))^2 / (1 - | w_i |)^2 \qquad \text{Eq. 22.16}$$

Equations 15 and 16 change equation 14 to:

$$F\_x_{N-m+i}(t) = (x_{N-m+i}(t) - X_i(t+1))/(1 - |w_i|)^2$$

$$+ (1 - w_i) * F\_x_i(t+1) + \sum_{j=i+1}^{m} F\_net_{N-m+j} * W_{N-m+j,N-m+i}$$

$$\text{Eq. 22.17}$$

The relaxation constants may be adapted in proportion to (minus):

$$F\_w_i = \sum_{t} F\_x_i(t) * (X_i(t) - x_{N-m+i}(t-1))$$

$$+ (1/2) \left\{ \sum_{t+1} (X_i(t) - X_i(t))^2 \right\} \frac{\partial}{\partial w_i} \frac{1}{(1 - |w_i|)^2} \qquad \text{Eq. 22.18}$$

Equation 16 was devised in order to minimize an ad hoc approximation to the standard errors in estimating the original weights. The details are given only in an unpublished final report to DARPA. Certainly there is room to improve upon equation 16, but it is at least a working place to start. One interesting possibility might be to eliminate *all* weights W', but allow some *intermediate* neurons to obey equation 15, with $X_i(t)$ replaced by the usual expression for $x_i(t)$.

Another possibility might be to minimize a more exact robust measure of standard errors such as that of Hjalmarsson and Ljung (1990).

22.6.5.   Example of a Special-Case Network Structure

Figure 22.8 represents more generality than most people actually use. Unfortunately, the needs of different applications vary greatly, and there is no *one* pattern of connections which is suitable for all problems. As in classical statistics (Box and Jenkins), one needs to use previous knowledge of the problem (where available) to suggest possible structures, and empirical tests to decide which structure fits the empirical data best. It is important to seek parsimony — to cut one's network down to a size where the training set is much larger than the number of weights in the network — in order to get performance in new situations, outside of the training set. (A beautiful example of a parsimonious network is given in Guyon et al., 1989.)

In order to make full use of parallel-processing neural network chips, most researchers prefer a "layered" structure, where networks in one layer do not take inputs from other neurons in the same layer. A simple structure on these lines — which is almost a special case of Equations 22.1–22.5 might be:

$$x_1(t) = w_1 X_1(t) + (1 - w_1) x_{N-m+1}(t-1) \qquad i = 1,...,m \qquad \text{Eq. 22.19}$$

$$x_{m+1}(t) = u_1(t) \qquad i = 1,...,n \qquad \text{Eq. 22.20}$$

$$net_1(t) = \sum_{j=1}^{m+n} W_u X_j(t) + W_1' \, net_1(t-1) \qquad i = m+n+1,\dots, N-m$$

Eq. 22.21

$$or = \sum_{j=m+n+1}^{N-m} W_u X_j(t) \qquad i = N-m+1,\dots, N$$

Eq. 22.22

$$x_1(t) = s(net_1(t)) \qquad i = m+n+1, N$$

Eq. 22.23

with the error function as defined in equation 16. In this structure, x_1 through x_{m+n} form an "input layer"; x_{m+n+1} through x_{N-m} form a "hidden layer"; and x_{N-m+1} through x_N form an "output layer."

One can go even further, by arbitrarily making some of the hidden layer neurons fit:

$$net_1(t) = net_1(t-1) + \sum_{j=1}^{M} W_u(X_j(t) - X_1(t))$$

Eq. 22.24

There is a slight but significant difference between equations 21 and 24, on the one hand, and the corresponding special case of the network in 22.6.1. (The feedback calculations are also changed in a very simple way, which follows directly from the chain rule for ordered derivatives.) The logic behind equation 21 is the idea that a weight of $w_i' = 1$ can really force a *stable* long-term memory. This formulation was inspired by the sticky neuron concept described in recent discussion with Houk and Barto regarding the Purkinje cells in the cerebellum; however, my formulation is slightly different, in order to reflect criticisms by David Robinson of Johns Hopkins based on his studies of the vestibulo-optic reflex. Equation 24 goes further, by strongly trying to *force* certain cells to represent slowly changing process parameters; estimates of these parameters should change when predictions based on the old parameters deviate from actuality. Both of these formulations are new, so far as I know.

There are many permutations of these ideas which one might try. As with the pure robust method, however, it may be crucial to *begin* adaptation with a loss function which downplays the long-term lag effects, and use the results as initial values to a more powerful but difficult formulation.

22.6.6 Biological Parallels and Future Research

Because the brains of mammals are far more complex than existing artificial systems, it would require more complex mathematics to understand them. Therefore, I will have to assume a full understanding of the preceding theory to comment seriously (if briefly) on such parallels.

It has been said that, "If there are five ways to compute something, the brain probably uses all five plus five more that no one has thought of." In that spirit, I hope that the existing neurocontrol designs are indeed relevant — if not wholly sufficient — to understanding organic brains.

The back-propagation of utility usually requires calculations which proceed backward in time, and make use of an exact memory of earlier time periods; therefore, adaptive critic designs — which learn in real-time — are far more plausible as models of the foresight capability of organic brains.

Werbos (1987a) cited a wide range of physiological experiments to support the idea that the "limbic system" or "rhinencephalon" acts as an adaptive critic, for the high-level long-term planning capabilities of the mammalian brain. Werbos (1987b) suggested that cells descended from the limbic system now located in the cerebral cortex (in mammals) could be acting as a local critic *within* the cerebral cortex, in order to replace back-propagation through time in system identification. (In other words, the feedback calculations of Section 22.6.2 can be replaced by an adaptive critic arrangement, as discussed on p.352 of Werbos (1988).) Unfortunately, the "wiring diagrams" of the brain published earlier in this decade are incomplete, and it may be years before really complete diagrams are available for engineering analysis; however, the physiological evidence related to these issues remains extensive.

Kawato (1990) has cited extensive experiments on human arm movement, suggesting that we do in fact minimize a cost function over time. However, the only known designs which work in doing this, in a general and flexible way, are the back-propagation of utility and adaptive critic designs. Houk et al. (1990) have suggested that fibers from the olive nucleus, in the brain stem, act as a "training signal" to the Purkinje cells in the cerebellum, controlling movement. In informal conversation, Barto has gone slightly further and speculated that the olive nucleus may serve as an adaptive critic for the (old) lower-level motor control system.

Schwaber of Dupont has speculated (after considerable experimentation) that the LNS of the brain stem — which controls the beating of the heart — may actually be a Model Reference Adaptive Control (MRAC) system. In this case, certain pacemaker cells would serve as the Reference Model which the heart is made to track. Narendra's current design for a neural MRAC requires the back-propagation of utility through time; however, it would be easy enough to set up an adaptive critic system to try to minimize the same cost measure which Narendra uses. In this connection, it is interesting that the LNS and the olive have extensive connections to each other; perhaps, the olive might serve as a critic for more than just the cerebellum.

To go beyond speculation and the lawyer-like collection of arguments, it would be important to do new experiments here, *informed* by the neurocontrol designs and issues discussed in this book. It would be important to test the engineering *capabilities* of sections of the brain, either in vivo or in living tissue extracts. (For example, to test if a system really is an MRAC, one might change or perturb what it is trying to control or its pacemakers to see if they stay in synch, under favorable but artificial conditions.) In brief, true reverse engineering is very difficult without an engineer taking an active role. Whether the biological research is truly useful to reverse engineering (or to true understanding), as opposed to the accumulation of raw data, will depend on future efforts to deepen the collaboration between engineers and biologists.

REFERENCES

Albus, J.S. (1971) A theory of cerebellar function. *Mathematical Bioscience*, No. 10, p. 25-61.

Barto, A., Sutton and Anderson (1983). Neuron-like adaptive elements that can solve difficult learning control problems. *IEEE Trans. SMC*. SMC-13, p. 834-846.

Bavarian, B. and E. Ranolli (1989), A modified 3-layer perceptron for control of robot manipulators. In *Proceedings of the 28th IEEE Conference on Decision and Control*. New York: IEEE.

Bhat, N. and McAvoy, T., Use of neural nets for dynamic modeling and control of chemical process systems. In *1989 American Control Conference*. New York: IEEE.

Box, G. and Jenkins, *Time-Series Analysis: Forecasting and Control*. San Francisco: Holden-Day.

Cotter, N. and P. Conwell (1990), Methuselah networks and optimal recurrent control. In *Proceedings of the International Joint Conference on Neural Networks* (IJCNN) (San Diego). New York: IEEE.

Ephremides, A. and Verdu (1989), Control and optimization methods in communication network problems. *IEEE Trans. AC*, Vol. 34, No. 9, September.

Franklin, J.A. (1988), Reinforcement of robot motor skills through reinforcement learning. In *Proceedings of the 27th IEEE Conference on Decision and Control*. New York: IEEE.

Freud, S. (1961), *The Ego and the Id* (Standard Edition). London: Hogarth.

Grossberg, S. and D. Levine (1987), Neural dynamics of attentionally modulated Pavlovian conditioning. *Applied Optics*, Vol. 26, No. 23.

Guez, A. and J. Selinsky (1988), A trainable neuromorphic controller. *Journal of Robotic Systems*. August, Volume 5, No. 4.

Guyon, I., Poujaud, Personnaz, Dreyfus, Denker and LeCun (1989), Comparing different neural network architectures for classifying handwritten digits. In *IJCNN Proceedings*. New York: IEEE.

Hjalmarsson,H. and L. Ljung, How to estimate model uncertainty in the case of undermodelling. *1990 American Control Conference*. New York: IEEE.

Houk, J.C., S. Singh, Fisher and Barto (1990), An adaptive sensorimotor network inspired by the anatomy and physiology of the cerebellum. In Miller, Sutton and Werbos (1990).

Jacobson, D. and Mayne (1970). *Differential Dynamic Programming*. New York: Elsevier.

Jordan, M. (1989), Generic constraints on underspecified target trajectories. In *IJCNN Proceedings*. New York: IEEE.

Jorgenson, C. and C. Schley (1990), A neural network baseline problem for control of aircraft flare and touchdown. In miller, Sutton and Werbos (1990).

Kawato, M. (1990), Computational schemes and neural network models for formation and control of multijoint arm trajectory. In Miller, Sutton and Werbos (1990).

Kraft, L.G. and D. Campagna (1990), A summary comparison of CMAC neural network and traditional adaptive control systems. In Miller, Sutton and Werbos (1990).

Kuperstein. M. (1987), Adaptive visual-motor coordination in multijoint robots using parallel architecture. In *Proceedings of the IEEE Conference on Robotics and Automation.* Washington, D.C.: IEEE Computer Society.

Lukes, G., B. Thompson and P.Werbos (1990), Expectation driven learning with an associative memory. In *IJCNN Proceedings* (Washington). Hillsdale, NJ: Erlbaum.

Mead, C. (1989). *Analog VLSI and Neural Systems.* Reading, Mass.: Addison-Wesley.

Miller, Sutton and Werbos (1990), *Neural Networks for Robotics and Control.* Cambridge, Mass.: MIT Press.

Narendra, K. and A. Annaswamy (1989). *Stable Adaptive Systems.* Englewood, NJ: Prentice-Hall.

Narendra, K. and J. Boskovic, A combined direct, indirect and variable structure method for robust adaptive control. In *1990 American Control Conference.* New York: IEEE.

Narendra, K. and K. Parthasarathy (1990), Identification and control of dynamical systems using neural networks, *IEEE Trans. Neural Networks,* March.

National Space Society (1989), Settling space: the main objective. In *America's Future in Space: A Briefing Book of the NSS Family of Organizations.* Washington D.C.: NSS.

Nguyen, D. and B.Widrow (1990) The truck backer-upper: An example of self-learning in neural networks. In Miller, Sutton and Werbos (1990).

Raiffa, H. (1968). *Decision Analysis: Introductory Lectures on Making Choices Under Uncertainty.* Reading, Mass.: Addison-Wesley.

Ungar, L. (1990), A bioreactor benchmark for adaptive neural-based process control. In Miller, Sutton and Werbos (1990).

Werbos, P. (1974). *Beyond Regression: New Tools for Prediction and Analysis in the Behavioral Sciences.* Ph. D. thesis to Harvard U. Committee on Applied Mathematics, November 1974.

Werbos, P. (1977). Advanced forecasting methods for global crisis warning and models of intelligence, *General Systems Yearbook,* 1977 issue.

Werbos, P. and J. Titus (1978), An empirical test of new forecasting methods derived from a theory of intelligence: the prediction of conflict in Latin America, *IEEE Transactions SMC,* September.

Werbos, P. (1986), Generalized information requirements of intelligent decision-making systems. In *SUGI-11 Proceedings.* Cary, NC: SAS Institute. A revised version, available from the author, is somewhat easier to read, and elaborates more on connections to humanistic psychology.

Werbos, P. (1987a), Building and understanding adaptive systems: a statistical/numerical approach to factory automation and brain research. *IEEE Trans. SMC,* March/April.

Werbos, P. (1987b). Learning how the world works: specifications for predictive networks in robots and brains. In *IEEE Proceedings SMC.* New York: IEEE.

Werbos, P. (1988), A generalization of back-propagation with application to a recurrent gas market model, *Neural Networks,* October.

Werbos, P. (1989), Maximizing long-term gas industry profits in two minutes in Lotus using neural network methods, *IEEE Transactions SMC,* March/April.

Werbos, P. (1990), Back-propagation through time: what it does and how to do it. *Proceedings of the IEEE,* August.

Werbos, P. (1990b), Energy and population: transitional issues and eventual limits. *NRG Forum,* August 1990. Teaneck, N.J.: NPG (201) 837-3555.

Werbos, P. (submitted), Stochastic modeling and representations of reality: a linear starting point. Submitted to *Neural Networks,* March 1990.

Widrow, B., N. Gupta and S. Maitra (1973), Punish/Reward: learning with a critic in adaptive threshold systems. *IEEE Trans. SMC.* Vol. SMC-3, No. 5, September.

Widrow, B. and F.W.Smith, Pattern-recognizing control systems. In *1963 Computer and Information Sciences (COINS) Symposium Proceedings.* Washington, D.C.: Spartan.

Wilhemsen and Cotter (1990), Neural-network based controller for a single degree of freedom robotic arm. In *IJCNN Proceedings* (San Diego). New York: IEEE.

23

APPLICATION OF NEURAL NETWORKS TO ROBOTICS

Craig T. Harston

23.0 OVERVIEW

The ultimate robot is an autonomous robot. Such a robot would understand diverse instructions and execute the appropriate tasks, sense the environment and move as needed, and adapt to various conditions and requirements to complete the task. The ideal robot would notice accidents, obstacles, unexpected interruptions, and recover from mistakes.

Many robot researchers and developers expect that true robotic autonomy will be achieved through neural network technology. It seems reasonable that modeling the nervous system, which functions autonomously, will provide the technology to build independent functional robots. No doubt, several sensory mediums such as sonar, light or radar along with tactile input will be integrated. These systems could be described as closed loop with extensive sensory-motor integration.

Robotic labor may be needed for most of our industrial and commercial activities. Situations do exist which require the presence of autonomous robots, such as: involvement in toxic areas, wartime situations, and space exploration. Unfortunately, fully independent robots are not available commercially nor do they exist in the research laboratory due to: lack of integration between technologies, inadequate computerized sensor processing, and extensive path planning and math modelling procedures.

Even though neural network control systems and robotic applications are undeveloped, their potential may equal or exceed that of the brain. If we are to evaluate different technologies for use with robots, we must keep an eye on the future. The question is not which technology facilitates robots most now, but which technology will take us furthest. Dead-end tracks must be avoided. It is entirely possible that increasing functionality with one approach will be accompanied by accelerating computational complexity and associated demands on computer resources.

23.1 NEUROLOGY APPLIED TO ROBOTICS

There are various neural network researchers who effectively tie neuroscience to computer simulations. For example, James Albus used simple but effective learning techniques to access table values (1975, 1979). His approach, patterned after the structure of the cerebellum, was called the *Cerebellar Model Articulation Controller* (CMAC). Andras Pellionisz (1987, 1988) developed a sensory motor model which involved linear tensor transformations. Another cerebellar based neural network robotic model has been reported (Niznik, Hoss & Watts, 1987).

23.2 NEURAL NETWORKS APPLIED TO ROBOTIC TASKS

The following is a discussion of tasks which are involved with robot control. Sensory processing involves pattern recognition and multisensory fusion. While these input processes are critical for successful robotics, this chapter focuses on planning and control concerns.

Task planning as encompassed by both path planning and path/trajectory resolution has been the focus of many articles. Path planning includes determining the optimal trajectory for single and multiple degree freedom robots. The coordination of multiple robots or two robot arms can be considered as an extension to multiple joint path planning. Apparently neural networks are successful because publications related to planning and the development of trajectories make up the largest group of robotic neural network topics.

The *control* of motors and real equipment is a problem of its own. Often there are irregularities, slippage, twisting or other component nonlinearities associated with the control of hardware. We call this topic the motor integration and dynamic control problem. While much work has been done on determining the path or inverse dynamics and kinematics for movement, comparatively few papers focus on managing the nonlinear control of motors and gears. On the other hand, several path planning neural networks have been integrated with robotic hardware, so it appears that these neural networks have learned how to control motor dynamics with some degree of success.

The *integration* of sensory systems with motor systems is critical for the development of independent, fully functional robots. The use of primary feedback and secondary feedback is of concern to many neural network researchers.

Obstacle detection and avoidance has been demonstrated by several neural systems. Typically these obstacles are fixed in the environment and little has been done to accommodate moving obstacles in potential robotic pathways. Probably detection of moving obstacles will be integrated with sensory-motor systems in the future. Accident detection and recovery is another area which needs more work.

Additional concerns involved with developing autonomous robots are *grasping and hand control.* There is growing concern about this robotic task and some research has been published on the topic. A particular problem is the adaptive response to variable payloads. Robots need to be able to grasp and carry objects without preprogramming the response. Objects as diverse as paper cups and steel bars are potential payloads for robots and automated machines need to be responsive to the tactile needs. Although little work has been done to compensate for responding to gravity from different directions, the need exists in space and underwater for robots to adapt to the pull of gravity. Many robotic neural network developers are conscious about jerk control.

23.3 TECHNOLOGICAL CONSIDERATIONS

There are several technological topics which should be examined: multiple neural network design, learning and teaching, generalization, computational requirements, and nonlinear mapping.

23.3.1 Multiple Neural Network Design

Early progress in the use of computer simulated neural networks for control problems resulted from the development of the *Adaptive Critic Element* (ACE) in conjunction with the *Associative Search Element* (ASE) [Barto, Sutton and Anderson, 1983]. Their ASE system was based on ideas from Hebb (1949), which were refined and called *reinforcement* learning. Reinforcement learning updates relevant weights defined by their involvement in ongoing activity. Klopf (1987) used this concept to develop an important and sophisticated neural model which simulates classical conditioning. Not only did Barto, et al., incorporate reinforcement into their ASE, but they reported that the ability of a neural network system to learn how to balance a pole mounted on a cart was enhanced by their ACE.

The ACE predicted the performance of the system based on mathematical models. This predictive information was incorporated and interfaced with the ASE neural network. The modeled information was used to update or reinforce neuronal unit weights as early as possible. The system learned how to balance the pole sooner with the ACE than without it. The reason was because the corrective information in the form of predicted results could be used to update the systems weights before corrective adjustments could be made in the non-ACE systems.

Another successful double neural network system was the truck backer-upper developed in Widrow's laboratory [Nguyen & Widrow, 1989]. They used one network which knew what commands to give along with another network which knew how to perform the commands. These two networks worked together to simulate a tractor-trailer backing up from several initial positions into a dock.

Successful multiple neural network control systems is apparent in research reported from Mitsuo Kawato's ATR Auditory and Visual Perception Research Laboratories at Osaka University. His laboratory developed robotic control systems which learn quickly (within 30 minutes), develop faster movements, execute different movements, and generalize learned movements [Miyamoto, Kawato, Setoyama & Suzuke, 1988]. These advancements appear to be based on at least two factors.

The first is the appreciation of the role of the adaptive critic in learning by neural network control systems. Two elements or subsystems are used which function some-what like adaptive critics. These adaptive critics are different than the early ACE of Barto, et al. They are neural networks which modify the function of a motor control neural network. As with the ACE, these networks are involved with predictive feedforward control of the motor system.

An important second influence is research from the field of neuroscience. Their model was patterned on evidence about how the brain works. They used neuroscience literature from Japan and western sources to develop subsystems which were integrated into their robotic control system.

The system not only learned how to move the robot's components but learned to coordinate smooth, efficient movements. The success of the system was partially based on the incorporation of sophisticated neurological research into the computerized simulations.

23.3.2 Learning and Teaching

Apparently, what neural network control systems do is dependent on what they are taught. This implies that the model used to generate training examples will limit the robot. If the neural network learns by copying human movements, then the system will be limited to human-like procedures. If a math model generates the training sets, then problems with the math model may show up in the robotic performance.

Typically, optimal movements are developed via a mathematical model with respect to some criteria such as the minimization of jerk, time, or distance. Kawato's group studied how humans move and improve while learning a manual task. They concluded that humans optimize based on minimization of torque. Thus, Kawato's group taught movements to their neural network system based on minimum torque optimization (Uno, Kawato & Suzuki, 1989).

23.3.3 Generalization

It is almost impossible to teach a robotic neural network every possible movement. In any robotic task there is an infinite variety of possible moves. The payload could be located anywhere, as well as the starting position of the robot. There are millions of possible start/goal combinations which the neural network would need to learn. The network would have to be impossibly large to learn all these possibilities. Fortunately, the neural network does not have to learn all these combinations. One of the best things neural networks do is to generalize. This ability to generalize is one of the most important and attractive features of neural network technology (Josin, 1988; Massone & Bizzi, 1989). Additionally, training should conform to the tasks which are expected from the robot. Careful planning and design of the training procedures may teach the robot and provide the needed results through generalization.

23.3.4 Computational requirements

When compared to traditional kinematic or geometric calculations required by today's robotic control systems neural networks use few computations. This reduction of computational workload reduces the computational resources required to control robots. If less computational resources are needed, then smaller and less expensive computers can be used for robots. If smaller computers can be used for robots, then robots are more likely to carry their computers on board and less space will be needed to carry the necessary electronics. Obviously, less is more in this case.

Additionally, neural network algorithms are completely compatible with parallel processing. To the extent that parallel computer chips or chips with parallel processing are available, then the neural network computations, few though they may be, can be quickly done. The faster these calculations can be done, the faster or more real time, neural network robots will be. Clearly, there are some powerful advantages to using neural network controlled robots.

23.3.5 Nonlinear mapping

Neural networks can learn to map nonlinear relationships. This capability is one of the most important advantages available from neural network technology. For robotics, we expect networks to compensate for the nonlinear dynamics necessary to control robotic hardware. Some hardware components flex or slip during operation. Variable force is necessary to manipulate different payloads. Payloads may get stuck or scrap, thus adding unpredictable friction which must be dynamically adjusted by a truly autonomous robot. The nonlinear capability of neural networks are a potential solution to these and other requirements.

23.4.2 Published examples of neural network robotic tasks

An examination of neural networks applied to robotics is in order. The question is which neural networks have solved the various robotic tasks. The fact is that different neural network types have been used for various robotic problems. There have been many neural network simulations; however, some have integrated their neural network control systems with hardware robots. The following section attempts to categorize and summarize the neural network robotic efforts as described in the public literature. There are many relevant topics so, with some difficulty, the list was limited to the following:

23.4 SUMMARY

Neural networks bring many important capabilities to robotics. Three of the most useful are the ability to nonlinearly transform information, generalize to novel conditions and reduce the dependency for massive computations on serial computers. Additional work is necessary to resolve current problems and prove the usefulness of neural network technology, however the promise and potential is clearly evident. The long-term potential of emulating the nervous system suggests that serious development is appropriate. While neurally controlled robots are not available for commercial applications, research and development is progressing well.

REFERENCES

Albus, J., (1975). A new approach to manipulator control: The cerebellar model articulation controller (CMAC), *Transactions ASME, Journal Dynamic Syst. Meas., Control*, 97: 220-227.

Albus, J.S., (1979). Mechanisms of planning and problem solving in the brain, *Mathematical Biosciences*, 45: 247-293.

Anderson, C.W., (1989). Learning to control an inverted pendulum using neural networks, *IEEE Control Systems Magazine*, April, 31-36.

Arteaga-Bravo, F.J., (1990). Multilayer Back-Propagation Network for learning the forward and inverse kinematics equations, *International Joint Conference on Neural Networks*, II-319-324.

Barto, A.G., Sutton, R.S. & Anderson,C.W., (1983). Neuronlike adaptive elements that can solve difficult learning control problems, *IEEE Transactions on Systems, Man, and Cybernetics*, SMC-13: 834-846.

Barhen, J., Gulati, S. & Zak, M., (1989). Neural Learning of constrained nonlinear transformations, *Computer*, 22: 67-76.

Bassi, D.F. & Bekey, G.A., (1989). High precision position control by cartesian trajectory feedback and connectionist inverse dynamics feedforward, *International Joint Conference on Neural Networks*, II-325-332.

Bullock, D. & Grossberg, S., (1987). A neural network architecture for automatic trajectory formation and coordination of multiple effectors during variable-speed arm movements, *IEEE 1st International Conference on Neural Networks*, IV: 559-566.

Bullock, D. & Grossberg, S., (1990). FLETE: An Opponent Neuromuscular design for factorization of length and tension, *International Joint Conference on Neural Networks*, II-209-212.

Eckmiller, R., (1987). Neural network mechanisms for generation and learning of motor programs, *IEEE 1st International Conference on Neural Networks*, IV: 545-550.

Eckmiller, R., (1989). Neural control of intelligent robots, *New developments in neural computing*, meeting on neural computing, London, 217-232.

Eckmiller, R., Bechmann, J., Werntges, H. & Lades, M., (1989). Neural kinematics net for a redundant robot ARM, *International Joint Conference on Neural Networks*, II-333-340.

Elsley, R.K., (1988). A learning architecture for control based on back-propagation neural networks, *IEEE International Conference on Neural Networks*, II-587-594.

Gardner, E.P. (1988). What the robot's hand should tell the robot's brain: Feature detection in a biological neural network, IJCNN, II-557-565.

Graf, D.H. & LaLonde, W.R., (1989). Neuroplanners for hand/eye coordination, *International Joint Conference on Neural Networks*, II-543-548.

Grossberg, S. & Kuperstein, M., (1986). *Neural Dynamics of Adaptive Sensory-Motor Control: Ballistic Eye Movements*, Amsterdam, Elsevier/North-Holland Press.

Guez, A. & Ahmad, A., (1988). Solution to the inverse kinematics problem in robotics by neural networks, *IEEE International Conference on Neural Networks*, II-617-624.

Guez, A. & Ahmad, A., (1989). Accelerated convergence in the inverse kinematics via multilayer feedforward networks, *International Joint Conference on Neural Networks*, II-341-344.

Guez, A., Eilbert, J.L. & Kam, M., (1988). Neural network architecture for control, *IEEE Control Systems Magazine*, April, 22-25.

Guez, A. & Selinsky, J., (1988). A neuromorphic controller with a human teacher, *IEEE International Conference on Neural Networks*, II-595-602.

Guo, J. & Cherkassky, V., (1989). A solution to the inverse kinematic problem in robotics using neural network processing, *International Joint Conference on Neural Networks*, II-299-304.

Harston, C.T., Maren, A.J. & Pap, R.M., (1989). Neural network sensory motor robotics application, *Proceedings of WESCON/89*, San Francisco, Nov. 14-15, 699-708.

Hebb, D.O. (1949). *The Organization of Behavior*, New York, Wiley.

Hering, D., Khosla, P. & Vijaya Kumar, B.V.K., (1990). The use of modular neural networks in tactile sensing, *International Joint Conference on Neural Networks*, II-355-358.

Hosogi, S., (1990). Manipulator control using layered neural network model with self-organizing mechanism. *International Joint Conference on Neural Networks*, II-217-220.

Jakubowicz, O. & Spina, R., (1990). An artificial neural network approach for solving autonomous navigation control problems, *International Joint Conference on Neural Networks*, II-367-370.

Jordan, M.I., (1989). Generic constraints on underspecified target trajectories, *International Joint Conference on Neural Networks*, I: 217-225.

Jorgensen, C.C., (1987). Neural network representation of sensor graphs for autonomous robot navigation, *IEEE 1st International Conference on Neural Networks*, IV: 507-515.

Jorgensen, C., Hamel, W. & Weisbin, C., (1986). Autonomous robot navigation, *BYTE*, 223-235, Jan.

Josin, G., (1988). Integrating neural networks with robots, *AI Expert*, August, 50-58.

Josin, G., Charney, D. & White, D., (1988). Robot control using neural networks, *IEEE International Conference on Neural Networks*, II-625-632.

Kawato, M., Isobe, M., Maeda, Y. & Suzuki, R., (1988). Coordinates transformation and learning control for visually-guided voluntary movement with iteration: A Newton-like method in a function space, *Biological Cybernetics*, 59: 161-177.

Kawato, M., Isobe, M. & Suzuki, R., (1988). Hierarchical learning of voluntary movement by cerebellum and sensory association cortex. In: Arbib, M.A., Amari, S. (eds), *Competition and cooperation in neural nets. II. Lecture notes in Biomathematics*, Springer, New York, 195-214.

Kawato, M., Setoyama, T. & Suzuki, R., (1988). Feedback error learning of movement by multi-layer neural network, *Abstracts of the first annual INNS meeting*, Boston, 1: 342.

Kawato, M, Uno, Y., Isobe, M. & Suzuki, R., (1987). A hierarchical model for voluntary movement and its application to robotics. *IEEE 1st International Conference on Neural Networks*, IV: 573-581.

Khoukhi, A., (1990). A multilevel neural architecture for robot dynamic control, *International Joint Conference on Neural Networks*, II-383-384.

Kitamura, S., Kurematsu, Y. & Nakai, Y., (1988). Application of the neural network for the trajectory planning of a biped locomotive robot, *Abstracts of the first annual INNS meeting*, Boston, 1: 344.

Klopf, A.H. (1987). A neuronal model of classical conditioning, Air Force Systems Command, AFWAL-TR-87-1139, & *Psychobiology*, 16 (2), 85-125, 1988.

Kumar, S.S. & Guez, A., (1990). Adaptive pole placement for neurocontrol, *International Joint Conference on Neural Networks*, II-397-400.

Kung, S. & Hwang, J., (1989). Neural network architectures for robotic applications. *IEEE Transactions on Robotics and Automation*, 5(5), 641-657.

Kuperstein, M., (1988). An adaptive neural model for mapping invariant target position, *Behav. Neurosci.*, 102: 148-162.

Kuperstein, M., (1988). Neural network model for adaptive hand-eye coordination for single postures, *Science*, 239: 1308-1311.

Kuperstein, M. & Rubinstein, J., (1989). Implementation of an adaptive neural controller for sensory-motor coordination, *International Joint Conference on Neural Networks*, II-305-310.

Kuperstein, M. & Wang, J., (1990). Neural controller for adaptive movements with unforeseen payloads, *IEEE Transactions on Neural Networks*, 1: 137-142.

Levin, E., Gewirtzman, R & Inbar, G.F., (1989). Neural network architecture for adaptive system modeling and control, *International Joint Conference on Neural Networks*, II-311-316.

Liu, H., Iberall, T. & Bekey, G.A., (1988). Building a generic architecture for robot hand control, *IEEE International Conference on Neural Networks*, II-567-574.

Martinetz, T.M., Ritter, H.J. & Schulten, K.J., (1989). 3D-neural-net for learning visuomotor-coordination of a robot arm, *International Joint Conference on Neural Networks*, II-351-355.

Martinez, O. & Harston, C., (1990). Research on robotic neural networks (unpublished).

Massone, L. & Bizzi, E., (1989). A neural network model for limb trajectory formation, *Biological Cybernetics*, 61: 417-425.

Massone, L. & Bizzi, E., (1989). Generation of limb trajectories with a sequential network, *International Joint Conference on Neural Networks*, II-345-350.

Massone, L. & Bizzi, E., (1990). On the role of input representations in sensorimotor mapping. *International Joint Conference on Neural Networks*, I-173-176.

Miller, W.T., (1987). Sensor-based control of robotic manipulator using a general learning algorithm, *IEEE J. Robotics Automat.*, RA-3: 157-165.

Miller, W.T., (1989). Real-time application of neural networks for sensor-based control of robots with vision, *IEEE Transactions on Systems, Man, and Cybernetics*, 19: 825-831.

Miller, W.T., Sutton, R.S. & Werbos, R.S., (1989). *Neural Networks for Robotics and Control*, Cambridge, MIT Press.

Miyamoto, H., Kawato, M., Setoyama, T. & Suzuki, R. (1988). Feedback-error-learning neural network for trajectory control of a robotic manipulator, *Neural Networks*, I, 251-265.

Narendra, K.S. & Parthasarathy, K., (1990). Identification and control of dynamical systems using neural networks, *IEEE Transactions on Neural Networks*, 1: 4-27.

Niznik, C.A., Hoss, W. & Watts, C., (1987). Modeling and mathematical formalism for robotic cerebellar neural network pathway information measures, *IEEE First International Conference on Neural Networks*, II, 583-592.

Nguyen, D. & Widrow, B., (1989). The truck backer-upper: An example of self-learning in neural networks. *International Joint Conference on Neural Networks*, II-357-364.

Pap, R.M., Harston, C.T., Parten, C.R., Maren, A.J., Rich, M.L. & Thomas, C., (1990). Application of Neural Networks for Telerobotics as applied to the space shuttle, *First Workshop on Neural Networks: Academic/Industrial/NASA/Defense (WNN-AIND 90)*, Auburn, Alabama; Feb. 5-6.

Park, J. & Lee, S., (1990). Neural computation for collision-free path planning, *International Joint Conference on Neural Networks*, II-229-232.

Pearlmutter, B.A., (1989). Learning state space trajectories in recurrent neural networks. *International Joint Conference on Neural Networks*, II-365-372.

Pellionisz, A.J., (1987). sensorimotor operations: A ground for the co-evolution of brain theory with neurobotics and neurocomputers, *IEEE First International Conference on Neural Networks*, IV, 593-600.

Pellionisz, A.J. (1988). Intelligent decisions and dynamic coordination: Properties of geometrical representation by generalized frames intrinsic to neural and robotic systems, IJCNN, II-603-610.

Porcino, D.P. & Collins, J.S., (1990). An application of neural networks to the guidance of free-swimming submersibles, *International Joint Conference on Neural Networks*, II: 417-420.

Psaltis, D., Sideris, A. & Yamamura, A., (1987). Neural Controllers, *IEEE 1st International Conference on Neural Networks*, IV: 551-558.

Ritter, H.J., Martinez, T.M. & Schulten, K.J., (1989). Topology-conserving maps for learning visuo-motor-coordination, *Neural Networks*, 2: 159-168.

Rosenblatt, J.K. & Payton, D.W., (1989). A fine-grained alternative to the subsumption architecture for mobile robot control, *International Joint Conference on Neural Networks*, II-317-324.

Rudolph, F., (1990). Locally optimizing neural networks in adaptive robot path planning, *International Joint Conference on Neural Networks*, II-425-428.

Sobajic, D.J., 7 Lu, J-J. & Pao, Y-H., (1988). Intelligent control of the Intelledex 605T robot manipulator, *IEEE International Conference on Neural Networks*, II-633-640

Tolat, V.V. & Widrow, B., (1988). An adaptive 'Broom Balancer' with visual inputs, *IEEE International Conference on Neural Networks*, II-641-647.

Trelease, R.B., (1988). Connectionism, Cybernetics, and the Cerebellum, *AI Expert*, August, 30-34.

Troudet, T & Merrill, W., (1989). Neuromorphic learning of continuous-valued mappings in the presence of noise: Application to real-time adaptive control, *International Joint Conference on Neural Networks*, II-621.

Tsutsumi, K., Katayama, K. & Matsumoto, H., (1988). Neural computation for controlling the configuration of 2-dimensional truss structure, *IEEE International Conference on Neural Networks*, II-575-586.

Tsutsumi, K. & Matsumoto, H., (1987). Neural computation and learning strategy for manipulator position control, *IEEE 1st International Conference on Neural Networks*, IV: 525-534.

Uno, Y., Kawato, M. & Suzuki, R., (1989). Formation and control of optimal trajectory in human multijoint arm movement, *Biological Cybernetics*, 61: 89-101.

Wang, H., Lee, T.T. & Gruver, W.A., (1989). A neuromorphic controller for a three-link biped robot, *International Joint Conference on Neural Networks*, II-624.

Widrow, B. & Smith, F.W., (1963). Pattern recognizing control systems. *Computer Information Sciences (COINS) Symposium*.

24

BUSINESS WITH NEURAL NETWORKS

Craig T. Harston

24.0 INTRODUCTION

Business is a diversified field with several general areas of specialization such as accounting or financial analysis. Almost any neural network application would fit into one business area or another. The general areas of marketing, scheduling and planning, personnel, financial analysis and auditing are discussed with regard to neural network applications.

There is some potential for using neural networks for business purposes, including resource allocation and scheduling. There is also a strong potential for using neural networks for database mining, that is, searching for patterns implicit within the explicitly stored information in databases. Most of the funded work in this area is classified as proprietary. Thus, it is not possible to report on the full extent of the work going on. Most work in applying neural networks to business applications involves use of well-known networks, such as the back-propagation network for pattern recognition and prediction, and the Hopfield-Tank network for optimization and scheduling. Thus, we cite only the most recent work as illustrative of applications in these areas.

24.1 MARKETING

There is a marketing application which has been integrated with a neural network system. The *Airline Marketing Tactician* (a trademark abbreviated as AMT) is a computer system made of various intelligent technologies including expert systems. A feedforward neural network is integrated with the AMT and was trained using back-propagation to assist the marketing and control of airline seat allocations. The adaptive neural approach was used because of the large number of interacting factors which were not amenable to rule expression. Additionally, the application's environment changed rapidly and constantly, which required a continuously adaptive solution. The system is used to monitor and recommend booking advice for each departure. Such information has a direct impact on the profitability of an airline and can provide a technological advantage for users of the system [Hutchison & Stephens, 1987].

While it is significant that neural networks have been applied to this problem, it is also important to see that this intelligent technology can be integrated with expert systems and other approaches to make a functional system. There is another significant technological contribution by the AMT system. Neural networks were used to discover the influence of undefined interactions by the various variables. While these interactions were not defined, they were used by the neural system to develop useful conclusions. It is also noteworthy to see that neural networks can influence the bottom line.

24.2 OPERATIONS MANAGEMENT

Operations management is a major business concern associated with some of the toughest problems known. Often sophisticated mathematical models attempt to determine the best solution. While these approaches are often successful, better solutions are always welcome. Scheduling and planning is discussed below, however, quality control, diagnostics, security and personnel concerns are also important. Significantly neural network technology has been applied to problems in all of these areas.

24.2.1 Scheduling and Planning

The classic exemplar of a scheduling problem is the *Traveling Salesman Problem* (TSP). Many of neural network applications relate to the traveling problem but are not reviewed here. We focus this review on scheduling neural networks related to specific business applications.

Hopfield described a neural network algorithm which has been the basis for solving the Traveling Salesman problem [Hopfield, 1984; Hopfield & Tank, 1985]. While many have used it for the TSP, others have identified practical applications such as computer task scheduling [Gulati, Iyengar, Toomarian, Protopopescu & Barhen, 1987]. Modified, the Hopfield network has been used for electric power distribution [Fukui & Kawakami, 1990].

Various other neural network approaches have been used to solve the TSP. A generalized neural network outperformed the Hopfield approach and performed comparably to conventional solutions to the traveling salesman problem [Xu & Tsai, 1990]. Kohonen's self-organizing feature maps have been used for TSP [Angeniol, de La Croix Vaubois & Texier, 1990], and to schedule data path operations into control steps. This Kohonen network performed on a par with other approaches for the data path scheduling [Hemani & Postula, 1990]. A Neocognitron was used to help plan cutting patterns from cloth. This stock cutting program used an expert system in combination with the neural approach. The neural network was used for pattern recognition, and the decisions were made by the expert system [Dagli, Ashouri, Leininger & McMillin, 1990]. The CAD design of integrated circuits with improved channel routing [planning] has been done with neural networks [Fujii, Tenorio & Zhu, 1989]. The pros and cons of using neural networks for scheduling satellite communications in broadcasting has been discussed

[Bourret, Remy & Goodall, 1990; Bourret, Goodall & Samuelides, 1989]. Using neural networks for airline and fast food crew scheduling has been recommended [Poliac, Lee, Slagle & Wick, 1987]. Clearly, the neural technology has been applied to various planning problems, sometimes in combination with expert system technology.

At times, the Hopfield approach fails because the program becomes trapped in local minima. When this happens, the algorithm will not converge to a solution. To avoid the trap, a third-order Boltzmann machine was used to decompose this problem for machine scheduling [Masti & Livingston, 1990]. The Boltzmann machine is an optimization technique used for neural networks so the neural network can successfully converge to an appropriate minimum.

24.2.2   Quality Control

Quality control is a leading concern of operations management in today's competitive environment. Statistical quality control measures have proven useful in Japan, the U.S. and throughout the world. Statistical quality techniques typically concentrate on monitoring one variable at a time. When a measure exceeds the statistically defined limits, corrective measures can be taken. While statistical controls are effective, neural network technology can offer assistance. As with expert systems, neural networks may be able to integrate with statistical quality control techniques to enhance their performance. For example, neural networks can evaluate the performance of several statistically monitored variables at once. Potentially dangerous performance by the combination of several variables can be evaluated and identified. While the records of all the variables, taken one at a time, can indicate no problem, there may be a combination of events which relates to a quality problem. Neural networks can evaluate the combination of events and identify problems caused by the interaction of two or more variables. This interaction need not be understood or previously defined for the neural network to function. Indeed, the neural network may be able to discover and use the interaction during the initial training. Such a use of neural networks for quality control would represent an important advance in the quality control technology.

As yet, neural technology has not been integrated with traditional statistical control techniques, however several inspection systems have been published. For example, a self training classifier neural network has been tested on the visual inspection of complex printed patterns [Beck, McDonald & Brzakovic, 1989]. Commercially applied systems are on the market. An industrial parts inspection system has been developed by Nestor on an AT microcomputer. The system only uses about 50K RAM and processes a pattern in about two seconds with 97 percent accuracy. A disk-drive head assemblies have been diagnosed by a Nestor neural network system. In this case, increasing or decreasing concavity or convexity patterns cut from stock material was detected and indicated when the process was beginning to drift out of tolerance. Training took less than 2 person-days and classified errors with 100 percent accuracy. The detection of quality deterioration reduced costs and allowed early corrective action.

A cost-competitive monitoring system was developed by Glover to examine soda bottles. This neural network system ensures that each bottle is capped and filled appropriately. Additionally, the system has examined sponges, dials, syrup and shampoo bottles. Glover has used back-propagation and counter propagation neural networks, and some of the published results suggested that back-propagation networks performed better. His system combined traditional technologies with neural network techniques. The system runs on an AT class microcomputer with a neurocomputer coprocessor board installed [Glover, 1988, 1989].

24.2.3 Diagnostics

Diagnostics is closely related to quality control and may become a major application for neural networks. A feedforward network was used for fault diagnosis of electronic equipment [Jakubowicz & Ramanujam, 1990]. Initially they trained the network to find one fault for one symptom and discussed the possibility of relating one symptom to several faults. Satellite communication networks have been diagnosed with feedforward neural networks [Casselman & Acres, 1990]. This system [DASA/LARS] is extensive and is deployed at Ft. Detrick, MD. Similar to diagnostic applications, sensor calibration has been done with feedforward and CMAC neural networks and compared favorably with polynomial curve fitting [Masory & Aguirre, 1989, 1990]. In their second paper on this topic [1990], the authors concentrated on the development of the feedforward network for this task.

24.2.4 Security and Signatures

Security systems can be like monitoring and diagnostic systems in that neural networks are used to provide surveillance. SIAC, a major player in the commercialization of neural network technology, developed an airline baggage explosive detection system. They analyzed gamma ray energy with a feedforward neural network and deployed the system in airports [Shea, 1989]. Neural networks have been recommended as an adaptive reference monitor for information security systems [Kellum, 1987].

Banks show a continuous interest in signature verification. Such a system is commercially available from Nestor and operates on an AT class microcomputer using about 45K RAM. Their neural system evaluates a pattern in a two to four seconds with 95 percent accuracy in their conservative mode. Nestor technology was also used by Richard Q. Fox of AEG to produce a signature verification system. This system had an accuracy of 92-98 percent detecting forgeries. Naturally, the banking industry is interested in more speed (which is always desirable when processing thousands of checks). No doubt, when neural systems are deployed on chips which incorporate extensive parallel processing, the speed of neural network systems will be acceptable. Another major concern in banking is the error rate which needs to be much less than 1 percent.

24.2.5 Personnel Management

Neural networks were used to classify data from individual interviews. Their data were categorized into binary form and analyzed by feedforward neural networks taught with back-propagation [Surkan, 1988]. Unfortunately, the results were inconclusive.

24.3 FINANCIAL ANALYSIS

One of the most exciting applications of neural network technology is that of financial analysis. The potential for improving forecasting, credit evaluation and securities trading is indicated in this review. There is relatively little work completed at this time; however, some success has been achieved. No doubt interest will increase as research data become available.

24.3.1 Forecasting

Paul Werbos developed an error back-propagation technique to adjust the weights of inner or hidden layers of neural processing units. In his mind, this approach was an effective method for determining the minimization of error. As such, it would be useful for predicting or forecasting economic models. While his early work was done as a thesis published in 1974, Paul has been fruitful in identifying many applications for his approach. He has recommended and encouraged many to use neural networks for robotics and controls among other tasks. He has developed and uses the back-propagation-of-error for a recurrent gas market model [Werbos, 1988]. The Department of Energy finds this approach important for forecasting energy through time-related markets.

Others also have used neural networks for forecasting; some have found that the networks performed better than regression techniques especially when applied to novel data [Dutta & Shekhar, 1988]. Others have found that neural networks forecast about as well as the Box-Jenkins technique [Sharda & Patil, 1990]. The performance of a mortgage delinquency prediction system marketed by Nestor seemed low. However, the results provided a potential savings by reducing delinquencies within the insurance agency by 12 percent [Ghosh, Collins & Scofield, 1988]. NeuralWare was able to develope a neural system which performed 5 percent better than professional analysts. The system did not work well until the cost choosing a bad stock was included. Although the margin of improvement was small for the Nestor and NeuralWare systems, these differences were meaningful and profitable.

24.3.2 Investment Screening

Records from a value-line database of fundamental financial data were extracted and taught to the enhanced associative neural network [Martinez & Harston, 1989, 1990 a & b]. Identification of each company was associated with categorical representation of the fundamental data. Following this associative learning, a request could be made of the network to identify companies with specified financial characteristics. The system could identify an exact match, if such could be recalled. Usually a company with the specified ideals did not exist in the database. With the typical database query system a new request would have been required to find additional information. Our modified associative network made the new request unnecessary because the system could find alternatives. When the M-H frequency value was adjusted, the network located close approximations to the original request. Each group of companies recalled had similar fundamental statistics, only differing on one or two items. Further modification of the frequency revealed additional groups of companies which were similar without redoing the request [Harston & Martinez, 1990]. As a result, reasonable investment opportunities could be identified according to our criteria.

24.3.3   Credit Evaluation

The HNC Company, founded by Robert Hecht-Nielsen, has developed several neural network applications. One of them is a credit scoring system which increased the profitability of the existing model up to 27 percent. The HNC neural systems were also applied to mortgage screening. A neural network automated mortgage insurance underwriting system was developed by the Nestor Company [Reilly, Collins, Scofield & Ghosh, 1990; Scofield, Collins & Ghosh, 1988]. This underwriting system was trained with 5,048 applications of which 2,597 were certified. The data related to property and borrower qualifications. In a conservative mode the system agreed with the underwriters on 97 percent of the cases. In the liberal mode the system agreed on 84 percent of the cases with the underwriters. This system ran on an Apollo DN 3000 system and used 250K memory while processing a case file in approximately one second.

24.3.4   Trading

If neural networks are such great pattern matchers and can be used for prediction and forecasting, then can they be used to predict the stock market? If so, we can all get rich. Naturally, such thinking is to be expected and someone tried to do it [White, 1988]. He used NLS and feedforward neural networks to predict daily IBM stock prices. He also used back-propagation for training. Unfortunately, the results were disappointing. In some ways, this result could have been expected. After all, neural networks can only process information, make data transformations and detect patterns. They cannot make up something from nothing. Where no information exists, neural networks cannot magically find meaning. Assuming that stock prices are nearly random on a day-to-day basis, then neural networks cannot be expected to predict the next day's stock price. However, negative results do not prove that the task of predicting stock prices cannot be done.

More realistically, neural networks may prove to be useful if more reasonable problems are worked on. For example, HNC used neural networks to analyze foreign currency trading. Their system was able to discover features in the data. They analyzed information about the Pound Sterling, Japanese Yen and Deutsch Mark. With this neural network system inexperienced traders could make profitable decisions. Others found that feedforward neural networks were substantially better than regression techniques when used for corporate bond rating [Dutta & Shekhar, 1988]. Using their neural network systems, Nestor Company developed a successful automated securities trading program. They correctly classified 75 percent of the patterns which were prescreened for unambiguously identified patterns. This result was good when compared to other automated traders which operate in the range of 50 percent to 60 percent accuracy. Their system was done on a DEC VAX and processed a pattern in about one second. A bond-trading system was developed by the Nestor Company which could make correct recommendations 72 percent of the time. A non-neural network system only gave correct signals 55 percent of the time. While the percent improvements were small, the profitability was significant.

It appears that neural networks can be used to an advantage for trading or market tasks. Possibly one of the secrets to successful market analysis, is to define and restrict the problem to one which is potentially solvable. Neural networks can not identify information which is not available in the original data. However, if there is information in the raw input, even if it is hidden, neural networks seem to be able to pull it out. The successful examples above suggest that neural networks can be used for profitable trading. Their success suggests that more consideration should be given to using neural networks for financial analysis.

24.4   WHERE IS ACCOUNTING–AUDITING?

Most accounting systems work just fine. There appears to be no need for sophisticated pattern detection programs for many accounting tasks. Little can be gained by spending time and money improving payroll, accounts payable or receivable systems with neural networks. On the other hand, auditing is a complex task which requires considerable skill and knowledge. Expert systems have been examined for auditing procedures and neural networks should be considered for application to the audit task. As yet, I have not found any publications directly related to auditing with neural technology. This may indicate that auditing is an area where important progress will be made by neural networks in the next few years.

The only neural network application similar to auditing, known to me, is that of a credit-card-fraud detection system advertised by the Nestor Company. Back-propagation was trained with 1,000 sample credit card transactions, some of which were fraudulent. The network system discovered patterns of fraudulent activity and identified these patterns early. It appears that successful financial monitoring and auditing can be done with neural network technology.

These successful financial applications are not surprising because the nature of financial data is compatible with neural-based processing techniques. Financial data is often noisy; patterns are commonly found and known to exist in financial data; there are many nonlinear and interactive relationships associated with financial problems. Neural networks are especially appropriate for use with these situations and characteristics. We can expect more useful applications of neural network technology to financial problems in the future.

24.5   SUMMARY

Clearly neural networks are useful for a wide variety of problems. Most of them appear to be successful, although some were marginal performers, and some did not perform much better than other approaches. Unfortunately, most of the results were not compared to other ways of doing things. Also, it was difficult to judge how much effort or investment these projects took. Given the state of the art, it is expected that a neural network application will likely be a custom development. It is significant that neural networks were used in a wide variety of business fields. Problems from fraud, investments, and operations have been approached with promising results. A reasonable conclusion of the evidence would recommend further development and consideration of neural network technology.

REFERENCES

Angeniol, B., de La Croix, Vaubois, G. & Texier, J.L. (1988). "Self-organizing feature maps and the travelling salesman problem," *Neural Networks*, 1, 289-293.

Beck, H., McDonald, D. & Brzakovic, D. (1989). "A self-training visual inspection system with a neural network classifier," *International Joint Conference on Neural Networks*, I, 307-311.

Bourret, P., Remy, F. & Goodall, S. (1990). "A special purpose neural network for scheduling satellite broadcasting times." *International Joint Conference on Neural Networks*, II, 535-538.

Bourret, P., Goodall, S. & Samuelides, M. (1989). "Optimal scheduling by competitive activation: Application to the satellite-antenna scheduling problem," *International Joint Conference on Neural Networks*, I, 565-572.

Casselman, F. & Acres, J.D. (1990). "DASA/LARS, a large diagnostic system using neural networks," *International Joint Conference on Neural Networks*, II, 539-542.

Collins, E., Ghosh, S. & Scofield, C.L. (1988). "An application of a multiple neural network learning system to emulation of mortgage underwriting judgments," *IEEE Int'l. Conf. on Neural Networks*, II, 459-466.

Dagli, C.H., Ashouri, M.R., Leininger, G. & McMillin, B. (1990). "Composite stock cutting pattern classification through neocognitron," *International Joint Conference on Neural Networks*, II, 587-590.

Dutta, S. & Shekhar, S. (1988). "Bond rating: A non-conservative application of neural networks," *IEEE International Conference on Neural Networks*, II, 443-450.

Fujii, R., Tenorio, M.F. & Zhu, H. (1989). "Use of neural nets in channel routing," *International Joint Conference on Neural Networks*, I, 321-325.

Fukui, C. & Kawakami, J. (1990). "Switch pattern planning in electric power distribution systems by Hopfield-type neural network, *International Joint Conference on Neural Networks*, II, 591-594.

Ghosh, S., Collins, E.A. & Scofield, C.L. (1988). "Prediction of mortgage loan performance with a multiple neural network learning system," *Abstracts of the First Annual INNS Meeting*, 1 (sup.1), 439.

Glover, D.E. (1988). "Neural nets in automated inspection," *Synapse Connection The Digest of Neural Computing*, 2(6), 1-13, 17.

Glover, D.E. (1989). "Optical processing and neurocomputing in an automated inspection system," *Journal of Neural Network Computing*, Fall, 17-38.

Gulati, S., Iyengar, S.S., Toomarian, N., Protopopescu, V. & Barhen, J. (1987). "Nonlinear neural networks for deterministic scheduling," *IEEE International Conference on Neural Networks*, II, 745-752.

Harston, C.T. & Martinez, O.M. (1990). "Neural network analysis of fundamental financial database," (In preparation)

Hemani, A. & Postula, A. (1990). "Scheduling by self organization," *International Joint Conference on Neural Networks*, II, 543-546.

Hopfield, J.J & Tank, D.W. (1990). "Computation of decisions in optimization problems," *Biol. Cybern.*, 52, 141-152.

Hopfield, J.J. (1984). "Neurons with graded response have collective computational properties like those of two-state neurons," *Proc. National Acad. Sci. USA, 81,* 3088-3092.

Hutchison, W.R. & Stephens, K.R. (1987). "The airline marketing tactician (AMT): A commercial application of adaptive networking," *Proc. First IEEE International Conference on Neural Networks*, San Diego, CA, II, 753-756.

Jakubowicz, O. & Ramanujam, S. (1990). "A neural network model for fault-diagnosis of digital circuits," *International Joint Conference on Neural Networks*, II, 611-614.

Kellum, C. (1987). "An adaptive reference monitor for information system security," *IEEE First International Conference on Neural Networks*, IV, 667-676.

Martinez, O.M. & Harston, C.T. (1989). "Recall in saturated associative neural networks," *International Joint Conference on Neural Networks*, II, 570.

Martinez, O.M. & Harston, C.T. (1990). "Interfacing database to find the best and alternative solutions to problems by obtaining the knowledge from the database," *International Joint Conference on Neural Networks*, II, 663.

Martinez, O.M. & Harston, C.T. (1990). "A system in control of its knowledge that provides alternative and different solutions from one input set," *International Joint Conference on Neural Networks*, I, 664.

Masory, O. & Aguirre, A.L. (1989). "Sensor calibration using artificial neural networks," *International Joint Conference on Neural Networks*, II, 577.

Masory, O. & Aguirre, A.L. (1990). "Neural Network calibrates a displacement sensor," *Sensors The Journal of Machine Perception*, 7(3), 48-56.

Masti, C.L. & Livingston, D.L. (1990). "Neural networks for addressing the decomposition problem in task planning," *International Joint Conference on Neural Networks*, II, 555-558.

Poliac, M.O., Lee, E.B., Slagle, J.R. & Wick, M.R. (1987). "A crew scheduling problem," *IEEE International Conference on Neural Networks*, II, 779-786.

Reilly, D. L., Collins, E., Scofield, C. & Ghosh, S. (1990). "Risk assessment of mortgage applications with a neural network system: An update as the test portfolio ages," *International Joint Conference on Neural Networks*, II-479-482.

Scofield, C., Collins, E.A. & Ghosh, S. (1988). "Prediction of mortgage loan performance with a multiple neural network learning system," *Abstracts of INNS*, 1, 439.

Sharda, R. & Patil, R.B. (1990). "Neural networks as forecasting experts: an empirical test," *International Joint Conference on Neural Networks*, II. 491-494.

Shea, P.M. (1989). "Detection of explosives in checked airline baggage using an artificial neural system," *International Joint Conference on Neural Networks*, II, 31-34.

Surkan, A.J. (1988). "Application of neural networks to classification of binary profiles derived from individual interviews," *IEEE International Conference on Neural Networks*, II, 467-472.

Werbos, P.J. (1988). "Generalization of back-propagation with application to a recurrent gas market model," *Neural Networks*, 1, 339-356.

White, H. (1988). "Economic prediction using neural networks: The case of IBM daily stock returns," *IEEE International Conference on Neural Networks*, II, 451-458.

Xu, X. & Tsai, W.T. (1990). "An adaptive neural algorithm for traveling salesman problem," *International Joint Conference on Neural Networks*, II, 716-719.

25

NEURAL NETWORKS FOR DATA COMPRESSION AND DATA FUSION

Alianna J. Maren

25.0 OVERVIEW

There are several useful neural networks methods for data compression and data fusion. Methods have been developed for both time-varying signal compression and for image data compression. Some involve finding a codebook, and others involve reducing the dimensionality of the expression for the data. Considerations of data recovery are important for data compression methods. So far, there are few studies comparing the efficacy of neural networks for data compression and data fusion with more traditional methods.

25.1 INTRODUCTION

There is an increasing need to develop methods for data compression, data dimensionality reduction, and data fusion in multiple applications areas. Some of the applications domains include, but are not limited to: compression and reconstruction of images from telemetry sources or for medical diagnosis, and data compression and reconstruction for sensor readings for aerospace component tests, power plant records, seismic records, and other areas. Many analyses (ranging from power plant or jet aircraft fault diagnosis to multi-electrode EEG recording) require that time-varying data from multiple sensors be correlated and sometimes fused. Similarly, many image analyses require correlation from multispectral or stereo images.

There are several neural network methods which have been developed to deal with data compression, data dimensionality reduction, and multisensor data correlation/fusion. The approaches which have been most explored include the *Learning Vector Quantization* (LVQ) network for data compression, and special architecture of the back-propagation network for data compression, dimensionality reduction, and sensor data fusion. The self-organizing Topology-Preserving Map (TPM) can be used for dimensionality reduction, and special architectures have been developed for multispectral, multisensor, and/or stereo image data correlation.

The key issue is to maximize data compression while retaining full information relevant to novel input. There are several traditional approaches to data compression, including data sampling and re-representation of the data in terms of segmented features. Both approaches can be used adaptively; e.g., the data can be adaptively sampled or some locally or globally adaptive criteria may be used to determine segmentation boundaries. Another approach is to characterize the data using pattern recognition. This approach is essentially that of finding of a different, more abstract (and hence compressed) data representation. We can frequently gain a great deal of data compression by going from one representation level to another. For example, when we represent a large number of data points as a weighted sum of sine waves, we have accomplished a change in representation level.

Another way to compress data is to reduce the dimensionality of the representation. This would work if, for example, data could be described by a large set of features, and we could find a smaller set of features which would work just as well. This implies hidden redundancy in the original feature set. This does not seem to be the case with the current data, as features such as relative curve amplitudes, phases, and frequencies are a compact representation.

In this chapter, we deal with strict *compression* and *dimensionality reduction* methods, as opposed to pattern recognition methods. The desired characteristics of these methods are:

- Robust discrimination of novel data from baseline,
- Potential for very rapid operation (both software andhardware optimization),
- Ability to specify loss involved in data compression and reconstruction, and
- High compression ratios to facilitate data storage.

25.2 NEURAL NETWORKS FOR DATA COMPRESSION AND DIMENSIONALITY REDUCTION

One of the most well-known approaches to data compression is to develop a codebook whose distribution of codewords approximates the probabilistic distribution of data which is to be represented. For simple systems, the accuracy (or degree of data fidelity on reconstruction) of codebook-based representation corresponds to the number of codewords used. Several researchers have used the Learning Vector Quantization network [Kohonen, 1989] for this purpose, where the exemplar vectors in the LVQ become the codewords.

Alston and Chau [1990] have proposed an interesting VLSI-implementable decoder for long constraint length convolutional codes. Their decoder has a highly regular topology with strictly local connections, which makes it attractive for scaling to large constraint length. The decoder has constant throughput rate. Similarly, the more advanced self-organizing Topology-Preserving Map has been used for dimensionality reduction. As an example, Ritter et al. [1988] used a TPM to represent a mapping from sets of seven-dimensional vectors (corresponding to two stereo x-y observations of a robot hand from two simulated cameras and to the location of the hand in a 3-dimensional space) in a 3-D topographic map representation.

Another approach to dimensionality reduction is to set up a back-propagation (or similar) network which has the same number of nodes in the input and output layers, and typically three hidden layers. The innermost hidden layer has substantially fewer nodes than in the input (output) layer. By training the network to regenerate itself in the output layer, and by forcing the input data through a feed-in, fan-out process in the hidden layers, the hidden layer nodes learn to represent the characteristic features of the input.

After training, the network structure is reconfigured into two networks; a compression network and a decompression network. The compression network contains the input layer, intermediary layer(s), and what was the innermost hidden layer (which becomes the new output layer). The decompression network contains a copy of the innermost hidden layer, any intermediary layers connecting it to the output, and the original output layer. The connection weight values are kept as they are.

To compress a pattern, it is presented to the compression (first) layer. The resulting activation in the output layer (which earlier was the innermost hidden layer) contains a dimensionally-reduced representation of this pattern. For decompression, the input layer to the decompression network (which has the same dimensionality as the output of the compression network) is activated with the compressed representation. It generates a decompressed pattern in the output layer. This method has been explored (with some enhancements) by Saund [1989], Myers [1987], and Kadaba [1990].

Lang and Hinton [1990] have done some interesting work in reducing the dimensionality of a time varying signal where only a small portion of the signal contained the significant distinguishing information; recognition of the speaker — independent spoken letters B,D,V, and E. Their approach was to form a cost function which allowed a back-propagation network to create complicated weight patterns only when they helped reduce overall network error. Their work illustrates how important it is to control the attention of a network (placing it on the region of spectral data which contains useful information), rather than just controlling the overall resolution of data representation. Also, they found that use of a time-delay neural network helped the network to actively discriminate between relevant and irrelevant portions of the training segments. A slightly more complex version of their network achieved 91% accuracy, as compared to 80% accuracy of a hidden Markov model system and 94% for human listeners.

25.3   NEURAL NETWORKS FOR IMAGE DATA COMPRESSION

Several different neural network approaches for image compression have been developed. However, not all methods have yielded highly accurate image reconstructions from the compressed data. Anthony et al. [1990] have shown that a back-propagation approach to image compression was inferior to principal component analysis. Sonehara et al. [1989] used a method like that described earlier (a back-propagation-based network for dimensionality reduction) for image data compression and reconstruction. Reconstruction of images (of faces) which were part of the training set was about 25dB, and for images which were not part of the training set, reconstruction was about 15dB.

Manikopoulos et al. [1990] have demonstrated an LVQ-based coding method for image sequences which yields up to 32dB PSNR. The process operates on a spatial domain of 16-dimensional vectors, representing segments of blocks above and to the left of the 16-element block being coded. The method combines an intraframe algorithm followed by an interframe algorithm, operating on a bundle of image frames. Both algorithms are based on a finite approach. Storage requirements for the finite states are reduced using the LVQ method.

Ahalt et al. [1989] have developed a similar LVQ-based coding mechanism which uses a competitive learning aspect to distribute the codebook vectors. They have compared their results to the well-known Linde-Buzo-Gray method for vector quantization (e.g., SNR of 33.95dB with a codebook size of 24, compared to 34.16dB SNR with a codebook of 23 vectors, respectively), and found that their approach offered comparable performance with higher potential for real-time implementation (in speech and image processing).

Naillon and Theeten [1989] have developed a vector quantizer using a Hopfield-type network for TV image codebook compression. They include "metastable state" as well as the induced stable states as attractors in their Hopfield network, and keep the number of metastable states usefully low with a minimal overlap of criterion.

The ideal for image compression would be a computational method which would store both a symbolic or high-level description of the structures in an image (especially patterns which might imply some sort of meaningful event, such as roads or rivers) and a highly compressed representation of the image itself, which could be reconstructed with high fidelity. It would be especially useful if the high-level description could indicate which portions of the image needed to be reconstructed to re-visualize salient areas. This would save on computational effort.

Earlier work conducted by a member of this team [Minsky and Maren, 1989] led to the development of a Hierarchical Scene Structure method which produced a high-level representation of an image. This representation is useful for extracting salient (perceptually interesting) information, which is represented near the top of a hierarchical data structure. By examining the data structure, it one can find out if the image contains the desired features. However, the method does not currently allow for facile reconstruction of the image.

One method which yields highly accurate image reconstruction (about 38dB SNR) uses prior knowledge of image segmentations. Okamoto et al. [1990] have demonstrated a model-based image compression which uses a large array of interacting filter banks. The line process of a Markov random field model represents image discontinuities. The image is reconstructed using stochastic relaxation (about 20 iterations) based on the acquired MRF model.

25.4 NEURAL NETWORK METHODS FOR MULTISOURCE INFORMATION CORRELATION/FUSION

25.4.1 Neural Networks for Correlating Multiple Streams of Time-Varying Data

There is some promising work indicating that neural networks might be useful for multisensor data integration and fusion (for reviews, see [Maren, 1989; Luo & Kay, 1989]. An example is the work done by Cain et al., which favorably compares the results of a back-propagation neural network for multisensor information fusion with the results of a combination of proprietary and statistical algorithms [Cain, 1989]. Similar work has been done by Kam et al. [1989], in developing a two-layer network of binary threshold units for adaptive sensor fusion.

25.4.2 Neural Networks for Correlating Multisensor, Multispectral, and Stereo Images

At least one instance of using a neural network (or rather, distributed parallel) approach has been useful in a specific aspect of image processing: that of correlating stereo images (Drumheller, 1986). This leads to a great improvement in processing speed for cooperative methods which were suggested much earlier (Marr & Poggio, 1976). Further, Grossberg (1987) has shown how his Boundary Contour System may be extended to yield stereo fusion.

There are issues in multispectral data correlation which are unique. These are that different regions with different features (e.g., size, shape, and location) may appear in images taken at different resolutions and wavelengths. Low-level pixel-to-pixel correlation is not always feasible, nor is it always desirable. Because of the difference in image characteristics, feature-to-feature correlation (using features extracted from segmented regions or edges) may not be possible either. High-level, structural descriptions, capturing significant perceptual relationships, might be a solid basis for multispectral and/or multisensor image correlation. The HSS method [Minsky and Maren, 1989; Maren et al. 1989(a) & (b)] would provide such a basis for correlation.

REFERENCES

Ahalt, S.C., Chen, P., & Krishnamurthy, A.K. (1989). "Performance analysis of two image vector quantization techniques," *Proc. First Int'l. Joint Conf. Neural Networks* (Washington, D.C., June 18-22, 1990), I-169–I-175.

Alston, M.D., & Chau, P.M. (1990). "A neural network architecture for the decoding of long constraint length convolutional codes," *Proc. Third Int'l. Joint Conf. Neural Networks* (San Diego, CA; June 17-21, 1990), I-121–I-126.

Anthony, D., Hines, E., Barham, J., & Taylor, D. (1990). "A comparison of image compression by neural networks and principal component analysis," *Proc. Third Int'l. Joint Conf. Neural Networks* (San Diego, CA; June 17-21, 1990), I-339–I-342.

Cain, M.P., Stewart, S.A., & Moore, J.B. (1989) "Object Classification Using Multi-spectral Data Fusion," *Proc. SPIE Technical Symposia on Aerospace Sensing, Sensor Fusion Session* (Orlando, FA; March 27-31, 1989).

Drumheller, M. (1986). "Connection machine stereomatching," *Proc. AAAI-86* (Philadelphia, PA; August, 1986), 748-753.

Grossberg, S. (1987). "Cortical dynamics of three-dimensional form, color, and brightness perception: II. Binocular theory," *Perception and Psychophysics, 41* (2), 117-158.

Kadaba, N., Nygard, K.E., Juell, P.L., & Kanga, L. (1990) "Modular back-propagation neural networks for large domain pattern classification," *Proc. Second Int'l. Joint Conf. on Neural Networks* (Washington, D.C., Jan. 15-19, 1990), II-551–II-554.

Kam, M., Naim, A., Labonski, P., & Guez, A. (1989). "Adaptive sensor fusion with netw of binary threshold elements," *Proc. First Int'l. Joint Conf. Neural Networks* (Washington, D.C.; June 18-22, 1989), II-57–II-64.

Kohonen, T. (1989). *Self-Organization and Associative Memory* (Springer-Verlag: Berlin).

Lang, K.J., & Hinton, G. (1990). "Dimensionality reduction and prior knowledge in E-set recognition," in D. Touretzky (Ed.), *Advances in Neural Information Processing Systems 2* (Morgan Kaufmann, San Mateo, CA), 178-185.

Luo, R.C., & Kay, M.G. (1989). "Multisensor integration and fusion in intelligent systems," *IEEE Trans. Systems, Man, & Cybernetics, SMC-19, 901-931.

Manikopoulos, C., Antoniou, G., & Metzelopoulou, S. (1990). "LVQ of image sequence source and ANS classification of finite state machine for high compression coding," *Proc. Third Int'l. Joint Conf. Neural Networks* (San Diego, CA: June 17-21, 1990), I-481–I-486.

Maren, A.J., Pap, R.M., & Harston, C.T., (1989a). "A hierarchical data structure representation for fusing multisensor information," *Proc. SPIE Technical Symposia on Aerospace Sensing, Sensor Fusion Section* (Orlando, FA; March 27-31, 1989).

Maren, A.J., Pap, R.M., & Harston, C.T. (1989b). "A hierarchical structure approach to multisensor information fusion," *Proc. Second Nat'l. Symposium on Sensor and Sensor Fusion* (Orlando, FA; March 27-31, 1989).

Maren, A.J. (1989). *Multisensor Information Fusion: An Annotated Bibliography, TCNEA AHCEL Team Technical Report 89-01,* University of Tennessee Space Institute.

Marr, D., & Poggio, T. (1976). "Cooperative computation of stereo disparity," *Science,* 194, 283-287.

Minsky, V., & Maren, A.J. (1989). "A multilayered, cooperative-competitive neural network for segmented scene analysis," *J. Neural Network Computing, 1,* Issue 3, 14-33.

Myers, M., Kuczewski, R., & Crawford, W., *Application of New Artificial Information Processing Principles to Pattern Classification,* Final Report, U.S. Army Research Office, Contract DAAG-29-85-C-0025, 1987.

Naillon, M., & Theeten, J.-B. (1989). "Neural approach for TV image compression using a Hopfield type network," in D. Touretzky (Ed.), *Advances in Neural Information Processing Systems 1* (Morgan-Kaufman, San Mateo, CA), 264-271.

Okamoto, T., Kawato, M., Inui, T., & Miyake, S. (1990). "Model based image compression and adaptive data representation by interacting filter banks," in D. Touretzky (Ed.), *Advances in Neural Information Processing 2* (Morgan Kaufmann, San Mateo, CA), 298-305.

Ritter, H.J., Martinetz, T.M., & Shulten, K.J. (1988). "Topology-preserving maps for learning visuo-motor coordination," *Neural Networks, 2,* 159-168.

Saund, E. (1989). "Dimensionality-reduction using connectionist networks," *IEEE Trans. Pattern Analysis and Machine Intelligence, 11,* 304-314.

Sonehara, N., Kawato, M., Miyake, S., & Nakane, K. (1989). "Image data compression using a neural network model," *Proc. First Int'l. Joint Conf. Neural Networks* (Washington D.C.; June 18-22, 1989), II-35-II-41.

26

DATA COMMUNICATIONS

Robert M. Pap

26.0   OVERVIEW

Neural networks are expected to be especially useful in the communications industry. Their promise as pattern recognizers and data correlators are naturally applicable to areas such as switching and queuing, transmission, error-correcting coding and data compression. They have demonstrated remarkable capabilities for performing optimization and resource allocation. One of the first practical uses of neural networks was the use of the ADALINE for adaptive equalization in modems for computer communications.

Systems built of neural networks can be applied to a broad range of uses in data communications including network management, satellite telemetry, switching and data interpretation. We will explore a number of these areas. Neural networks used with fiber optics in optical implementations is a potential method for expanding data throughput while maintaining the advantage of the optical signal in the fiber.

A special issue of the *IEEE Communications* [Posner, 1989] magazine investigated the basics of the technology. Posner points out some reasons for the interest in neural networks as applied to the communications industry. They have demonstrated the ability to generalize and optimize quickly and with less computation and memory resources than current techniques. Neural networks seem to be able to perform the pattern recognition and optimization tasks used in transmission, switching, memory, data compression and the man/machine interface. He points out that the potential for learning reduces the system life-cycle cost by reducing sustaining engineering efforts and the costs involved in customization of systems and networks. Posner's own research team has been working on increasing the capacity of Hopfield associative memory, important to hardware implementations of optimization systems [McEliece, Posner, Rodemich and Venkatesh, 1988].

26.1   NETWORK MANAGEMENT

Optimal or efficient routing of data communications traffic is a major area of concern. Many different technologies have been tried in attempts to solve routing problems. Rauch and Winarske [1988] and Pap et al. [1989] have explored various ways of implementing neural networks for network management improvements. Rauch and Winarske simulated traffic flow on a 16 node network linked by two satellites and five earth stations to determine optimal traffic routing. This involved finding the routing, from a given origin node to a defined destination node, which minimizes the loss function. Their simulation uses a modification of the Hopfield and Tank [1985, 1974, 1985] traveling salesman algorithm and shows reasonable convergence after 250 iterations. Rauch and Winarski raise the question of whether this model will work when scaled up to a 1000 node network. Some of the concepts that should be considered in network management problems include the questions raised by Wilson and Pauley [1988] and addressed by Szu [1988].

Satellite communication networks as well as local area and wide area network systems are potential applications for neural network based communications network management systems. Attempts to develop such systems using expert systems have not been perfected. The combination of expert systems with neural networks or fuzzy logic [Kershenbaum and Rubinson, 1989] are being explored for network planning and design.

Network management is essential in trying to maximize network efficiency when a satellite communications network is under stress due to overload or failure. Ansari and Chen [1990] have proposed using a neural network derived from a Kohonen Self-Organizing Model for use on a Time Division Multiple Access digital satellite communications network. The first layer of their three-layer neural network consists of a pattern recognition network that selects the exemplar map which most resembles the input data. Layer 2 modifies the chosen exemplar map and discerns if there is a discrepancy with the input data. A new map is generated by the third layer. By using the small but powerful neural network they were able to simulate the viability of their algorithm in a ten link satellite communications network with a capacity of a thousand channels.

Bourret, Remy, and Goodall [1990] have shown a concept for scheduling satellite broadcasting times using a competitive activation function instead of a Hopfield network [Bourret, Goodall and Samuelides, 1989]. The concepts of Hopfield and Tank [1985] were not attempted with the modifications suggested by Szu [1988] and did not take into account the stability problem as reported by Wilson and Pauley [1988]. This example of the optimization problem was developed with modifications to the network and the activation functions to keep the activation function competitive.

Frankel, Shacham and Mathis [1988] discuss the idea of new generation distributed communication network management. The concepts outlined in Frankel's self-organizing network could use the Kohonen method for topology mapping. The routing hierarchy will develop into an organizational structure similar to a neural network. He calls out the desire to have a system that is robust, actively adapts to accommodate its configuration and is able to modify its training.

26.2 ISDN COMMUNICATIONS NETWORK CONTROL

One practical application being developed is an *Asynchronous Transfer Mode* (ATM) network for the data transmission layer of the Broadband Integrating Digital Services Network [Dayton,1989]. Hiramatsu [1989, 1990] of NTT is developing a network controller which can manage traffic. Hiramatsu discusses a learning control method for service quality control using neural networks. Some of the areas he addresses in his design include the use of neural networks for classification of transmission patterns, window size control, regulation and control of calls, optimal routing and dynamic assignment of trunk capacity as well as call regulation based upon congestion. He projects the use of an expert system for prediction of traffic and network design assistance, and has outlined the requirement of a network controller that would use neural networks in a variety of areas to assure service qualities and traffic characteristics.

26.3 NETWORK SWITCHING

The concept of using a Hopfield neural network for control of a crossbar switch was proposed by Marrakchi and Troudet [1989] and Troudet and Waters from Bellcore. This application requires real-time switching of packets at very high rates with maximum throughput. Bellcore has developed a VLSI implementation of an 8X8 neural network controller in 2-micron CMOS technology.

Brown [1989] reports on the concepts of using the parallelism for switching applications. One conclusion is that the crosspoint count of a switch is often used as a measure of its complexity. He considers not only large multistage switching networks, but also the ability for neural networks to do routing. He explores multiple problems with both feedforward and feedback neurons as well as the Winner Take All circuit.

26.4 DATA ROUTING

While outlining concepts for channel routing on a CAD system, Fujii, Tenario, and Zhu [1989] have developed some concepts that can be incorporated into the layout of a data communications network. These can evolve into a method for improving traffic throughput. Kitayama and Ito [1989] used an associative memory concept incorporating a Hopfield model to control multiple fiber optics to demonstrate that the global interconnection between two-dimensional units can be easily achieved.

26.5 DATA INTERPRETATION

Alston and Chau [1990] present a hardware model for a neural network that can decode block-encoded digital data. This model is designed to be part of a forward error-correcting digital communications system.

Information encoding using a hardware pulse coding neuron was reported by Canditt and Eckmiller [1990]. This hardware concept encodes the data at an instantaneous impulse rate while being analog and asynchronous. The synaptic weights of the neurons are stored in 8-bit digital form.

Chesmore [1989] has investigated a design for the design of phase-shift modulated signals in his application of pulse processing neural networks.

26.6   OPTICAL IMPLEMENTATIONS

Optics has emerged as a promising technology for neural network implementations because its capabilities match the requirements of most neural network models. As noted by Psaltis et al. [1989], the primary advantage lies in an optical system's ability to provide the massive interconnections between processors that are required in most neural network models. In electronic systems, signals must travel on physical wires that are subject to interference and crosstalk, while optical signals can propagate through free space and even pass through each other without interacting.

Psaltis is working with holographic media, which he feels can simultaneously provide the massive physical connectivity and the large memory required to specify the connections. In the near term, however, it is possible to implement a complete neural computer that incorporates both optic and electronic technology that is available today. He describes two implementations, a hybrid optoelectronic multilayer feedforward neural network and an optical autoassociative memory, that use optical disk and analog-integrated circuit technologies. The weakness in optical implementations is that optics does not provide complex nonlinear processing as easily as electronics.

Optical implementations have a natural use in fiber optic systems. Neugebauer, Agranat and Yariv [1990] have demonstrated an opto-electronic network that computes 1000 connections with 5-6 bit accuracy in less than 10 microseconds in low illumination, giving a computational rate of greater than 100,000,000 connections per second. This is based upon an integrated circuit containing a 32 × 32 array of synapses and 32 decision functions. The concept of an inverse XOR operation was used to demonstrate the capability.

Oita et al. [1989] of Mitsubishi reported on an opto-electronic associative neural network with 32 neurons, in which eight state vectors can be stored for time division multiplexing.

The importance of neural networks for wide area networks using fiber optics is essential to the proposed high-performance computing network [Gore, 1989] which specifically identifies neural networks as an important technology to be developed for high speed data communications.

26.7   ADAPTIVE FILTER

The use of neural networks as adaptive filters is one of the earliest practical implementations. In an earlier chapter, we explored the ADALINE and MADALINE neural networks, which apply to adaptive signal processing. Widrow and his students Glover and McCool developed this technology into an adaptive noise filter for use in modems for computer data communication. The concept was reported in a series of papers by Widrow et al. [1975] and Widrow, Glover and McCool [1977] which detail the technique.

Broomhead, Jones et al. [1989] explore a parallel architecture for nonlinear adaptive filtering and pattern recognition that consider the Perceptron as well as the linear least squares method for recognizing data. They conclude that implementation of a radial basis function network will be of use where processing of high data rates is of concern and the ability to retrain without shutting down the system is required.

In Siu, Gibson, and Cowan [1989] the concept of decision feedback equalization using a multilayer perceptron is explored. This paper expanded on their earlier work [Gibson, Siu and Cowan [1989] by applying this technology to be an adaptive equalizer for data communications. The results show better *bit error rate* (BER) performance than the standard LMS decision feedback equalizers concept.

26.8   QUADRATURE AMPLITUDE MODULATION

The *Quadrature Amplitude Modulation* (QAM) is a two-dimensional scheme that allows two double-sideband suppressed carrier modulated waves to occupy the same transmission bandwidth and still allow for the separation of the signals at the receiver output. The same basic concept is used in the CCITT V.29 constellation. Limiting the linear distortion of the signal constellation is important to the quality of the data.

Kohonen, Raivio, Venta and Henriksson [1990] propose to use a competitive neural network, a Self-Organizing Map, to produce localized responses to the input signal. This discrete signal identification scheme will work with the QAM as outlined in the paper and can also be applied to the CCITT V.29 data. It uses the active topology preserving property of the learning technique. As the received signal is quantized, only certain combinations of data can occur. Kohonen et al. show that if two-dimensional transmissions were distorted, a neural network as the adaptive signal identification mechanism can recover from possible errors such as noise and distortion.

26.9   LOCAL AND WIDE AREA NETWORKS

Barga and Melton [1990] reported on the interconnection of an ANSkit to a large network for simulation. Traditional LAN (ISO layer) concepts have not taken advantage of neural network techniques for autonomous configuration and collision detection and resolution. Recent progress in high speed Local Area Network (LAN) communications technology was reported by Pap [1989] in "Using Neural Networks in Local Area Network Planning and Design". This detailed the concepts needed for very high performance (approaching 3 gigabit) data transmission. These neural network pattern recognition implementations consisting feedforward neural nets can be applied to the requirements of distributed communication networks and to the capabilities for both FDDI and ISDN. Neural networks have a distinct advantage over today's decoding techniques in the amount of time needed for pattern recognition and addressing the data as well as in the speed of the processor board for monitoring and recognizing the data in its transit around the communications network. This is also significant to the future of large wide area networks such as NSFNET, and is directly related to the needs of the proposed High Performance Supercomputing Act [Gore, 1989].

Current approaches using artificial neural network concepts have not been applied to the true LAN and FDDI bandwidth problem. The real problem is technology limited at the speed of the receiving and transmitting terminations in the LAN and FDDI standards [(*Carrier Sense Multiple Access with Collision Detection (CSMA/CD)*, ANSI/IEEE Standard 802.3-1985, and Draft International Standard ISO/DIS 8802/3, 1985] and (Draft Proposed American National Standard, *FDDI Token Ring Physical Layer Protocol (PHY)*, ANSC X3T9.5, 1986; McCook, 1988) and (*The Ethernet. A Local Area Network. Data Link Layer and Physical Layer Specification*, 1982]. Neural network hardware for pattern recognition can decode delimiters and interpret transaction length better than conventional technology [Pap et al., 1990]. The key criteria for evaluating the performance issues are outlined in Bux [1984]. It appears that the neural network concept will satisfy the requirements of a LAN based upon the learning and fault tolerance.

The following is a concept proposed to be developed by Pap et al. [1990]. The characteristics of neural network systems would include adaptability, fault tolerance and possibly learning from observation. These neural nets would understand commands yet operate independently to perform the necessary tasks. Compatibility through parallel processing is a significant advantage of neural network technology and allows data comparison and transformations. This will eliminate the need for massive LAN software overhead that resides on each node. Capacities such as learning and generalization would facilitate control interpolation for unplanned variations in task requirements and performance. These learning systems are capable of controlling nonlinear functions and could control concurrent functions to manipulate multiple data paths in a coordinated fashion.

An approach to the time dependencies of interpreting asynchronous bit patterns across time would be backpropagation through time [Werbos, 1974, 1987, 1988]. This is a variation of a feedforward neural network which includes information from different time periods to control the output. Learning is based on the modification of adaptive weights for each time period. In a LAN or WAN applications, backpropagation through time might be able to perform in real time to predict conflicts, detect collisions or control a sequence of movements.

A modification of backpropagation through time might allow the network to react correctly to patterns before their time dependent sequence is completed. Incoming data would be assessed continuously (in the hardware configuration) putting out responses even as the data comes in to the network. To be sure, the best answer for network diagnosis would result from processing following the completion of the input sequence; however, useful classifications could result long before the input sequence finished.

REFERENCES

Alston, M.D. and Chau, P.M., "A Decoder for Block-Coded Forward Error Correcting Systems," *Proceeding of the International Joint Conference on Neural Networks*, (Washington, D.C., January 15-19, 1990], Vol 2, pp.II-302 to II-305.

Ansari, N. and Chen, Y., "Dynamic Digital Satellite Communications Network Management by Self-Organization," *Proceeding of the International Joint Conference on Neural Networks*, (Washington, D.C., January 15-19, 1990], Vol 2, pp.II-567 to II-570.

Barga, R.S. and Melton, R.B., "Framework for Distributed Artificial Neural System Simulation,"*Proceeding of the International Joint Conference on Neural Networks*, (Washington, D.C., January 15-19, 1990], Vol 2, pp.II-94 to II-97.

Bourret, P., Remy, F., and Goodall, S., "A Special Purpose Neural Network For Scheduling Satellite Broadcasting Times," *Proceeding of the International Joint Conference on Neural Networks*, (Washington, D.C., January 15-19, 1990], Vol 2, pp.II-535 to II-538.

Bourret, P., Goodall, S. and Samuelides, M., "Optimal Scheduling by Competitive Activation: Application to the Satellite Scheduling Problem," *Proceedings of International Joint Conference on Neural Network* (Washington, DC: June 18-22, 1989], Vol. I, pp. I-565-572, 1989.

Broomhead, D.S., Jones, R., McWhirter, J.G. and Shepherd, T.J., "A Parallel Architecture for Nonlinear Adaptive Filtering and Pattern Recognition," *Proceeding of the First IEE International Conference on Artificial Neural Networks* (London, England; 16-18 October, 1989], Conference Publication 313, pp. 265-269.

Brown, T.X., "Neural Networks for Switching" from Posner, Edward C., guest editor, "Special Issue on Neural Networks in Communications," *IEEE Communications Magazine*, November 1989, pp 72-81.

Bux, W., "Performance Issues in Local Area Networks," *IBM Systems Journal*, Vol 23, No 4, pp. 351-374, 1984.

Carrier Sense Multiple Access with Collision Detection (CSMA/CD), ANSI/IEEE Standard 802.3-1985, and Draft International Standard ISO/DIS 8802/3. IEEE, New York, 1985.

Canditt, S. and Eckmiller, R., "Pulse Coding Hardware Neurons that Learn Boolean Functions," *Proceeding of the International Joint Conference on Neural Networks*, (Washington, D.C., January 15-19, 1990], Vol 2, pp.II-102 to II-105.

Chesmore, E.D., "Application of Pulse Processing Neural Networks in Communications and Signal Processing," *Proceeding of the First IEE International Conference on Artificial Neural Networks* (London, England; 16-18 October, 1989], Conference Publication 313, pp. 337-341.

Dayton, R.L., *Guide to Integrating Digital Services*, McGraw Hill, New York, 1989

Draft Proposed American National Standard, *FDDI Token Ring Physical Layer Protocol (PHY)*, ANSC X3T9.5, Oct. 10, 1986.

Frankel, M., Shacham, N. and Mathis, J.E., "Self Organizing Networks," *Future Generation Computer Systems*, Vol 4, pp. 95-115, 1988.

Fujii, R., Tenorio, M.F., and Zhu, H., "Use of Neural Nets in Channel Routing," *Proceedings of International Joint Conference on Neural Network* (Washington, DC: June 18-22, 1989], Vol. I, pp. I-321-326, 1989.

Gibson, G.J., Siu, S. and Cowan, C.F., "Multi-layer Perceptron Structures Applied to Adaptive Equalizers for Data Communications," *Proceedings of ICASSP* (Glasgow, Scotland: May, 1989], pp. 1183-1186.

Gore, Senator A., Jr., "Proposed National High Performance Computing Act," U.S. Senate Resolution, 1067.

Hiramatsu, Atsushi, "ATM Communications Network Control by Neural Networks," *IEEE Transactions on Neural Networks*, Vol.1, No.1. pp.122-130, March, 1990.

Hiramatsu, A. "ATM Communications Network Control by Neural Networks," *Proceedings of International Joint Conference on Neural Network* (Washington, DC: June 18-22, 1989], Vol. I, pp. I-259-266, 1989.

Hopfield, J.J., "Neurons with Graded Response have Collective Computational Properties like those of Two State Neurons," *Proceeding of the National Academy of Science USA*, Vol 81, pp. 3088-3092, 1984.

Hopfield, J.J., "Neural Network and Physical Systems with Emergent Collective Computational Abilities," *Proceeding of the National Academy of Science USA*, Vol 79, pp. 2554-2558, 1982.

Hopfield, J.J. and Tank, D.W., "Neural Computation of Decisions in Optimization Problems," *Biological Cybernetics*, Vol 52, pp. 141-152, 1985.

Kershenbaum, A. and Rubinson, T.C., "Fuzzy logic and expert system for network planning and design," in *Proceeding of Interface '89*, March 13-16, 1989.

Kitayama, K. and Ito, F., "Multiple Fiber Coupler Associative Memory with Bit Significance Retrieval," in *Proceedings of the International Joint Conference on Neural Networks*, (Washington, D.C.: June 18-22, 1989], Vol. II pp. 633, 1989.

Kleinrock, L., *Queuing Systems*, V.1, pp. 140, John Wiley & Sons, New York, 1975.

Kohonen, T., Raivio, K., Venta, O. and Henriksson, J., "An Adaptive Discrete-Signal Detector Based on Self-Organizing Maps," *Proceeding of the International Joint Conference on Neural Networks*, (Washington, D.C., January 15-19, 1990], Vol 2, pp. II-249 to II-252.

Marrakchi, A. and Troudet, T. "A Neural Network Arbitrator for Large Crossbar Packet Switches," in *IEEE Transactions on Circuits and Systems*, Vol 36, No 7, pp 1039-1041, July, 1989.

McCool, J., "A Look at the Emerging FDDI Standard," Preprint from Electro 88 (New York), 1988.

McEliece, R.J., Posner, E.C., Rodemich, E.A., & Venkatesh, S.S., "The Capacity of Hopfield Associative Memory," in *IEEE Transactions of Information Theory*, Vol. IT-33, pp. 461-482, July, 1988.

Neugebauer, C.F., Agranat, A. and Yariv, A., "Optically Configured Phototransistor Neural Networks," *Proceeding of the International Joint Conference on Neural Networks*, (Washington, D.C., January 15-19, 1990], Vol 2, pp. II-64 to II-67.

Oita, M., Tai,S., Ohta, J., Hara,K., Kyuma,K. and Nakayama, T., "An Opto-Electronic Implementation of the Associative Neural Network Using Time Division Multiplexing Techniques," in *Proceedings of the International Joint Conference on Neural Networks*, (Washington, D.C.: June 18-22, 1989], Vol. II pp. 634, 1989.

Pap, R.M., "Using Neural Networks for Network Planning & Design," in *Proceedings of Interface '89*, March 13-16, 1989.

Pap, R. M., Maren, A. J., Harston, C. T. and Parten, C. R., "Application Of Neural Networks To High Performance Data Communications Networks," Submitted to *The Journal of Communications Technology*, 1990.

Posner, E. C., "Neural Networks in Communications," Letter to *IEEE Transactions on Neural Networks*, Vol.1, No.1. pp.145-147, March, 1990.

Posner, E. C., guest editor, "Special Issue on Neural Networks in Communications," *IEEE Communications Magazine*, November 1989.

Psaultis, D., Yamamura, A.A., Hsu, K., Lin., Gu, X-G., Brady, D., "Optoelectronic Implementations of Neural Networks," Posner, Edward C., guest editor, "Special Issue on Neural Networks in Communications," *IEEE Communications Magazine*, November 1989, pp.37-40, 71.

Rauch, H.E. and Winarske, T., "Neural Networks for Routing Communications Traffic," *IEEE Control Systems Magazine*, pp 26-31, April, 1988.

Siu, S., Gibson, G.J. and Cowan, C.F.N., "Decision Feedback Equalization using Neural Network Structures," *Proceeding of the First IEE International Conference on Artificial Neural Networks* (London, England; 16-18 October, 1989], Conference Publication 313, pp. 125-128.

Szu, H., "Fast TSP Algorithm Based on Binary Neuron Output and Analog Neuron Input Using the Zero-Diagonal Interconnect Matrix and Necessary and Sufficient Constraint of the Permutation Matrix" in *Proceedings of the Second IEEE Neural Network Conference* (San Diego, 1988], Vol II, pp. 259-266.

Tank, D.W. and Hopfield, J.J., "Simple Neural Optimization Networks : An A/D Converter, Signal Decision Circuit and a Linear Program Circuit," *IEEE Transactions on Circuits and Systems*, Vol. CAS-33, No. 5, pp. 533-541, 1986.

The Ethernet. A Local Area Network. Data Link Layer and Physical Layer Specification, DEC Corp., Intel Corp., and Xerox Corp., Ver. 2.0, Nov. 1982.

Werbos, P.J., "Generalization of backpropagation with application to a recurrent gas market model," *Neural Networks*, Oct. 1988.

Werbos, P.J., "Building and Understanding Adaptive Systems: A Statistical/Numerical Approach to Factory Automation and Brain Research," *IEEE Transaction on Systems, Man and Cybernetics*, pp. 7-20, 1987.

Werbos, P.J., "Beyond Regression: New Tools for Prediction and Analysis in the Behavioral Sciences," Harvard Ph.D. thesis, Committee on Applied Mathematics, Nov. 1974.

Widrow, B., Glover, J.R. and McCool, J., "Reply to Comments by D.F. Tuft on Adaptive Noise Canceling: Principles and Applications," *Proceedings of the IEEE*, Vol. 65, No. 1, pp. 171-173, January, 1977.

Widrow, B., Glover, J.R., McCool, J., Kaunitz, J., Williams, C.S., Hearn, R.H., Zeidler, J.R., Dongle, E. and Goodlin, R.C., "Adaptive Noise Canceling: Principles and Applications," *Proceedings of the IEEE*, Vol. 63, No. 12, pp. 1692-1716, December, 1975.

Wilson, G.V., Pawley, G.S., "On the Stability of the Traveling Salesman Problem Algorithm of Hopfield and Tank," *Biological Cybernetics*, Vol 58, pp. 63-70, 1988.

27

NEURAL NETWORKS FOR MAN/MACHINE SYSTEMS

Alianna J. Maren

27.0 OVERVIEW

In recent years, there has been a growing awareness of the importance of man/machine interfaces. Many systems for monitoring, diagnosis, and control of complex processes (e.g., power plants, oil refineries, flying high-performance aircraft) are becoming more difficult to manage. This is due to the combined factors of increased flow rate and volume per unit time of sensor-based information, the complexity of the processes, and the rapidity at which changes can sometimes occur. On a broader social scale, as more and more segments of the population interact with computers, there will be an increased need to make the man/machine interface as comfortable as possible.

There are several ways in which neural networks can play a role in enhancing man/machine systems. They are:

- User-specific adaptive interfaces
- Adaptive aiding for operators
- Modeling human performance
- Bioengineering

27.1 ADAPTIVE INTERFACES

One of the most rapidly growing areas in the man/machine systems area is that of designing and developing intelligent and/or user adaptive interfaces. User interfaces are important in many system monitoring and control applications, ranging from piloting high performance aircraft and managing air traffic control to managing large, complex powerplants. There is a growing need for appropriate interfaces in intelligent tutoring systems. There are also numerous daily tasks which involve interaction with computers. Good interface design is important in maximizing productivity in these tasks.

Recently, the concept of user-adaptive interfaces has emerged [Norcio & Stanley, 1989; Noah & Halpin, 1986; Rouse, 1988; Holynski, 1988]. The basis for this concept is the realization that the performance of an integrated man-machine system can be improved if the user interface is adapted to the individual user. There are two major ways in which the interface can be adapted. The first is to a baseline distinction which distinguishes between different user characteristics. The second is to the time-varying nature of a man-machine participatory task. In order to label this distinction, we call the first aspect adaptive interface design, and the second adaptive aiding.

The first type of user adaptation embodies the realization that individual users vary in terms of cognitive style and perceptual style. For example, different users might naturally prefer to think in terms of concrete or abstract terms, to use linguistic or visual mental representations, or vary in the degree to which they want to reflect on or experiment with new information. These are examples of varience in cognitive style. Similarly, different users might prefer (and optimally use) graphical or command-line interfaces, with differing degrees of screen complexity, information content, and other perceptual design characteristics. User performance will vary significantly when information is presented in a manner that is in accord with or differs from their cognitive style. (See, e.g., [Sein & Bostrum, 1989] for results on a study of differences in learning when two different models were used for teaching, one analogic, the other abstract.) Once these issues have been clarified, an intelligent system can be designed to account for differences among individual users. Because of the possible complexity of the number of classes of users (resulting from measurements of user cognitive style along many variables), the adaptation of the interface to the user will be a complex task. It will involve recognizing and classifying the particular style of a user (a classification task), and creating the appropriate interface (a control/mapping task).

Neural networks are an appropriate technology for developing such baseline adaptive capabilities in an intelligent interface. As an example, Heger and Koen [under review] have developed a simple neural network-based adaptive interface to a database. The interface acquires knowledge through its interactions with both the user and the database. It uses this knowledge to personalize the database retrieval process and to induce new queries.

27.2   ADAPTIVE AIDING

A second major form of interface adaptation is to the time-varying nature of the task [Rouse, 1988]. For example, Morris et al. [1988] explored how computer aiding influenced human performance on concurrent tasks of visual target identification and tracking. They found that several factors influenced human performance, including the immediate terrain being searched and the terrain of the area that had just previously been searched. When the intelligent system was responsible for initiating appropriate aid, the overall man-machine system performed better in search tasks over a simulated water environment (which had a greater area to search for the target vessels). This is the type of pattern recognition (proportional amounts of water versus land, and scene complexity) for which neural networks are well suited. Further, this task exemplifies the type of scenario in which task responsibilities, information flow, and other factors should adaptively change in response to changing task and/or environmental conditions. Of the available technologies, neural networks appear most well-suited for creating either real-time or long-term system adaptations.

27.3 NEURAL NETWORKS TO EMULATE HUMAN PERFORMANCE

One aspect of designing adaptive man/machine interfaces is building a useful model of human behavior. This is a very complex model-building task. A useful model of human performance will facilitate adaptive aiding, and be useful in training human operators. Rouse et al. [1989] argue convincingly for building a structural (computer) model which captures human skills and knowledge in terms of input/output relations. They advocate developing algorithmic models for data which can be captured by signals (stimuli and responses). Models are necessarily limited in capturing phenomena; ideally, a system must be able to deal with approximations of performance.

Recent neural network models of human behavior show that they are able to capture and model complex processes more effectively than can traditional artificial intelligence systems. For example, Fix [1990] used a back-propagation network to model the performance of different human operators controlling simulated cars on a circular racetrack. Operators gained points by passing other cars and lost points when they were bumped by other cars. The four other cars were computer controlled and had strategies to keep the operator from passing them. The operator was able to switch lanes and adjust speed. Fix trained three three-layer back-propagation networks, each using data from a different operator. In each case, the network successfully learned to mimic the operators performance, including their stylistic differences.

McMahon [1990] has similarly developed an neural network model which selected offensive and defensive maneuvers for the domain of air combat manuevering. He compared the performance results of the neural network system against those of a rule-based expert system. When tested on a series of 40 airspace scenarios (evaluated by expert fighter pilots), the neural network agreed with expert pilot maneuver selection over 2.5 times more than the rule-based production system.

This type of model can be extended to simulate the performance of a human-operated system, such as an intelligent adversary in training military pilots and other personnel. This approach could also be used for modeling operations where multiple repetitions of a situation are not possible (making it difficult to use standard statistical methods). The amount of labor involved in creating such a model is much less than for other techniques. This would facilitate building models of complex human behavior in applications where they are desirable, but have not been developed because of the high cost.

As evidenced by McMahon's work, a special advantage of the neural network system over the rule-based system is that neural networks, operating on signal rather than symbolic data, are able to extract information from the spatio-temporal configuration of patterns (hostile aircraft positions). This allows them to operate on a richer information base than is readily available to expert systems.

Other neural networks have been developed which mimic human performance. Kraiss and Ku ttelwesch [1990] use a functional link network for vehicle control and path planning. They found that the neural network did not copy the human operator exactly, but rather extracted the consistent aspects of the operator's behavior and learned to mimic those. Josin [1990] has described a reinforcement-learning network which can take on some of the tasks of an autopilot for a high-performance aircraft.

27.4 NEURAL NETWORKS FOR BIOENGINEERING

An important new application of neural networks is just beginning to emerge: that of bioengineering. This takes the concept of an adaptive interface to a new level, where the interface between the human and the machine becomes much more intimate, and the precise boundaries between man and machine begin to dissolve. Most work in this area is highly preliminary.

There are several exciting and novel ways in which this type of bioengineering, leading to true bionic systems , can develop. One is to perform pattern recognition on sensors attached to the skin. Another would be to interpret EEG readings. Yet a third, the most innovative and daring, is to interpret the readings from probes directly inserted into the neural system (either brain or afferent neurons).

Enhancements to the DataGlove and similar tactile, force-feedback, and position sensors [Foley, 1987] will involve more sensor responses. The postion of the glove could be best determined by interpretation of the positions of a large number of sensors in an interdependent manner. Neural networks are well-suited for this task.

A great deal of basic work in interpreting EEG readings has already been done, some involving neural network applications. Gevins et al. [1987] have reported that brain electrical patterns before accurate performance differ from the patterns observed before inaccurate performance. Using 26 electrodes and several signal-enhancing procedures, they found that brief spatially distributed patterns over distinct cognitive, somesthetic-motor, and integrative motor areas were observed prior to performance and may be essential precursors of accurate visuomotor performance. It is possible to use neural networks to facilitate this analysis.

Gevins and Morgan [1988] have developed and tested a system of interacting neural networks to evaluate 64-channel Event-Related Potentials (ERPs) as people play a special video game. About a gigabyte of information is recorded from each experiment. They have been able to identify spatio-temporal ERPs associated with subject's preparation to react to a number on the video screen. When subjects do not exhibit these special ERP patterns, their performance on the upcoming task is inaccurate. As a result of training, the neural network system was able to choose features which in combination would predict subsequent performance accuracy two-thirds of the time. They also found that certain changes in these patterns preceded deterioration of performance as subjects performed difficult concentration tasks for many hours. In one subject, they identified leading indicators which distinguished between fully alert and incipiently impaired performance 81% of the time. Currently, all their analyses are performed off-line.

Hiraiwa et al. [1990] have described a neural network system which evaluates ERPs in real-time in the context of potential use as a man/machine interface. They found that the ERP's generated before pronouncing syllables contained some information on those syllables, and the ERP's observed before the subjects moved a joystick are related to the directionality in which they moved the joystick.

Bernard Widrow has described a major step towards faciliting a direct interface between the human nervous system and a physical (prosthetic) device, which is reported in Wan et al. [1990]. They have interposed silicon-based interfaces, perforated by small via holes (holes about the diameter of an axon) between two cut ends of severed peripheral nerve fascicles. The axons regenerate, grow through the holes, and connect with each other in a random manner. During this process, they become physically isolated and spatially fixed relative to microelectrodes attached to the silicon interface. At this stage of research, passive neural interfaces (with microelectrodes but without active microelectronic circuitry) have been fabricated and implanted in the nerves of rats. Their preliminary studies show that both recording and stimulating the neurons with the interface is possible.

Wan, Widrow, and their colleagues Kovacs and Rossen project two potential uses of this type of direct interface. The first would be for nerve repair. When a nerve is cut and regrown together, the individual axonal connections become scrambled. A bidirectional pattern-recognizing artificial neural network mediating the reconnection should be able to electronically intercept and reroute motor and sensory neural impulses, providing for increased recovery after a nerve injury.

Wan et al. also foresee the possibility of using such a direct neural interface to provide a means for neuronal control of a prosthetic device. This would require feature extraction, adaptive clustering of features into groups, and then forming an average demodulated signal for each feature signal type. This information would be routed to a controller for the prosthetic device. Afferent information would be routed back to the neural bundle. In sum, a system of neural networks would act as the bridge between the neural signals and the actual prosthetic device, such as a robotic hand. Its function would be to interpret microelectrode signals (generated both biologically and by the device) to make use of the prosthetic transparent to the patient.

27.5 SUMMARY

There are many aspects of the man-machine interface to which neural networks could be applied, including adaptive interfaces, adaptive aiding, human performance models, and bioengineered direct man-machine interfaces. Work in all of these areas is highly preliminary, but is exciting. There is a reasonable potential for using neural networks to replace some aspects of user modeling or automatic aiding that are currently being done by expert systems. There is strong potential for using neural networks to model human performance. In such cases, a network or system of networks may learn to emulate specific operators who were used for training. They may provide the ability to generate more complex automated training systems, such as cockpit simulations of enemy aircraft. There is a long-term but fascinating potential of using neural networks to facilitate direct man-machine interfaces, either by interpreting sensor data provided by a human or by creating some form of sensor data input (e.g., an artificial cochlea or retina). At the furthest element on the time-line, artificial neural networks may facilitate direct connection between the human nervous system or brain and a physical device, such as a prosthesis.

REFERENCES

Fix, E. (1990). "Modeling human performance with neural networks," *Proc. Third Int'l. Joint Conf. Neural Networks* (San Diego, CA; June 17-21, 1990), I-247–I-252.

Foley, J.D. (1987). "Interfaces for advanced computing," *Scientific American* (October), 127-135.

Gevins, A.S., Morgan, N.H., Bressler, S.L., Cutillo, B.A., White, R.M., Illes, J., Greer, D.S., Doyle, J.C., & Zeitlin, G.M. (1987), "Human neuroelectric patterns predict performance accuracy," *Science, 235,* 580-585.

Gevins, A.S., Morgan, N.H. (1988). "Application of neural-network (NN) signal processing in brain research," *IEEE Trans. Acoustics, Speech and Signal Processing, 36,* 1152-1166.

Heger, A.S., & Koen, B.V. (1991). "KNOWBOT: An adaptive database interface," *Nuclear Science and Engineering Journal, 107,* #2. (To appear in February issue.)

Hiraiwa, A., Shimohara, K., & Tokunaga, Y. (Preprint: Expected publication in 1990). "EEG topography recognition by neural networks," Preprint from Human Interface Laboratories, Nippon Telegraph and Telephone Corp., Japan.

Holynski, M. (1988). "User-adaptive computer graphics," *Int. J. Man-Machine Studies, 29,* 539-548.

Hunt, E. (1975-1978). "Cognitive theory applied to individual differences," in W. K. Estese (Ed.), *Handbook of Learning and Cognitive Processes* (Hillsdale, NJ: Lawrence Erlbaum), 81-110.

Josin, G.M. (1990). "Development of a neural network autopilot model for a high performance aircraft," *Proc. Second Int'l. Joint Conf. Neural Networks* (Washington, D.C., Jan. 15-19, 1990), II-547–II-550.

Kraiss, K.F., & Küttelwesch, H. (1990). "Teaching neural networks to guide a vehicle thorugh an obstacle course by emulating a human teacher," *Proc. Third Int'l. Joint Conf. Neural Networks* (San Diego, CA; June 17-21, 1990), I-333–I-337.

McMahon, D.C. (1990). "A neural network trained to select aircraft maneuvers during air combat: A comparison of network and rule based performance," *Proc. Third Int'l. Joint Conf. Neural Networks* (San Diego, CA; July 17-21), I-107–I-111.

Morris, N.M., Rouse, W.B., & Ward, S.L. (1988). "Studies of dynamic task allocation in an aerial search environment," *IEEE Trans. Systems, Man, & Cybernetics, 18,* 376-389.

Noah, W.W., & Halpin, S.M. (1986). "Adaptive user interfaces for planning and decision aids in C3I systems," *IEEE Trans. Systems, Man, & Cybernetics, SMC-16,* 909-918.

Norcio, A.F., & Stanley, J., "Adaptive human-computer interfaces: A literature survey and perspective," *IEEE Trans. Systems, Man, & Cybernetics, 19* (March/April, 1989), 399-408.

Rouse, W. (1988). "Adaptive aiding for human/computer control," *Human Factors, 30,* 431-443.

Rouse, W.B., Hammer, J.M., & Lewis, C.M. (1989). "On capturing human skills and knowledge: Algorithmic approaches to model identification," *IEEE Trans. Systems, Man, & Cybernetics, 19,* 558-573.

Sein, M.K., & Bostran, R.P. (1989). "Individual differences and conceptual models in training novice users," *Human-Computer Interaction, 4,* 197-229.

Snow, R.E., "Toward assessment of cognitive and conative structures," *Educational Researcher* (1989-90).

Snow, R.E., "Cognitive-conative aptitude interactions in learning," in R. Kanfer, P.L. Ackerman, & R. Cudeck (Eds.), *Abilities, Motivation, and Methodology* (Hillsdale, NJ: Lawrence Erlbaum, expected 1989 publication date).

Snow, R.E., "Individual differences and the design of educational programs," *American Psychologist* (October, 1986), 1029-1039.

Sternberg, R.J., "Domain-generality versus domain-specificity: The life and impending death of a false dichotomy," *Merill-Palmer Quarterly, 35* (January, 1989), 115-130.

Sternberg, R.J., "Mental self-government: A theory of intellectual styles and their development," *Human Development, 31* (1988), 197-224.

Wan, E.A., Kovacs, G.T.A., Rosen, J.M., & Widrow, B. (1990). "Development of neural network interfaces for direct control of neuroprostheses," *Proc. Second Int'l. Joint Conf. Neural Networks* (Washington, D.C., Jan. 15-19, 1990), II-3-21.

28

CAPTURING THE FUTURE: NEURAL NETWORKS IN THE YEAR 2000 AND BEYOND

Aliann J. Maren

28.0 INTRODUCTION

In contrast to the pragmatic and specific considerations of earlier chapters in this book, this chapter opens a door to the future, to the possibilities, to what can — and possibly will — happen.

Because gazing into the future is somewhat like gazing into a crystal ball, we give our descriptions of future possibilities as "predictions." However, these predictions are not just guesses. Each prediction rests on some sort of evidence or established trend which, with extrapolation, clearly takes us into a new realm. Some of these realms are fascinating. Some are frightening. All of them, in one way or another, clearly depict a reality which is different from the one which we inhabit now.

28.1 PREDICTION 1

Neural networks will soon lose their "sex appeal." Neural networks are currently viewed as something of a glamour topic. They're hot, they're sexy, they're the "in thing" for scientists and engineers of all types. Our first prediction is that this will change.

This change will be for a positive reason. As neural networks proliferate, are applied to more and more applications, and inserted into more and more real-life, working systems, they will lose their special, "out-there" status as they become everyday tools. Familiarity will not necessarily breed contempt, but it will make commonplace what was once a glamour item. The more neural networks are accepted, the greater this tendency will be.

Corollary to the First Prediction: Scientists, engineers, and technicians in a wide range of fields will gain a wide working knowledge of neural networks as a natural and useful tool for their work. Neural networks will take on a "tool" status, much the way many scientists and engineers use their working knowledge of other fields, such as statistics, computer programming, and use of electronic devices.

28.2 PREDICTION 2

Neural networks will rapidly "infiltrate" everyday life. Neural networks can be much easier to insert into existing systems than traditional, rule-based "AI" (artificial intelligence) systems. To be useful and effective, AI systems have typically required large numbers of rules, encoding the knowledge painfully acquired from experts by "knowledge engineers." This is a painstaking, long-term effort. It tends to draw a lot of attention to itself. It is hard to introduce (or insert) an AI system without being rather obvious.

On the other hand, neural networks can be small and simple, and still useful. They can be designed as modules which can be readily inserted into existing systems. In these systems, they can take over a few functions, or many. Certain networks which can learn from "experience" (i.e., a good training set), can be taught off-line and inserted without too much fuss and bother. This ease of insertability, and their versatility for many pattern recognition, classification, and control tasks makes them ideal candidates for step-by-step system upgrades. Slowly but surely, neural networks will infiltrate existing systems.

Neural networks can be introduced with a psychological "soft-sell." There are some environments where regulations, pressures, or just plain old-fashioned conservatism make it difficult to introduce new technologies. But we don't have to call a neural network a "new technology" if we don't want to. A given network can simply be an "adaptive system" which just happens to perform better than the existing system. This will make it easy to insert them within the context of traditional computing systems. After all, Widrow's adaptive filters have been in use for over 20 years!

Further, neural network chips will make it possible to easily insert networks into physical devices. The rapid development of neural network chips, and the ability to put a useful and functioning neural network onto a single chip, will make network insertion a much easier and more natural process than was ever possible with AI systems.

28.3 PREDICTION 3

Neural networks will enable simple forms of "non-living intelligence" to carry out various tasks. Some of the most engaging research reports to appear have dealt with "non-living intelligence" — artificial "bugs" which search for food, experience boredom and thus seek experiences, and experience danger and thus balance their explorations with a tendency to stay in known safe areas [Scanlon & Johnson, 1988; Johnson & Scanlon, 1987a, 1987b, Goldberg, 1988]. At least two researchers are suggesting neural systems which need to "dream" in order to consolidate learned information! [Hutton & Sigillito, 1990; Johnson, 1990]. Although primitive, these constructs may be the basis for a new form of robotic engineering. Tasks for which these simple robots would be well-suited would include cleaning or scavenging, especially in areas too dangerous for humans.

These non-living intelligences may be built on a very small scale. A neural controller on a chip will be able to incorporate simple sensory feedback and direct the motions of a small robot. The result will be the first order of autonomous systems which will exist in society. These will not be the huge robotic automations envisioned during the 1950s, but simpler gadgets — almost like bugs. They will do simple tasks, such as cleaning the insides of pipes and the sides of boats. They will be cheap, disposable, and — by the year 2000 — common.

28.4   PREDICTION 4

Neural networks can greatly affect business practices such as direct marketing. During the 1980s, direct marketing took on a new form. Retailers realized that catalogs could be targeted to highly specialized markets. Innovations such as round-the-clock access, video catalogs, and special databases of client lists became commonplace. The widespread use of direct marketing has had two major implications. The first is that very specialized "stores," whether for gardening supplies or for computer peripherals, can reach markets which extend far beyond their geographic area. Companies whose special niches might have allowed them to operate only in high-density urban areas can now reach a large client population via effective direct mail. Second, in order to maximize their effectiveness, they must precisely target their mailings. The more closely they can match up a catalog with the persons who are likely to buy the products described, the more likely they are to succeed in making sales.

This is where neural networks, combined with advances in database technology, are likely to have a very distinct impact. A direct marketing firm will be able to use neural networks to identify patterns of purchasing trends within a database of customer purchases. Once a database has been analyzed to extract certain "types" of customers, a relatively small number of purchases by a given customer will suffice to type him or her within a certain category. This category — one far more specific than those used by current catalogs — can be directly marketed to the customer.

By creating sets of offerings based on the patterns of previous purchases, marketers can achieve higher sales volumes. As an example, suppose a music store kept records of all purchases by each customer, and wanted to create a service of "special offers" for each customer by advising them of new releases. Instead of relying on customer indications of generic type (e.g., "hard rock," "classical," etc.) the previous purchases themselves would provide the basis for the establishing the most likely purchase areas. By developing an adequate descriptor base using multiple features to describe each piece of music (e.g., "hard rock with sexy lyrics," or "romantic works performed on cello, violin, and harps"), and identifying the common features of the main body of purchases by each individual, it will be possible to identify what would interest that person the most in the future. The specific and exciting aspect of neural networks will be to extract patterns which are not just subsets of those clusters already identified, but which spread across previously identified categories.

This type of technology will enable individuals or small businesses to act as intermediaries or go-betweens between large numbers of individuals and large product inventories. By acting as a "broker" for purchases, it should be possible to make arrangements with the manufacturers, leading to the possibility of operating with low or no in-house inventories. Their sole service would be the brokerage, or coupling of buyer with product. In addition to facilitating sales in direct marketing (and similar operations), neural networks will allow the appearance of personalized service, while maintaining a large client base. This will have a dramatic impact on the way certain businesses operate. More than ever, the business of buying and selling will become an information-based operation.

There is a potential advantage in this for individuals as well as companies. This type of profiling may make it easier for individuals to access certain products which they wish to buy. There is also an equally strong downside possibility. If information-sifting technologies are coupled with covert psychological manipulation (e.g., advertisements cued to whatever may be detected as an individual's insecurities, fears, or fantasies), then a form of "subliminal seduction" which would be very hard to resist could evolve. This would be abusive of human rights. Thus, there needs to be some societal attention on how a company's knowledge of patterns of individual behavior are used in dealings with these people. There certainly needs to be some safeguards against proliferating data about individuals to other organizations.

28.5   PREDICTION 5

Neural networks will become a prime tool for manipulating information systems. There will be extensions of the previously described database mining to other aspects of business and society. In terms of using neural networks to sift through large amounts of data — we have only just begun. Nestor was one of the first companies to apply neural networks to this area, focusing first on mortgage loan evaluation (discussed in Part IV). Much of this work, as discussed earlier, is proprietary.

This type of neural network application can have widescale societal impact. By examining any large database for implicit patterns, especially causally-related patterns or patterns of temporal activity, it will be possible to extract far more information from data resources than has previously been possible. We can expect that this type of process will be used in business (predicting future trends), in finance (analyzing credit histories, assessing investment opportunities), and in security. This type of information-processing may also be done by political or governmental organizations. While there can be an upside — improved technologies for accessing and sifting and using information, there is an associated downside. It may be more difficult to maintain privacy. The possibility of being "analyzed" for political, business, or security purposes open a dangerous path for abuse. Thus, convergent with the development of this type of technology, we need to address issues of privacy and individual rights.

28.6   PREDICTION 6

Neural networks will facilitate breakthroughs with some of the technical "nuts" that have been toughest to crack — speech understanding, computer vision, adaptive robotic control — all areas which lead to intelligent man/machine interfaces and autonomous systems. There have been major difficulties in enabling each of these areas; speech understanding, image understanding, robotic motion. For a while longer, there will continue to be difficulties. However, each of these areas has been plagued by problems at the low end, in dealing with the raw data, that have been exacerbated by algorithmic rigidity. Inadequate segmentations, inabilities to associate segments with appropriate neighbors using context, difficulties in using context for interpretation ... these are all areas in which neural network technology has already demonstrated the ability to surpass obstacles where other technologies fall short.

Neural networks will probably enable the breakthroughs that have been the subject of so much science fiction and popular press stories. These breakthroughs will in their turn affect the reality in which we live by enabling much more sophisticated and varied man-machine interactions. This leads to the following prediction.

28.7   PREDICTION 7

Neural networks will facilitate new forms of man-machine interfaces.

There have been a lot of exciting developments over the past decade in the area of man-machine interfaces. Heads-up cockpit displays, DataGloves, and other "novel" interfaces are growing in importance and usefulness. But more subtle, more sophisticated, more personal interfaces are yet to come. Neural networks will be an essential tool for any interface which attempts to map an aspect of external, "physical" reality onto a person's somatosensory, "experienced" reality. Further, neural networks may make possible direct man-machine interfaces, as discussed in the previous chapter.

On what time scale can we expect such developments? Perhaps the best way to extrapolate is to turn to science fiction, which in the latter half of this century has become the best predictor of future developments which we might have. Many of us reading this book will have cut our teeth (figuratively) on some of the early science fiction; The Moon is a Harsh Mistress, by Heinlein, I, Robot, by Asimov, and others. The period in which this type of heavily technical, space-exploration science fiction began to emerge was in the 1940s, with a peak in the 1950s through the mid-1960s. As a corresponding real technical milestone, we landed on the moon on July 20,1969. Our current space program is not progressing nearly at the course that we would like if we wanted to fulfill our science-fiction fantasies of moon colonization, interplanetary travel, and interstellar adventures. However, we have achieved some milestones in an arena that has fired our imaginations. The time interval between popularization of a vision of future achievement (space travel) and the first significant milestone in that achievement was about 20-30 years — enough for the high-school students who read the stories to go to college, join NASA, and contribute to the missions. (We note that the science-fiction emphasis on hard technology and lunar and interplanetary space travel died just shortly before our first lunar landing.)

Beginning in the late 1960s, we began to see more stories about technologies involving extension of human capabilities. As an example, we have a passage from a book published in the late 1960s:

"She was born a thing and as such would be condemned if she failed to pass the encephalograph test required of all newborn babies ... The electro-encephalogram was entirely favorable ... There was the final, harsh decision: to give their child euthanasia or permit it to become an encapsulated "brain," a guiding mechanism in any one of a number of curious professions.

"On the anniversary of her 16th year, Helva was unconditionally graduated and installed in her ship, the XH-834 ... The neural, audio, visual, and sensory connections were made and sealed. Her extendibles were diverted, connected, or augmented and the final, delicate-beyond-description brain taps were completed while Helva remained anesthetically unaware of the proceedings. When she woke, she *was* the ship."

(From *The Ship Who Sang*, by Anne McCaffrey, Ballantine, 1969, pp. 1 and 7. © 1961, 1969 by Anne McCaffrey; reprinted from the story and book, both of the same name; by permission of the author and the author's agent)

Right now, this type of fiction is at its high-point. The "cyberpunk" genre of science fiction emphasizes refined forms of man-machine interfaces. An example from a recent novel illustrates the point:

"Cowboy forces a grin and gives him the finger, and then closes the dorsal hatch. He strips naked and sticks electrodes to his arms and legs, and then runs the wires from the electrodes to collars on his wrists and ankles... Next he plugs jacks into the sockets in his temples, the silver-chased sockets over each ear, the fifth socket at the base of his skull. He pulls his helmet on over them, careful not to stress the laser-optic wires coming out of his head. He tastes rubber and hears the hiss of anesthetic, loud here in the closed space of the helmet...

"Neurotransmitters awaken the five studs in his head and Cowboy watches the insides of his skull blaze with incandescent light, the liquid-crystal data matrices of the panzer molding themselves to the configuration of his mind. His heart beats faster, he's living in the interface again, the eye-face, his expanded mind racing like electrons through the circuits, into the metal and crystal heart of the machine."

(From *Hardwired,* by Walter Jon Williams. © 1986 by Walter Jon Williams and Tor Books, New York. Reproduced by permission of the publisher.)

If we take this as epitomizing the current heyday of the cyberpunk genre, we can probably expect to see real milestones achieved in this area within the 20 — 30 years of its initiation in the late 1960s; that is, by the year 2000. The type of research described in the previous chapter is likely to lead to this type of capability. Let's note the societal implications, however. Few of the cyberpunk stories portray a world in which many of us would wish to live. The integration of a human with a machine system, while possibly leading to greater abilities on the part of the hybrid man/machine system, is often linked (in current science fiction) with a society milieu in which the individual is devalued and survival is hard.

28.8   PREDICTION 8

Neural networks will facilitate user-specific systems for education, information processing, and entertainment. "Alternate realities," produced by comprehensive environments, are attractive in terms of their potential for systems control, education, and entertainment. This is not just a far-out research trend, but is something which is becoming an increasing part of our daily existence, as witnessed by the growing interest in comprehensive "entertainment centers" in each home. Current (legally sanctioned) tools for creating "alternative realities" include sophisticated, room-surround audio, big-screen TVs, and even some recent products which introduce specialized scents. There is a growing tendency toward personalized control with tools such as home video recording of movies and other broadcasts, programmable CD players, and remote control systems. Thus, there is a substantial basis just within the home entertainment industry for a certain degree of "artificial reality" experience. This does not begin to tap the potential for use in education or in enabling human control of complex systems.

One goal of such systems is to make the experienced reality more "real," which means that the different systems need to be more all-encompassing and more integrated. Ultimately, they should be able to give kinesthetic and tactile stimulus as well as visual, auditory, and olfactory. Another goal is to fine-tune the system to the individual, specifically to the individual's perceptual and cognitive style preferences (such as discussed in the preceding chapter). Also, it may be useful — especially in educational situations — to present information in certain special ways. (This is an area which has been opened up [Ostlander and Schroeder, 1979], but which requires substantial research to validate.) With neural networks, we could develop comprehensive control systems for integrating audio, visual, kinesthetic, and olfactory stimuli to produce environments containing specific cues.

There is some evidence [Robbins, 1986] which suggests that people respond to specific types and sequences of cues (audio, visual, kinesthetic). These "cue sequences" can purportedly arouse specific states in their target persons. These "states" might be love, anxiety, motivation, aversion. If we were able to learn our own state-specific "access sequences," we could use an audio/visual/somatic/olfactory environment to generate the sequence leading to a specific state. We could, in effect, learn to "program" ourselves. This is not a new idea [Lilly, 1967], but until recently, has not been technically viable.

This "programming" would require feedback from the user in order to be effective, but simple and "passive" sensors (e.g., fingertip sensors, gloves, or wristbands to sense pulse, blood pressure, skin ionization, etc.) could provide effective feedback into a neural control system. This could be achieved, for example, with sensors that would detect pulse, blood pressure, skin ionization, and other variables which the system could learn to correlate with a person's overall response state.

A neural network system could be developed to monitor a person's physiological state and to generate sensory-specific cues in proper order, and with appropriate temporal spacing. The temporal spacing would vary from session to session, depending on the physiological state of the person entering the session. In order to generate appropriate temporal spacing, the network system would use passive sensing of the person's physiological state. A training period (for both the network and the user) would be necessary, but after that, the network should be able to work effectively with only minor adaptations.

There would be tremendous societal implications from such developments. It would be easy to stress the positive — such as ways to accelerate learning, or make learning easier or more effective. It would be easy to point to a utopia of individual user-centered entertainment realms. However, there is also a possible downside. This is already a part of our societal mythology, as exemplified by the "Feelies" in Brave New World:

> "'Take hold of those metal knobs on the arms of your chair,' whispered Lenina. 'Otherwise you won't get any of the feely effects.'

> "'The Savage started. That sensation on his lips! He lifted a hand to his mouth; the titillation ceased; let his hand fall back on the metal knob; it began again."

(Excerpt from *Brave New World* by Aldous Huxley. ©1932. ©1960 by Aldous Huxley. Reprinted by permission of Harper-Collins Publishers.)

Given the current societal trend towards "cocooning," the possibility of an all-encompassing user-specific entertainment system which offers alternative realities might not be completely in the best interests of either the individual or of society.

28.9   PREDICTION 9

Neural networks will allow us to explore new realms of human capability realms previously available only with extensive training and personal discipline.

Back in the 1960s, we began to have some interesting experiments that used physical systems as feedback devices to help humans produce and maintain useful or interesting attentional or arousal states. The most well-known result of this period of investigation was the use of biofeedback monitors to help people cultivate "alpha" state (a term descriptive of the type of EEG readouts observed). This was thought to be useful to help people relax and deal with stress. This work was being done at about the same time that there was an upsurge in societal interest in other methods for inducing different states of awareness [Lilly, 1967; LaBerge, 1986, & whoever wrote the Relaxation Response, Tart, 1983].

Research in these areas — at least sponsored, journal-reported research — tapered off during the 1970s and 1980s. More recently, though, there have been developments in technologies to measure physiologically observed variables correlated with different human activities and different states of arousal/attention. Some of this work was cited in the previous chapter.

Turning once again to science fiction as an indicator of what may be to come, there may be some resurgence of interest in exploring ways in which humans can use different states of attention or arousal in order to accomplish different tasks, or to facilitate new forms of man/machine interface. The following excerpt is taken from a recent best-seller:

"Theta-sine-alpha indicated that James was relaxed, but it was a much deeper level of relaxation, more neurological, much more than ordinary muscle relaxation. The ability to get to theta-sine-alpha had taken months of training. They called it biofeedback when psychologists would hook a patient up to a mini-EEG or polygraph that would beep whenever a beta wave would be detected, indicating stress or irregular muscular or nervous activity. The idea was to relax the body or control nerve activity until the beeping stopped. James had to go far beyond such muscle relaxation — he had to relax his mind, open it, create a window into the subconscious.

"For Kenneth Francis James, the window to his mind did not open like a door or a window — it opened like a hot, rusty knife ripping though pink flesh. But that was the nature of the Advanced Neural Transfer and Response System that linked the brain with a digital computer. James had gone far beyond Carmichael's lectures. This was the real thing, the link-up between the computer on the plane and the suit...

"Once ANTARES was open it would transmit a complex series of preprogrammed questions to various conscious and subconscious areas of James' mind. The questions, programmed months earlier by countless hours in a simulator-recording unit, would match the existing brainwave patterns of each level encountered. After scanning, recognizing and matching the patterns, ANTARES would then over-power the particular neural function, force the original pattern to a compatible subconscious level and allow the ANTARES computer to control that level ...

"ANTARES could collect and transmit digital data signals to James' conscious mind, and James could receive that information as if it came from his own five senses. But James no longer had five senses — he had hundreds, thousands of them. The radar altimeter was a sense. The radar was a sense. So was the laser range-finder. Dozens of thermometers, aneroids, gallium-arsenide memory chips, limit switches, logic circuits, photocells, volt-meters, chronometers — the list was endless and never changing."

(From *Day of the Cheetah,* © 1989 by Dale F. Brown, Incorporated. Reprinted by permission of the publisher, Donald I. Fine, New York, N.Y.)

The distinction between this and the usual cyberpunk literature is that it suggests that a specific state of consciously induced neurophysiologically observable awareness is a necessary ingredient in facilitating a man-machine system interface. It is difficult to identify the extent to which technological forecasting, no matter how far-out, has a real potential, and to what extent it is simply used as a stage-prop to make believable an instantiation of a superhero — another cultural archetype. (In this story, the superhero is the man-plane combination, not one or the other separately.) Nevertheless, this story makes manifest an archetype which has a place in our society. Specifically, this kind of possibility has a place in the expectations for future possibilities in the minds of the young men and women who will someday be society's leading scientists and engineers.

28.10 PREDICTION 10

Neural networks, integrated with other artificial intelligence technologies, methods for direct culture of nervous tissue, and other exotic technologies such as genetic engineering, will allow us to develop radical and exotic life-forms, whether man, machine, or hybrid. The concept of humans creating non-human sentient life has been an archetypical theme throughout human history. There are many stories which manifest this theme, including such notables as the Greek story of Pygmalion and Mary Shelley's Frankenstein. With the advent of the computer age, and particularly the development of artificial intelligence systems, the possibility of artificial but sentient life — in both its positive and negative aspects — has permeated science fiction, our window into the future. Asimov's series of books on the "sentient robot" theme has epitomized this type of concern. So have movies and books which depict a conflict between intelligent and autonomous robots or AIs (artificial intelligences) and humanity. (Recent examples include the movies Robocop and The Terminator.)

The widespread societal interest in "future studies" began with publication of Alvin Toffler's *Future Shock* [1970], through to the *Third Wave* [1980], and moving on with Naisbitt's *Megatrends* [1982] and Naisbitt and Aburdene's *Megatrends 2000* [1990]. Yet none of these popular books (representing the common cultural expectations for the future) are dealing at all adequately with the impact of real autonomous systems and enhanced interfaces for man/machine systems.

For the most gripping explorations, we again turn to science fiction. Science fiction is often the medium through which we paint our imaginings of what the future might be, and explore the possible outcome of different alternatives. It is also the way in which we explore in our minds what the enabling technologies for creating these futures. In the context of creating artificial sentient life (as distinct from the artificial but insentient life described earlier), we are seeing more and more inclusion of neurally-based technologies. This is exemplified in the following extract:

"'TOKUGAWA is a child. Naturally, we hope it's a very bright child — for example it's responding well to spoken input in both English and Japanese, and can converse in a simple manner in both languages via a speech synthesizer; unlike our human child, TOKUGAWA doesn't have to achieve a certain level of neuromuscular control before it can speak, and there *are* tidbits of knowledge we can feed directly into it. But the fact remains that making TOKUGAWA fully operational involves a process more of education than programming.'

"From the head of the table Yoshimitsu Akaji beamed like a father pleased with the progress of a bright offspring. 'So you are ready to commence what we might call TOKUGAWA's vocational training, then, Doctor?'

"'Yes, Yoshimitsu-san.' She adjusted her glasses. 'But I consider that a relatively unimportant part of the educational process.'

"Suzuki raised an eyebrow. 'Oh? And what *do* you consider to be the important part of the process, Dr. O'Neill?'

"'Teaching TOKUGAWA how to be a human being.'"

(From *Cybernetic Samurai*, by Victor Milán. © 1990. Reprinted by permission of William Morrow and Company, Inc.)

As this potential becomes more and more real, we have many issues which will confront us both as individuals and as a society. Questions will come up, such as: If human or animal neural tissue (or complete neural systems) are used to build an autonomous system, then do we have a person or animal, or a machine? What are the legal — and moral — aspects of using human and/or animal tissue for experimentation with man/machine or animal/machine hybrids? Even if an autonomous machine is built entirely from man-made components, at what point to we consider it to be a sentient life? If we do, then what are its rights?

Many scientists are becoming actively involved in consideration of such issues, but these issues need to be addressed by society — an educated society — as a whole. Each of us, at some level, will have to contribute a response — and a responsibility — to creating the reality in which we and our children will live.

This depiction of future scenarios leads from the prosaic to the extraordinary. Many technologies will play a role in creating these futures. To paraphrase scientist Stephen Hawkin, we will most likely experience a reality that is not only stranger than we imagine but stranger than we can imagine. Neural networks will help mediate a transfer of what we can learn about our own minds and brains into external environments, and then back again.

REFERENCES

Asimov, I. (1984, 1950). *I, Robot,* Ballantine, New York.

Benson, H. (1975). *The Relaxation Response,* Morrow, New York.

Butler, F. (1978). *Biofeedback: A Survey of the Literature,* Plenum, New York.

Goldberg, M. (1988). "JSB: An AL Simulation," *AI Expert* (February), 63-65.

Heinlein, R. (1967). *The Past Through Tomorrow,* (collection of stories written between 1939 and 1958), G.P. Putnam's Sons, New York.

Hutchinson, M. (1986). *Megabrain,* Beech Tree (Wm. Morrow), New York.

Hutton, L., & Sigillito, V. (1990). "Experiments on constructing a cognitive map: A neural network model of a robot that daydreams," *Proc. Second Int'l. Joint Conf. Neural Networks* (Washington, D.C., Jan. 15-19, 1990), I-223–I-227.

Johnson, M., & Scanlon, R. (1987a). "Experiences with a feeling-thinking machine," *Proc. First IEEE Int'l. Conf. Neural Networks* (San Diego, CA; June 21-24, 1987), II-71–II-77.

Johnson, M., & Scanlon, R. (1987b). *Non-Living Intelligence,* ARCCB-TR-87009. Watervliet, NY: US Army ARDEC.

Johnson, R.C. (1990). "Speak, Memory," *OMNI* (February), 28.

LaBerge, S. (1986). *Lucid Dreaming,* Ballantine, New York.

Lilly, J. (1977). *The Deep Self,* Human Software, Inc., New York.

Lilly, J. (1987, 1972, 1967). *Programming and Meta-Programming the Human Biocomputer,* 1987 ed. published by Crown, New York.

McCaffrey, A. (1969). *The Ship Who Sang,* Ballantine, New York.

Milán, V. (1985). *Cybernetic Samurai,* Ace, New York.

Naisbitt, J. (1982). *Megatrends,* Warner Books, Inc, New York.

Naisbitt, J., & Aburdene, P. (1990). *Megatrends 2000,* William Morrow & Co., Inc., New York.

Ostlander, S., & Schroeder, L. (1979). *Superlearning,* Delacorte, New York.

Robbins, A. (1986). *Unlimited Power,* Fawcett Columbine, New York.

Scanlon, R.D., & Johnson, M. (1988). "Design of a feeling-thinking machine," *Proc. Fifth Army Conf. Applied Mathematics and Computing* (West Point, NY, June 15-18, 1988), ARCCB-TR-88013, Watervliet, NY: US Army ARDEC.

Tart, C. (1983, 1975). *States of Consciousness,* Psychological Processes, El Cerrito. (1975 publication by Dutton, New York.)

Toffler, A. (1971, 1970). *Future Shock,* Bantam, New York. (1970 publication by Random House, New York.)

Toffler, A. (1980). *The Third Wave,* Bantam, New York.

Williams, W. J. (1986). *Hardwired,* Tor Books, New York.

INDEX

ABOUT THE AUTHORS

Dr. Alianna J. Maren is a Visiting Associate Professor of Neural Network Engineering at the University of Tennessee Space Institute, a founding member of the Tennessee Center for Neural Engineering and Applications, and Director of Emerging Technologies with Accurate Automation Corporation. Since joining UTSI in 1987, Dr. Maren and her students have developed multilayered cooperative/competitive methods for improved image understanding and novel approaches to databases. In 1989, she founded AHCEL, the Advanced Human Capabilities Engineering Laboratory, which is exploring neural networks and other technologies for adaptive man-machine systems. She is Technical Head of developing neural networks for sensor fusion on a Phase II Accurate Automation Corporation contract with the Naval Ocean Systems command, and has participated in neural networks contracts funded by the Department of Energy, the Department of Transportation, the National Aeronautics and Space Agency, and other organizations. Dr. Maren is an editor and regular columnist for the Journal of Neural Network Computing, and has given many short courses, tutorials, seminars, and lectures in neural networks.

Dr. Craig T. Harston is Founder and President of Computer Applications Service in Chattanooga, TN. He has investigated the effect of pharmaceuticals on animal behavior and brain morphology and chemistry. He has been in the data processing/computer science field since 1981, with an emphasis on developing neurally-based approaches for sensory-motor robotic control. He has consulted on neural networks projects with the National Science Foundation, the National Aeronautics and Space Agency, the U.S. Navy, General Dynamics, and other organizations. Dr. Harston is currently developing neural-based approaches for material consistency analysis, mental health triage, and object identification in microscopic images.

Mr. Robert M. Pap is Founder and President of Accurate Automation Corporation, a Chattanooga-based firm. Mr. Pap is the Principal Investigator for the U.S. Navy on a project applying neural networks to radar processing, and is also Principal Investigator on a project to do neural network-based diagnosis of the National Aerospace Plane. His firm, Accurate Automation, has won additional contracts from such organizations as the National Aeronautics and Space Agency, the National Science Foundation, the Department of Transportation, and General Dynamics. Mr. Pap's technical interests also include process control systems, avionics, and data communications. Mr. Pap is an editor for the Journal of Neural Network computing, and is a frequent speaker at conferences such as WESCON and INTERFACE.

Dr. Stanley P. Franklin, a topologist turned computer scientist, became involved with neural networks via artificial intelligence and cognitive science. His neural netowrk papers and prior book chapters are concerned with the global dynamics of neural networks and with neural computability. Current research projects include a general purpose (programmable) neurocomputer, a medical diagnostic system, and training neural networks via genetic logorithms on parallel computers.

Mr. Robert L. Gezelter is Founding Principal of Gezelter Consultants, Inc. In this capacity, Mr. Gezelter acts as a consultant on a wide ange of technical areas including operating systems internals, networks, and related matters. His practise places a heavy emphasis on architectures and design issues. His clients range from major corporations listed in the Fortune 20 to small businesses, both nationally and internationally. In addition to his consulting practise, Mr. Gezelter serves as a Contributing Editor for Digital News. He is a frequent mmember of technical panels and clinics concerning areas covered by his practise. Mr. Gezelter holds M.S. and B.A. degrees in Computer Science from New York University.

Dr. Dan Jones, M.D., has a diversified background leading to his current work in using neural networks for medical diagnosis. Dr. Jones has a bachelor's degree in chemistry, electronics schooling with the U.S. Navy, a medical degree, ten years of medical practise, and five years of experience in working with neural networks and software development. Frustrated with the black-box model of psychology, he discovered neural networks in 1985, and immediately recognized their potential when coupled with ongoing neurophysiological research to eventually unravel the mechanisms of the mind. Dr. Jones has programmed networks for character recognition, speech recognition, and medical diagnosis. He recently founded Neural Arts, a company dedicated to the development of neural networks to assist in medical diagnosis.

Mr. Stephen G. Morton is the Founder and President of Oxford Computer, Inc., in Oxford, CT. His company is developing a family of memory-plus-processor chips, or "Intelligent Memory Chips," for applications in digital signal processing, pattern recognition, and computer graphics. He has a Bachelor's degree in Electrical Engineering (1971) and a Master of Science in Electrical Engineering (1972) from the Massachusetts Institute of Technology in Cambridge, MA. He has been awarded more than ten U.S. patents in the fields of chip design and parallel processing. He has more than twenty years of hands-on experience in analog circuit design, digital signal processing, image processing, computer architecture, parallel processing, fault-tolerant chip design, computer systems design, telecommunications and computer software. He has published many articles in technical journals and the electronic engineering press about parallel processors, chip design and neural networks.

Mr. Patrick K. Simpson is the principal investigator of the neural networks internal research and development project at General Dynamics Electronics Division, specializing in the application of neural networks, fuzzy logic, and artificial intelligence to defense-related problems in the areas of sonar and radar signal processing. Mr. Simpson has recently completed a book entitled Artificial Neural Systems: Foundations, Paradigms, Applications and Implementations, which provides a broad overview of neural network technology and succinctly describes 27 neural network paradigms with applications potential. Mr. Simpson is the program chair of the 1991 Workshop on Neural Networks for Ocean Engineering and teaches neural network courses at UCSD Extension, CLA Extension, and the Applied Technology Institute. Mr. Simpson is the Treasurer of the IEEE Neural Networks Council and a member of IAKE, INNS, and SPIE. Mr. Simpson has a B.A. in Computer Science from the University of California at San Diego.

Dr. Harold L. Szu is with the Naval Research Laboratory, and is Adjunct Professor with American University. He played a key role in organizing the International Neural Network Society, and was the first Treasurer of the Society. He is the developer of the Cauchy machine neural network. Dr. Szu is a frequent session chair, panelist, and speaker at international conferences on neural networks.

Dr. Paul J. Werbos is Program Director for Neuroengineering and Emerging Technology Initiation at the National Science Foundation (NSF) and Secretary of the International Neural Network Society (INNS). In June, 1990, in a unanimous action of the Governing Board, he was nominated for the Presidency of the INNS. Dr. Werbos developed the back-propagation method in 1974 for his doctorate in Applied Mathematics at Harvard University. His three other degrees from Harvard and the London School of Economics emphasized mathematical physics, international policy economy, and economics. While an Assistant Professor at the University of Maryland, Dr. Werbos developed the advanced adaptive critic designs or neurocontrol. Before joining NSF in 1989, he worked nine years at the Energy Information Administration of the Department of Energy, where he variously held lead responsibility for evaluating long-range forecasts (under Charles Smith), and for building models of industrial, transportation, and commercial demand using back-propagation and other methods. In previous years, he was regional Director and Washington Representative of the L-5 Society, predecessor to the National Space Society, and an organizer of the global Futures Roundtable. He has worked on occasion with the National Space Society, the Global Tomorrow Coalition, the Stanford Energy Modeling Forum, and the Adelphi Friends Meeting. He also retains an active interest in fuel cells for transportation and in the foundations of physics.